INTRODUCTION TO PROBABILITY

AND ITS APPLICATIONS

THE DUXBURY ADVANCED SERIES IN STATISTICS AND DECISION SCIENCES

Applied Nonparametric Statistics, Second Edition, Daniel
Applied Regression Analysis and Other Multivariable Methods, Second Edition, Kleinbaum, Kupper, and Muller
Classical and Modern Regression with Applications, Second Edition, Myers
Elementary Survey Sampling, Fourth Edition, Scheaffer, Mendenhall, and Ott
Introduction to Contemporary Statistical Methods, Second Edition, Koopmans
Introduction to Probability and Its Applications, Scheaffer
Introduction to Probability and Mathematical Statistics, Bain and Engelhardt
Linear Statistical Models, Bowerman and O'Connell
Probability Modeling and Computer Simulation, Matloff
Quantitative Forecasting Methods, Farnum and Stanton
Time Series Analysis, Cryer
Time Series Forecasting: Unified Concepts and Computer Implementation, Second Edition, Bowerman and O'Connell

THE DUXBURY SERIES IN STATISTICS AND DECISION SCIENCES

Applications, Basics, and Computing of Exploratory Data Analysis, Velleman and Hoaglin
A Course in Business Statistics, Second Edition, Mendenhall
Elementary Statistics, Fifth Edition, Johnson
Elementary Statistics for Business, Second Edition, Johnson and Siskin
Essential Business Statistics: A Minitab Framework, Bond and Scott
Fundamentals of Biostatistics, Third Edition, Rosner
Fundamentals of Statistics in the Biological, Medical, and Health Sciences, Runyon
Fundamental Statistics for the Behavioral Sciences, Second Edition, Howell
Introduction to Probability and Statistics, Seventh Edition, Mendenhall
An Introduction to Statistical Methods and Data Analysis, Third Edition, Ott
Introductory Statistics for Management and Economics, Third Edition, Kenkel
Mathematical Statistics with Applications, Fourth Edition, Mendenhall, Wackerly, and Scheaffer
Minitab Handbook, Second Edition, Ryan, Joiner, and Ryan
Minitab Handbook for Business and Economics, Miller
Operations Research: Applications and Algorithms, Winston
Probability and Statistics for Engineers, Third Edition, Scheaffer and McClave
Probability and Statistics for Modern Engineering, Second Edition, Lapin
Statistical Experiments Using BASIC, Dowdy
Statistical Methods for Psychology, Second Edition, Howell
Statistical Thinking for Behavioral Scientists, Hildebrand
Statistical Thinking for Managers, Second Edition, Hildebrand and Ott
Statistics: A Tool for the Social Sciences, Fourth Edition, Ott, Larson, and Mendenhall
Statistics for Business and Economics, Bechtold and Johnson
Statistics for Management and Economics, Sixth Edition, Mendenhall, Reinmuth, and Beaver
Understanding Statistics, Fifth Edition, Ott and Mendenhall

INTRODUCTION TO PROBABILITY

AND ITS APPLICATIONS

Richard L. Scheaffer

University of Florida

THE DUXBURY ADVANCED SERIES IN STATISTICS AND DECISION SCIENCES

PWS-KENT Publishing Company

Boston

PWS-KENT
Publishing Company

20 Park Plaza
Boston, Massachusetts 02116

PWS-KENT Publishing Company is a division of Wadsworth, Inc.

Library of Congress Cataloging-in-Publication Data

Scheaffer, Richard L.
Introduction to probability and its applications/Richard L. Scheaffer
p. cm.
ISBN 0-534-91970-7
1. Probabilities. I. Title.
QA273.S357 1989
519.2—dc20 89-16194
CIP

Editor: Michael Payne
Production Editor: Susan L. Krikorian
Manufacturing Coordinator: Margaret Sullivan Higgins
Interior and Cover Designer: Catherine L. Johnson
Interior Illustrator: Keyword Publishing Services
Compositor: The Universities Press Ltd.
Text Printer and Binder: The Maple-Vail Book Manufacturing Group

Printed in the United States of America

1 2 3 4 5 6 7 8 9—93 92 91 90 89

PREFACE

This textbook is designed to be covered in a one-semester or one-quarter course in probability for students who have a solid background in basic differential and integral calculus. Probability is presented as an area of mathematics that has many and varied applications to the solutions of real-world problems. Therefore, an axiomatic development of the theory is given, including a thorough discussion of random variables and the commonly used probability distributions. This is mixed with discussions of practical uses of probability, some involving real data. The text should provide a sound background for those students going on to more advanced courses in probability or statistics and, at the same time, provide a working knowledge of probability for those who must apply it in engineering or the sciences.

Many exercises are provided, most for practice in the calculation of probabilities, but some for allowing the student to extend the theory. It has been this author's experience that many probability problems appear deceptively easy, and students must be encouraged to practice the calculations in a variety of settings in order to really master the subject. Thus, students should work through most of the problems at the end of each section, along with a sampling of those from the supplementary set at the end of each chapter.

Some sections contain material of a more advanced nature, much of which might appear in a course in stochastic processes. These sections (3.11, 3.12, 5.8, 5.9, 5.10, 6.6, 6.7, and 7.6) can be skipped with no loss of continuity in the remaining sections.

In this age of technology, much of the applied work in probability is done with the aid of a computer, especially in cases involving nonstandard distributions. Thus, activities for the computer are presented and guidelines are given for generating probability distributions by use of random numbers. These sections should be especially useful for those who must simulate the properties of complex systems. Except for those sections designated for the computer, knowledge of computing is not required for completing the text.

I would like to thank the following individuals for their work in reviewing the manuscript for this text: Bernice L. Auslander, University of Massachusetts, Boston; Jay Devore, California Polytechnical State University, San Luis Obispo; Paul T. Holmes, Clemson University; David Robinson, St. Cloud State University, Minnesota; and Joseph J. Walker, Georgia State University.

I extend wholehearted thanks to Chris Franklin for providing the activities for the computer and other helpful comments on the material, and to Monroe Crews for a superb job of typing the manuscript.

Richard L. Scheaffer
Gainesville, Florida

CONTENTS

CHAPTER 3

DISCRETE PROBABILITY DISTRIBUTIONS 54

CHAPTER 4

CONTINUOUS PROBABILITY DISTRIBUTIONS 129

CHAPTER 5

MULTIVARIATE PROBABILITY DISTRIBUTIONS 202

CHAPTER 6

FUNCTIONS OF RANDOM VARIABLES 251

CHAPTER 7

SOME APPROXIMATIONS TO PROBABILITY DISTRIBUTIONS: LIMIT THEOREMS 280

INTRODUCTION TO PROBABILITY

AND ITS APPLICATIONS

CHAPTER 1

PROBABILITY IN THE WORLD AROUND US

1.1 WHY STUDY PROBABILITY?

We live in an information society. We are confronted—in fact, inundated—with quantitative information at all levels of endeavor. Charts, graphs, rates, percentages, averages, forecasts, and trend lines are an inescapable part of our everyday lives. They affect our decisions on health, citizenship, parenthood, jobs, financial concerns, and many other important matters. Today, an informed person must have some facility for dealing with data and making intelligent decisions based on quantitative arguments that involve uncertainty or chance.

We live in a scientific age. We are confronted with arguments that demand logical, scientific reasoning even if we are not trained scientists. We must be able to clearly see our way through a maze of reported "facts" in order to separate credible conclusions from specious ones. We must be able to intelligently weigh such issues as the evidence on the causes of cancer, the effects of pollutants on the environment, or the results of a limited nuclear war.

We live amidst burgeoning technology. We are confronted with a job market that demands scientific and technological skills, and students must be trained to deal with the tools of this technology in productive, efficient, and correct ways. Much of this new technology is concerned with information processing and dissemination; and proper use of this technology requires probabilistic skills. These skills are in demand in engineering, business, and computer science for jobs involving quality control, reliability, product development and testing, market research, business management, data management, and economic forecasting, just to name a few.

Few results in the natural or social sciences are known absolutely. Most results are reported in terms of chances or probabilities: the chance of rain tomorrow, the chance that you get home from school or work safely, the chance that you live past sixty years of age, the chance of getting a certain disease (or recovering from it), the chance of inheriting a certain trait, the chance of your annual income exceeding $30,000 in two years, the chance of winning an election. Students must obtain some knowledge of probability and be able to tie this concept to real scientific investigations if they are to understand science, and the world around them.

This book provides an introduction to probability that is both mathematical, in the sense that a theory is developed from axioms, and practical, in the sense that applications to real-world problems are discussed. It is intended that the material will provide a strong basis in probability for those who may go on to deeper studies in statistics, mathematics, engineering, or the physical and biological sciences, and, at the same time, provide a basis for practical decision making in the face of uncertainty.

1.2

DETERMINISTIC AND PROBABILISTIC MODELS

It is essential that we grasp the difference between theory and reality. Theories are ideas proposed to explain phenomena in the real world and, as such, are approximations or models for reality. These models, or explanations of reality, are presented in verbal form in some less quantitative fields and as mathematical relationships in others. Whereas a theory of social change might be expressed verbally in sociology, the theory of heat transfer is presented in a precise and deterministic mathematical manner in physics. Neither gives an accurate and unerring explanation for real life. Slight variations from the mathematically expected can be observed in heat-transfer phenomena and other areas of physics. The deviations cannot be blamed solely on the measuring instruments, the explanation that one often hears, but are due in part to a lack of agreement between theory and reality. Anyone who believes that the physical scientist now completely understands the wonders of this world need only look at history to find a contradiction. Theories assumed to be the "final" explanation for nature have been superseded in rapid succession during the past century.

In this text, we shall develop certain theoretical models of reality; we shall attempt to explain the motivation behind such a development and the uses of the resulting models. At the outset, we discuss the nature and importance of model building in the real world so that the reader will have a clear idea of the meaning of the term model and the types of models generally encountered.

Suppose that we wish to measure the area covered by a lake that, for all practical purposes, appears to have a circular shoreline. Since we know the area, A, is given by $A = \pi r^2$, where r is the radius, we would attempt to measure the radius (perhaps by averaging a number of measurements taken at various points) and substitute it into the formula. The formula $A = \pi r^2$, as used here, is called a *deterministic model.* It is *deterministic* because, once the radius is known, the area is assumed to be known. It is a *model* of reality because the border of the lake will have some irregularities and not form a true circle. Even though the plane figure in question is not exactly a circle, the model provides a useful relationship between the area and the radius, which makes approximate measurements of area easily obtained. Of course, the model becomes poorer and poorer as the figure deviates more and more from a circle until, eventually, it is no longer of use and a new model takes over.

Another deterministic model is Ohm's Law, $I = E/R$, which states that electric current is directly proportional to the voltage and inversely proportional to the resistance in a circuit. Once the voltage and resistance are known, the current is determined. If many circuits with identical voltages and resistances are investigated, the current measurements may differ by small amounts from circuit to circuit, owing to inaccuracies in the measuring equipment or other uncontrolled influences. Nevertheless, the discrepancies will be negligible, and for all practical purposes Ohm's Law provides a useful deterministic model of reality.

Contrast the two preceding situations with the problem of tossing a balanced coin and observing the upper face. No matter how many measurements we may make on the coin before it is tossed, we cannot predict with absolute accuracy whether the coin will come up heads or tails. However, it is reasonable to assume that if many identical tosses are made approximately one-half will result in the outcome of heads. That is, we cannot predict the outcome of the next toss, but we can predict what will happen in the long run. We sometimes convey this long-run information by saying that the "chance" or "probability" of heads on a single toss is 1/2. This probability statement is actually a formulation of a *probabilistic model* of reality. Probabilistic models are useful in describing experiments that give rise to random, or chance, outcomes. In some situations, such as tossing an unbalanced coin, preliminary experimentation would have to be conducted before realistic probabilities could be assigned to the outcomes; but it is possible to construct fairly accurate probabilistic models for many real-world phenomena. Such models are useful in varied applications, such as describing the movement of particles in physics (Brownian motion) and predicting the profits for a corporation during some future quarter.

1.3

APPLICATIONS OF PROBABILITY

We shall now illustrate two uses of probability theory. Both involve an underlying probabilistic model, but the first hypothesizes a model and then uses it for practical purposes, whereas the second deals with the more basic question of whether the hypothesized model in fact is a correct one.

Suppose that we attempt to model the random behavior of the arrival times and lengths of service for patients at a medical clinic. Such a mathematical function would be useful in describing the physical layout of the building and deciding how many physicians would be required to service the facility. Thus, this use of probability assumes that the probabilistic model is known and is a good characterization of the real system. The model is then employed to infer the behavior of one or more variables. The inferences will be correct, or nearly correct, if the assumptions that governed the construction of the model were correct.

The problem of choosing the correct model is the second use of probability theory, and this use reverses the reasoning procedure just described. Assume that the scientist does not know the probabilistic mechanism governing the behavior of arrival and service times. He or she might then observe an operating clinic and acquire a sample of arrival and service times. Using this sample data, inferences could be drawn concerning the nature of the underlying probabilistic mechanism, a type of application known as *statistical inference*. This book deals mostly with problems of the first type, but, on occasion, will use data as a basis for model formulation. Ideally, most students will go on to a formal course in statistical inference.

More specifically, consider the problem of replacing the light bulbs in a particular socket in a factory. A bulb is to be replaced either at failure or at a specific age, T, whichever comes first. Suppose that the cost, c_1, of replacing a failed bulb is greater than the cost, c_2, of replacing a bulb at age T. This may be true because in-service failures disrupt the factory, whereas replacements may not. A simple probabilistic model allows us to conclude that the average replacement cost per unit time, in the long run, is approximately

$$\frac{1}{\mu}[c_1 \text{ (probability of an in-service failure)} \\ + c_2 \text{ (probability of a planned replacement)}]$$

where μ denotes the average service time per bulb. The average cost is a function of T and, if μ and the indicated probabilities can be obtained from the model, a value of T can be chosen to minimize this function. Problems of this type are discussed more fully in Chapter 7.

Biological populations are often characterized by birth rates, death rates, and a probabilistic model that relates the size of the population at a given time to these rates. One simple model allows us to show that a population has a high probability of becoming extinct even if the birth and death rates are equal. Only if the birth rate exceeds the death rate might the population exist indefinitely.

Again referring to biological populations, models have been developed to explain the diffusion of a population across a geographic area. One such model concludes that the square root of the area covered by a population is linearly related to the length of time the population has been in existence. This relationship has been shown to hold reasonably well for many varieties of plants and animals.

Probabilistic models like those mentioned give scientists much information for explaining and controlling natural phenomena. Much of this information is intuitively clear, such as the fact that connecting identical components in series reduces the expected life length compared to that of a single component, whereas parallel connections increase expected life length. But many results of probabilistic models give new insights into natural phenomena, such as the fact that, if a person has a 50:50 chance of winning on any one trial of a gambling game, the excess of wins over losses will tend to stay either positive or negative for long periods of time, given that the trials are independent. (That is, the difference between number of wins and number of losses does not fluctuate rapidly from positive to negative.)

1.4

A BRIEF HISTORICAL NOTE

Probability has its origins in games of chance, which have been played throughout recorded history and, no doubt, since prehistoric times. Game boards have been found in excavations of Egyptian tombs and gaming was so popular in Roman times that laws had to be passed to regulate it. With this long history of games of chance it is somewhat surprising that formal study and development of probability theory did not take root until the middle of the seventeenth century.

It is generally agreed that a major impetus to the formal study of probability was provided by the Chevalier de Mere when he posed a gambling question to the famous mathematician Pascal (1623–1662). The question was along the following lines. A gambler is to throw a six with a die and has eight throws in which to do it. If he has no success on the first three throws, and the game is then ended, how much of the stake is rightfully his. Pascal cast this problem in probabilistic terms and engaged in extensive correspondence with another French mathematician, Fermat (1608–1665), about its solution. This correspondence began the formal mathematical development of probability theory.

Scientists in the eighteenth century (such as James Bernoulli and De Moivre) continued the development of probability theory and recognized its usefulness in solving important problems in science. The normal curve was introduced in this time period. Their work was continued by Gauss and Laplace in the nineteenth century, when the use of probability in data analysis began as a forerunner of modern statistics. In the twentieth century, probability theory became a major branch of mathematical research; and applications of the theory have spread to virtually every corner of scientific research.

1.5 A LOOK AHEAD

This text is concerned with the theory and applications of probability as a model of reality. We shall postulate theoretical frequency distributions for populations and develop a theory of probability in a precise mathematical manner. The net result will be a theoretical or mathematical model for the acquisition and utilization of information in real life. It will not be an exact representation of nature, but this should not disturb us. Like other theories, its utility will be measured by its ability to assist us in understanding nature and in solving problems in the real world. Such is the role of the theory of heat transfer, the theory of strengths of materials, and the other models of nature.

CHAPTER 2

PROBABILITY

2.1

AN ELEMENTARY DEFINITION OF PROBABILITY

If one flips a balanced coin into the air, what is the chance that it will land heads up? Most people would say that this chance, or probability, should be .5, or something very close to .5. Upon further questioning, one can determine that to most people the meaning of the .5 is that approximately one-half of the time, in repeated flipping, the coin should land heads up. From there on, the reasoning gets more fuzzy. Will 10 tosses result in exactly 5 heads? Will 50 tosses result in exactly 25 heads? Well, probably not. So, the .5 is assumed to be a long-run relative frequency, or limiting relative frequency, as the number of flips gets large.

To see what might happen in actual practice, we will look at coin-flipping data actually generated by John Kerrich, a mathematician interned in Denmark during World War II. Kerrich actually tossed a coin 10,000 times, keeping a tally of the number of heads. After 10 tosses he had 4 heads, a relative frequency of .4, after 100 tosses he had 44 heads (.44), after 1000 tosses he had 502 heads (.502), and after 10,000 tosses he had 5067 heads (.5067). The relative frequency of heads remained very close to .5 after 1000 tosses (although the figure at 10,000 is further from .5 than the figure at 1000).

If n is the number of trials of an experiment (like the number of flips of a coin), then it seems that one might define the probability of an event, E, (such as observing a head) by

$$P(E) = \lim_{n\to\infty} \frac{\text{number of times } E \text{ occurs}}{n}.$$

But, will this limit always converge? If so, can we ever determine what it will converge to without actually conducting the experiment many times? For these and other reasons, this is not an acceptable mathematical definition of probability, but it is a property that should hold, in some sense. So, another definition of probability must be found that allows such a limiting result to be proved as a consequence. This will be done in Section 2.3.

2.2

A BRIEF REVIEW OF SET NOTATION

Before going into a formal discussion of probability, it is necessary to outline the set notation we will use. Suppose we have a set S consisting of points labeled 1, 2, 3, and 4. We denote this by $S = \{1, 2, 3, 4\}$. If $A = \{1, 2\}$ and $B = \{2, 3, 4\}$, then A and B are subsets of S, denoted by $A \subset S$ and $B \subset S$ (B is "contained in" S). We denote the fact that 2 is an element of A by $2 \in A$. The union of A and B is the set consisting of all points that are in either A or B or both. This is denoted by $A \cup B = \{1, 2, 3, 4\}$. If $C = \{4\}$, then $A \cup C = \{1, 2, 4\}$. The intersection of two sets A and B is the set consisting of all points that are in both A and B, denoted by $A \cap B$, or merely AB. For the example $A \cap B = AB = \{2\}$ and $AC = \emptyset$, where $\emptyset$ denotes the null set, or the set consisting of no points.

The complement of A, with respect to S, is the set of all points in S that are not in A, denoted by $\bar{A}$. For the specific sets just given, $\bar{A} = \{3, 4\}$. Two sets are said to be mutually exclusive, or disjoint, if they have no points in common, as in A and C.

Venn diagrams can be used to portray effectively the concepts of union, intersection, complement, and disjoint sets, as in Figure 2.1.

We can easily see from Figure 2.1 that

$$\bar{A} \cup A = S$$

for any set A. Other important relationships among events are the *distributive*

FIGURE 2.1 Venn diagrams of set relations.

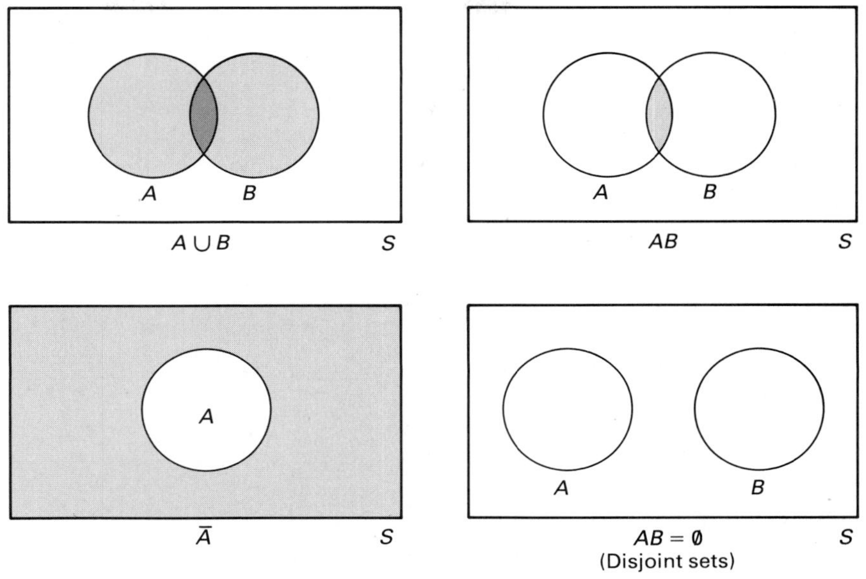

laws:

$$A(B \cup C) = AB \cup AC$$
$$A \cup (BC) = (A \cup B)(A \cup C)$$

and *DeMorgan's Laws:*

$$\overline{A \cup B} = \bar{A}\bar{B}, \quad \text{or} \quad \left(\overline{\bigcup_{i=1}^{n} A_i}\right) = \bigcap_{i=1}^{n} \overline{A_i},$$

$$\overline{AB} = \bar{A} \cup \bar{B}, \quad \text{or} \quad \left(\overline{\bigcap_{i=1}^{n} A_i}\right) = \bigcup_{i=1}^{n} \overline{A_i}.$$

It is important to be able to relate descriptions of sets to their symbolic notation, using the symbols given above, and to list correctly or count the elements in sets of interest. The following example illustrates the point.

Example 2.1 Twenty electric motors are pulled from an assembly line and insepected for defects. Eleven of the motors are free of defects, 8 have defects on the exterior finish, and 3 have defects in their assembly and will not run. Let A denote the set of motors having assembly defects and F the set having defects on their finish. Using A and F, write a symbolic notation for

(a) the set of motors having both types of defects

(b) the set of motors having at least one type of defect
(c) the set of motors having no defects
(d) the set of motors having exactly one type of defect.

Then give the number of motors in each set.

Solution

(a) The motors with both types of defects must be in A and F; thus, this event can be written AF. Since only 9 motors have defects, whereas A contains 3 and F contains 8 motors, 2 motors must be in AF (see Figure 2.2).

FIGURE 2.2 Venn diagram for Example 2.1 (numbers of motors shown for each set).

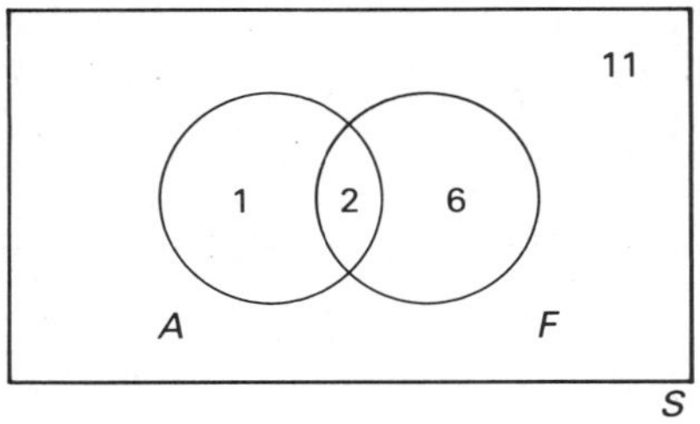

(b) The motors having at least one type of defect must have either an assembly or finish defect. Hence, this set can be written $A \cup F$. Because 11 motors have no defect, 9 must have at least one defect.

(c) The set of motors having no defects is the complement of the set having at least one defect, and is written $\overline{A \cup F} = \bar{A}\bar{F}$ (by DeMorgan's Law). Clearly, 11 motors fall into this set.

(d) The set of motors having exactly one type of defect must be in A and not in F or in F and not in A. This set can be written $A\bar{F} \cup \bar{A}F$; and 7 motors fall into the set.

EXERCISES

2.1 Of 25 microcomputers available in a supply room, 10 have circuit boards for a printer, 5 have circuit boards for a modem, and 13 have neither board. Using P to denote those that have printer boards and M to denote those that have modem boards, symbolically denote the following sets, and give the number of microcomputers in each set.

(a) the set of all microcomputers that have both boards
(b) the set of all microcomputers that have neither board

(c) the set of all microcomputers that have printer boards only

(d) the set of all microcomputers that have exactly one of the boards

2.2 There are 5 applicants (Jim, Don, Mary, Sue, and Nancy) available for 2 identical jobs. A supervisor selects 2 applicants to fill these jobs.

(a) List all possible ways in which the jobs can be filled. (That is, list all possible selections of 2 applicants from the 5.)

(b) Let A denote the set of selections containing at least 1 male. How many elements are in A?

(c) Let B denote the set of selections containing *exactly* 1 male. How many elements are in B?

(d) Write the set containing 2 females in terms of A and B.

(e) List the elements in $\bar{A}$, AB, $A \cup B$, and $\overline{AB}$.

2.3 Use Venn diagrams to verify the distributive laws.

2.4 Use Venn diagrams to verify DeMorgan's Laws.

2.3

A FORMAL DEFINITION OF PROBABILITY

Suppose a regular six-sided die is tossed onto a table and the number on the upper face is observed. This is a probabilistic situation because the number occurring on the upper face cannot be determined in advance. We will analyze the components of this experimental situation and arrive at a definition of probability that will allow us to mathematically model what happens in die tosses, as well as in many similar situations.

First, we might toss the die once, or several times, to collect data on possible outcomes. This data-generation phase is called an *experiment.*

Definition 2.1

An **experiment** is the process of making an observation.

Any experiment can result in a number of possible outcomes. A list, or set, of all possible outcomes for an experiment is called a *sample space.*

Definition 2.2

A **sample space**, S, is a set that includes all possible outcomes for an experiment, listed in a mutually exclusive and exhaustive manner.

Mutually exclusive means that the elements of the set do not overlap, and *exhaustive* means that the list contains all possible outcomes.

For the die toss, we could write a sample space as

$$S_1 = \{1, 2, 3, 4, 5, 6\}$$

where the integers indicate the possible numbers of dots on the upper face, or as

$$S_2 = \{\text{even}, \text{odd}\}.$$

Both S_1 and S_2 satisfy the Definition 2.2, but S_1 seems the better choice because it gives all the necessary details. S_2 has 3 possible upper-face outcomes in each listed element, whereas S_1 has only 1 possible outcome per element.

As another example, suppose a nurse is measuring the height of a patient. (This measurement process constitutes the experiment). A sample space could be listed as

$$S_3 = \{1, 2, 3, \ldots, 50, 51, 52, \ldots, 70, 71, 72, \ldots\}$$

if the height is rounded to the closest integer number of inches. On the other hand, an appropriate sample space could be

$$S_4 = \{x \mid x > 0\},$$

which is read "the set of all real numbers, x, such that $x > 0$." Whether S_3 or S_4 should be used in a particular problem depends on the nature of the measurement process. If decimals are to be used, we need S_4. If only integers are to be used, S_3 will suffice. The point is, then, that sample spaces for a particular experiment are not unique and must be selected to provide all pertinent information for a given situation.

Let us go back to our first example, the toss of a die. Suppose player A can have first turn at a board game if he or she rolls a six. Therefore, the *event,* "roll a six" is important to that player. Other possible events of interest in the die-tossing experiment are "roll an even number," "roll a number greater than four," and so on.

Definition 2.3

An **event** is any subset of a sample space.

Definition 2.3 holds as stated for any sample space that has a finite or countable number of elements. Some subsets must be ruled out if a sample space covers a continuum of real numbers, as S_4 given earlier, but any subset likely to occur in practice can be called an *event.*

From this definition of events, we see that an event is a collection of

FIGURE 2.3 Venn diagram for a die toss.

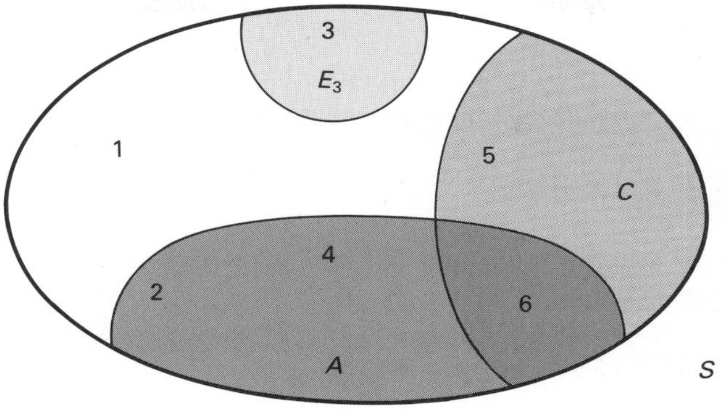

elements from the sample space. For the die-tossing experiment we can see how this works, for example, by defining

A to be "an even number"

B to be "an odd number"

C to be "a number greater than 4"

E_1 to be "observe a 1" and, in general,

E_i to be "observe integer i."

Then, if $S = \{1, 2, 3, 4, 5, 6\}$,

$$A = \{2, 4, 6\}$$
$$B = \{1, 3, 5\}$$
$$C = \{5, 6\}$$
$$E_1 = \{1\} \quad \text{and} \quad E_i = \{i\}, \qquad i = 1, 2, 3, 4, 5, 6.$$

Sample spaces and events often can be conveniently displayed in Venn diagrams. Some events for the die-tossing experiment are shown in Figure 2.3. For an experiment, we now know how to establish a sample space and list appropriate events. The next step is to define a probability for these events. We have seen already that the intuitive idea of probability is related to relative frequency of occurrence. A regular die should show an even number, when tossed, about 1/2 of the time and a three about 1/6 of the time. So, probabilities should be fractions between 0 and 1. One of the integers 1, 2, 3, 4, 5, or 6 must occur every time the die is tossed, and so the total probability associated with the sample space must be 1. In repeated tosses of the die, if a one occurs 1/6 of the time, then a one or two must occur $1/6 + 1/6 = 1/3$ of the time. Because relative frequencies for mutually exclusive events can be added, so must the probabilities. These considerations lead to the following definition.

Definition 2.4

Suppose that an experiment has associated with it a sample space S. Let A and B denote any two mutually exclusive events in S. A **probability** is a numerically valued function that assigns a number $P(A)$ to every event A so that the following axioms hold:

1. $P(A) \geq 0$.
2. $P(S) = 1$.
3. If $A_1, A_2, \ldots,$ is a sequence of mutually exclusive events (i.e., $A_iA_j = \emptyset$ for any $i \neq j$), then

$$P\left(\bigcup_{i=1}^{\infty} A_i\right) = \sum_{i=1}^{\infty} P(A_i).$$

From axiom 3 it follows that, if A and B are mutually exclusive events,

$$P(A \cup B) = P(A) + P(B).$$

This is similar to the addition of relative frequencies in the die tossing example discussed earlier.

Two elementary properties of probability now follow immediately. First, if $A \subset B$, then $P(A) \leq P(B)$. To see this, write

$$B = A \cup \bar{A}B$$

so that

$$P(B) = P(A \cup \bar{A}B) = P(A) + P(\bar{A}B).$$

Because $P(\bar{A}B) \geq 0$ by axiom 1, it follows that $P(A) \leq P(B)$. In particular, because $A \subset S$, for any event A, and $P(S) = 1$, then $P(A) \leq 1$.

Second, we can show that $P(\emptyset) = 0$. Because S and $\emptyset$ are disjoint with $S \cup \emptyset = S$:

$$1 = P(S) = P(S \cup \emptyset) = P(S) + P(\emptyset).$$

The definition of probability tells us only the axioms such a function must obey; it does not tell us what numbers to assign specific events. The actual assignment of numbers usually comes about from empirical evidence or from careful thought about the experiment. If a die is balanced, we could toss it a few times to see if the upper faces all seem equally likely to occur. Or we could simply assume this would happen and assign a probability of 1/6 to each of the six elements in S; that is, $P(E_i) = 1/6$, $i = 1, 2, \ldots, 6$. Once we have done this, the model is complete because, by axiom 3, we can now find the probability of any event. For example, for events defined on page 13 and in Figure 2.3:

$$\begin{aligned} P(A) &= P(E_2 \cup E_4 \cup E_6) \\ &= P(E_2) + P(E_4) + P(E_6) \\ &= \frac{1}{6} + \frac{1}{6} + \frac{1}{6} = \frac{1}{2} \end{aligned}$$

and

$$\begin{aligned} P(C) &= P(E_5 \cup E_6) \\ &= P(E_5) + P(E_6) \\ &= \frac{1}{6} + \frac{1}{6} = \frac{1}{3}. \end{aligned}$$

It is important to remember that Definition 2.4 along with the actual assignment of probabilities to events provide a probabilistic model for an experiment. If $P(E_i) = 1/6$ is used in the die-tossing experiment, the model is good or bad depending on how close the long-run relative frequencies for each outcome actually come to the numbers suggested by a theory. If the die is balanced, the model should be good; it tells us what we can expect to happen. If the die is not balanced, then the model is bad and other probabilities should be substituted for the $P(E_i)$. Throughout the remainder of this book, we will develop many specific models based on this underlying definition and discuss practical situations in which they work well. None is perfect, but many are adequate for describing real-world probabilistic phenomena.

Example 2.2 A purchasing clerk wants to order supplies from 1 of 3 possible vendors, which are numbered 1, 2, and 3. All vendors are equal with respect to quality and price, so the clerk writes each number on a piece of paper, mixes the papers, and blindly selects 1 number. The order is placed with the vendor whose number is selected. Let E_i denote the event that vendor i is selected $(i = 1, 2, 3)$, B the event that vendor 1 or 3 is selected, and C the event that vendor 1 is *not* selected. Find the probabilities of events E_i, B, and C.

Solution

Events E_1, E_2, and E_3 correspond to the elements of S because they represent all the "single possible outcomes." Thus, if we assign appropriate probabilities to these events, the probability of any other event is easily found.

Because 1 number is picked at random from the 3 available, it should seem intiutively reasonable to assign a probability of 1/3 to each E_i. That is,

$$P(E_1) = P(E_2) = P(E_3) = \frac{1}{3}.$$

In other words, we find no reason to suspect that one number has a higher chance of being selected than any of the others. Now,

$$B = E_1 \cup E_3$$

and by axiom 3 of Definition 2.4:

$$P(B) = P(E_1 \cup E_3) = P(E_1) + P(E_3) = \frac{1}{3} + \frac{1}{3} = \frac{2}{3}.$$

Similarly,

$$C = E_2 \cup E_3$$

and thus,

$$P(C) = P(E_2) + P(E_3) = \frac{2}{3}.$$

Note that different probability models could have been selected for the sample space connected with this experiment, but only this model is reasonable under the assumption that the vendors are all equally likely to be selected. The terms *blindly* and *at random* are interpreted as imposing equal probabilities on the finite number of points in the sample space.

The examples seen so far assign equal probabilities to the elements of a sample space, but this is not always the case. If you have a quarter and a penny in your pocket and you pull out the first one you touch, the quarter may have a higher probability of being chosen because of its larger size.

Often the probabilities assigned to events are based on experimental evidence or observational studies that yield relative frequency data on those events of interest. The data provide only approximations to the true probabilities, but these approximations often are quite good and usually are the only information we have on the events of interest. Example 2.3 illustrates the point.

Example 2.3 Figure 2.4 shows data on the marital status of the unemployed in the United States.

Suppose you met an unemployed worker in 1956. What is the approximate probability that the unemployed person would be

(a) a married woman?
(b) single?
(c) married?

Answer parts a, b, and c also for an unemployed worker met in 1982.

FIGURE 2.4 Marital status of the unemployed.

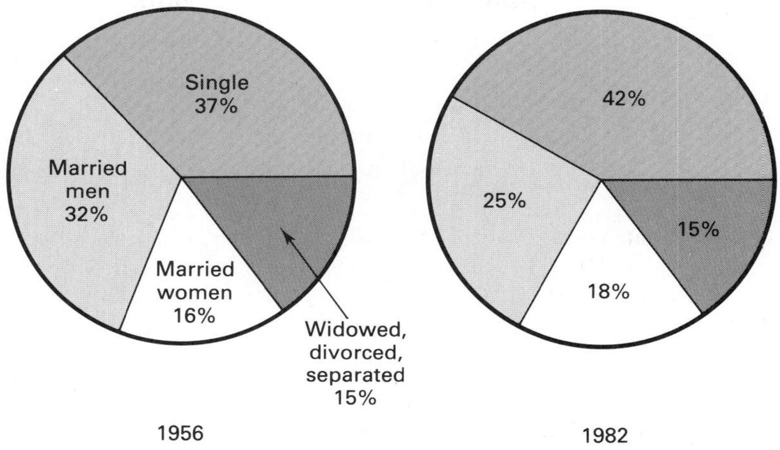

Source: U.S. Department of Labor, Bureau of Labor Statistics, "Workers Without Jobs: A Chartbook on Unemployment," BLS Bulletin 2174, July 1983, p. 21.

Solution

We assume nothing else is known about the unemployed person that you met. (The person is, in effect, randomly chosen from the population of unemployed workers.) Then we see directly from the chart for 1956 that

$$
\begin{aligned}
P(\text{married woman}) &= .16 \\
P(\text{single}) &= .37 \\
P(\text{married}) &= P(\text{married woman}) + P(\text{married man}) \\
&= .16 + .32 + .48.
\end{aligned}
$$

From the 1982 chart, we have

$$
\begin{aligned}
P(\text{married woman}) &= .18 \\
P(\text{single}) &= .42 \\
P(\text{married}) &= .18 + .25 = .43.
\end{aligned}
$$

Note that we *cannot* answer some potentially interesting questions from this chart. For example, we cannot find the probability that the unemployed person is a single woman.

EXERCISES

2.5 A vehicle arriving at an intersection can turn left, turn right, or continue straight ahead. If an experiment consists of observing the movement of one vehicle at this intersection, do the following:

(a) List the elements of a sample space.

(b) Attach probabilities to these elements if all possible outcomes are equally likely.

(c) Find the probability that the vehicle turns, under the probabilistic model of part b.

2.6 A manufacturing company has two retail outlets. It is known that 30 % of the potential customers buy products from outlet 1 alone, 50 % buy from outlet 2 alone, 10 % buy from both 1 and 2, and 10 % of the potential customers buy from neither. Let A denote the event that a potential customer, randomly chosen, buys from 1 and B the event that the customer buys from 2. Find the following probabilities:

(a) $P(A)$

(b) $P(A \cup B)$

(c) $P(\bar{B})$

(d) $P(AB)$

(e) $P(A \cup \bar{B})$

(f) $P(\bar{A}\bar{B})$

(g) $P(\overline{A \cup B})$.

2.7 For volunteers coming into a blood center, 1 in 3 gave O^+ blood, 1 in 15 have O^-, 1 in 3 have A^+, and 1 in 16 have A^-. What is the probability that the first person who shows up tomorrow to donate blood has

(a) type O^+ blood?

(b) type O blood?

(c) type A blood?

(d) either type A^+ or O^+ blood?

2.8 Information on modes of transportation for coal leaving the Appalachian region is shown on the following chart.

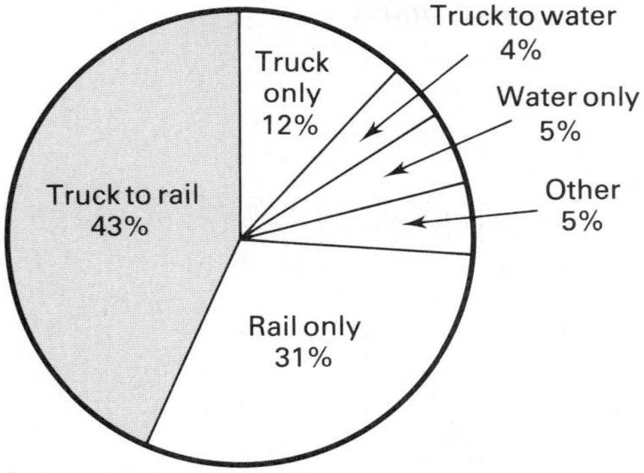

Source: G. Elmes, "Coal Transportation: An Undervalued Aspect of Energy Modeling," *Transportation Research* 18A, no. 1 (1984): 19. Used by permission.

If coal arriving at a certain power plant comes from this region, find the probability that it was transported out of the region

(a) by truck to rail.

(b) by water only.

(c) at least partially by truck.
(d) at least partially by rail.
(e) by modes not involving water. (Assume *other* does not involve water.)

2.9 Hydraulic assemblies for landing gear coming from an aircraft rework facility are inspected for defects. History shows that 8 % have defects in the shafts alone, 6 % have defects in the bushings alone, and 2 % have defects in both shafts and bushings. If a randomly chosen assembly is to be used on aircraft, find the probability that it has
(a) a bushing defect.
(b) a shaft or bushing defect.
(c) only one of the two types of defects.
(d) no defects in shafts or bushings.

2.4

COUNTING RULES USEFUL IN PROBABILITY

Let us look at the die-tossing experiment from a slightly different perspective. Because the 6 outcomes should be equally likely for a balanced die, the probability of A, an even number, is

$$P(A) = \frac{3}{6} = \frac{\text{number of outcomes favorable to } A}{\text{total number of equally likely outcomes}}.$$

This "definition" of probability will work for any experiment resulting in a finite sample space with *equally likely* outcomes. Thus, it is important to be able to count the number of possible outcomes for an experiment. The number of outcomes for an experiment easily can become quite large, and counting them is difficult unless one knows a few counting rules. Four such rules are presented as theorems in this section.

Suppose a quality control inspector examines 2 manufactured items selected from a production line. Item 1 can be defective or nondefective, as can item 2. How many outcomes are possible for this experiment? In this case, it is easy to list them. Using D_i to denote that the ith item is defective and N_i to denote that the ith item is not defective, the possible outcomes are

$$D_1D_2, \quad D_1N_2, \quad N_1D_2, \quad N_1N_2.$$

These four outcomes could be placed on a two-way table, as in Figure 2.5. This table helps us see that the four outcomes arise from the fact that the first item has 2 possible outcomes and the second item has 2 possible outcomes, and hence, the experiment of looking at both items has $2 \times 2 = 4$ outcomes. This is an example of the multiplication rule, given as Theorem 2.1.

FIGURE 2.5 Possible outcomes for inspecting two items (D_i denotes that the ith item is defective; N_i denotes that the ith item is not defective).

		2nd Item	
		D_2	N_2
1st Item	D_1	D_1D_2	D_1N_2
	N_1	N_1D_2	N_1N_2

Theorem 2.1 If the first task of an experiment can result in n_1 possible outcomes and, for each such outcome, the second task can result in n_2 possible outcomes, then there are n_1n_2 possible outcomes for the 2 tasks together.

The multiplication rule extends to more tasks in a sequence. If, for example, 3 items were inspected and each could be defective or not defective, then there would be $2 \times 2 \times 2 = 8$ possible outcomes.

Tree diagrams are also helpful in verifying the multiplication rule and for listing outcomes of experiments. Suppose a firm is deciding where to build 2 new plants, 1 in the East and 1 in the West. Four eastern cities and 2 western cities are possible locations. Thus, there are $n_1n_2 = 4(2) = 8$ possibilities for locating the 2 plants. Figure 2.6 lists these possibilities on a tree diagram.

FIGURE 2.6 Possible locations for two plants (*A, B, C, D* denote eastern cities; *E, F* denote western cities).

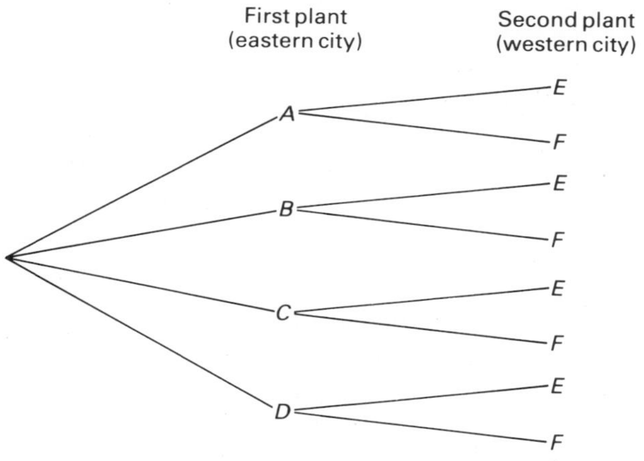

The multiplication rule (Theorem 2.1) helps only in finding the number of elements in a sample space for an experiment. We must still assign probabilities to these elements to complete our probabilistic model. This is done for the site selection problem in Example 2.4.

Example 2.4 Referring to the firm that plans to build two new plants, the eight possible locations are shown in Figure 2.6. If all eight choices are equally likely (i.e., one of the pairs of cities is selected at random), find the probability that city E is selected.

Solution

City E can get selected in four different ways, because there are four possible eastern cities to pair with it. Thus,

$$(E \text{ gets selected}) = (AE) \cup (BE) \cup (CE) \cup (DE).$$

Each of the eight outcomes has probability 1/8, because the eight events are assumed to be equally likely. Because these eight events are mutually exclusive,

$$\begin{aligned} P(E \text{ gets selected}) &= P(AE) + P(BE) + P(CE) + P(DE) \\ &= \frac{1}{8} + \frac{1}{8} + \frac{1}{8} + \frac{1}{8} = \frac{1}{2}. \end{aligned}$$

Example 2.5 Five motors (numbered 1 through 5) are available for use, and motor number 2 is defective. Motors 1 and 2 come from supplier I and motors 3, 4, and 5 come from supplier II. Suppose two motors are randomly selected for use on a particular day. Let A denote the event that the defective motor is selected and B the event that at least one motor comes from supplier I. Find $P(A)$ and $P(B)$.

Solution

We can see from the tree diagram in Figure 2.7 that there are twenty possible outcomes for this experiment, which agrees with our calculation using the multiplication rule. That is, there are twenty events of the form $\{1, 2\}$, $\{1, 3\}$, and so forth. Since the motors are randomly selected, each of the twenty outcomes has probability 1/20. Thus,

$$\begin{aligned} P(A) &= P(\{1, 2\} \cup \{2, 1\} \cup \{2, 3\} \cup \{2, 4\} \cup \{2, 5\} \cup \{3, 2\} \\ &\quad \cup \{4, 2\} \cup \{5, 2\}) \\ &= \frac{8}{20} = .4 \end{aligned}$$

FIGURE 2.7 Outcomes for experiment of Example 2.5.

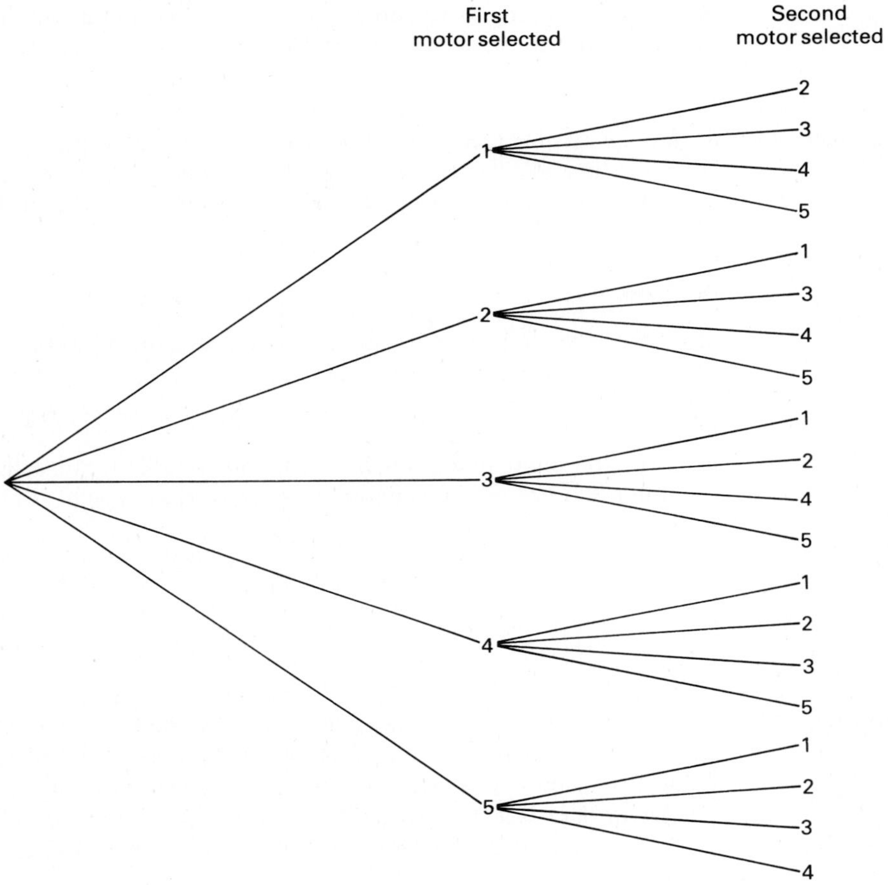

because the probability of the union is the sum of the probabilities of the events in the union.

The reader can see that B contains fourteen of the twenty outcomes and, hence, that

$$P(B) = \frac{14}{20} = .7.$$

The multiplication rule is often used to develop other counting rules. Suppose that, from three pilots, a crew of two is to be selected to form a pilot–copilot team. To count the number of ways this can be done, observe that the pilot's seat can be filled in three ways and the copilot's in two ways (after the pilot is selected), so that there are $3 \cdot 2 = 6$ ways of forming the team. This is an example of a *permutation,* for which a general result is given in Theorem 2.2.

Theorem 2.2 The number of ordered arrangements, or permutations, of r objects selected from n distinct objects ($r \leq n$) is given by

$$P_r^n = n(n-1)\cdots(n-r+1) = \frac{n!}{(n-r)!}.$$

Proof

The basic idea of a permutation can be thought of as filling r slots in a line, with one object in each slot, by drawing these objects one at a time from a pool of n distinct objects. The first slot can be filled in n ways, but the second in only $(n - 1)$ ways after the first is filled. Thus, by the multiplication rule, the first two slots can be filled in $n(n - 1)$ ways. Extending this reasoning to r slots, the number of ways of filling all r slots is

$$n(n-1)\cdots(n-r+1) = \frac{n!}{(n-r)!} = P_r^n.$$

Hence, the theorem is proven. ■

We illustrate the use of Theorem 2.2 with the following examples.

Example 2.6 From among ten employees, three are to be selected for travel to three out-of-town plants, A, B, and C, one to each plant. Since the plants are in different cities, the order of assigning the employees to the plants is an important consideration. In how many ways can the assignments be made?

Solution

Because order is important, the number of possible distinct assignments is

$$P_3^{10} = \frac{10!}{7!} = 10(9)(8) = 720.$$

In other words, there are 10 choices for plant A, but then only 9 for plant B, and 8 for plant C. This gives a total of 10(9)(8) ways of assigning employees to the plants.

Example 2.7 An assembly operation in a manufacturing plant involves four steps, which can be performed in any order. If the manufacturer wishes to compare experimentally the assembly times for each possible ordering of the steps, how many orderings will the experiment involve?

Solution

The number of orderings is the permutation of $n = 4$ things taken $r = 4$ at a time. (All steps must be accomplished each time.) This turns out to be

$$P_4^4 = \frac{4!}{0!} = 4! = 4 \cdot 3 \cdot 2 \cdot 1 = 24$$

because $0! = 1$ by definition. (In fact, $P_r^r = r!$ for any integer, r.)

Sometimes order is not important, and we are interested only in the number of subsets of a certain size that can be selected from a given set.

Theorem 2.3 The number of distinct subsets, or combinations, of size r that can be selected from n distinct objects ($r \leq n$) is given by

$$\binom{n}{r} = \frac{n!}{r!\,(n-r)!}.$$

Proof

The number of ordered subsets of size r, selected from n distinct objects, is given by P_r^n. The number of unordered subsets of size r is denoted by $\binom{n}{r}$. Since any particular set of r objects can be ordered among themselves in $P_r^r = r!$ ways, it follows that

$$\binom{n}{r} r! = P_r^n$$

or

$$\binom{n}{r} = \frac{1}{r!} P_r^n = \frac{n!}{r!\,(n-r)!}. \qquad \blacksquare$$

Example 2.8 In Example 2.6, suppose that three employees are to be selected from the ten to go to the same plant. In how many ways can the selection be made?

Solution

Here order is not important, and we merely want to know how many subsets of size $r = 3$ can be selected from $n = 10$ people. The result is

$$\binom{10}{3} = \frac{10!}{3!\,7!} = \frac{10 \cdot 9 \cdot 8}{1 \cdot 2 \cdot 3} = 120.$$

Example 2.9 Refer to Example 2.8. If two of the ten employees are women and eight are men, what is the probability that exactly one woman gets selected among the three?

Solution

We have seen that there are $\binom{10}{3} = 120$ ways to select 3 employees from the 10. Similarly, there are $\binom{2}{1} = 2$ ways to select 1 woman from the 2 available and $\binom{8}{2} = 28$ ways to select 2 men from the 8 available. If selections are made at random (that is, all subsets of 3 employees are equally likely to be chosen), then the probability of selecting exactly one woman is

$$\frac{\binom{2}{1}\binom{8}{2}}{\binom{10}{3}} = \frac{2(28)}{120} = \frac{7}{15}.$$

Example 2.10 Five applicants for a job are ranked according to ability with applicant number 1 being best, number 2 second best, and so on. These rankings are unknown to an employer, who simply hires two applicants at random. What is the probability that this employer hires exactly one of the two best applicants?

Solution

The number of possible outcomes for the process of selecting two applicants from five is

$$\binom{5}{2} = \frac{5!}{2!\,3!} = 10.$$

If one of the two best is selected, the selection can be done in

$$\binom{2}{1} = \frac{2!}{1!\,1!} \doteq 2 \text{ ways.}$$

The other selected applicant must come from among the three lowest ranking applicants, which can be done in

$$\binom{3}{1} = \frac{3!}{1!\,2!} = 3 \text{ ways.}$$

Thus, the event of interest (hiring one of the two best applicants) can come about in $2 \cdot 3 = 6$ ways. The probability of this event is then $6/10 = .6$.

Theorem 2.4 The number of ways of partitioning n distinct objects into k groups containing $n_1, n_2, \ldots, n_k$ objects, respectively, is

$$\frac{n!}{n_1!\, n_2! \cdots n_k!}$$

where

$$\sum_{i=1}^{k} n_i = n.$$

Proof

The partitioning of n objects into k groups can be done by first selecting a subset of size n_1 from the n objects, then selecting a subset of size n_2 from the $n - n_1$ objects that remain, and so on until all groups are filled. The number of ways of doing this is

$$\binom{n}{n_1}\binom{n-n_1}{n_2}\cdots\binom{n-n_1-\cdots-n_{k-1}}{n_k}$$

$$= \frac{n!}{n_1!\,(n-n_1)!} \cdot \frac{(n-n_1)!}{n_2!\,(n-n_1-n_2)!} \cdot \cdots \cdot \frac{(n-n_1\cdots-n_{k-1})!}{n_k!\,0!}$$

$$= \frac{n!}{n_1!\,n_2!\cdots n_k!}. \qquad \blacksquare$$

Example 2.11 Suppose that 10 employees are to be divided among three jobs with three employees going to job I, four to job II, and three to job III. In how many ways can the job assignment be made?

Solution

This problem involves partitioning the $n = 10$ employees into groups of size $n_1 = 3$, $n_2 = 4$, and $n_3 = 3$; and this can be accomplished in

$$\frac{n!}{n_1!\,n_2!\,n_3!} = \frac{10!}{3!\,4!\,3!} = \frac{10 \cdot 9 \cdot 8 \cdot 7 \cdot 6 \cdot 5}{3 \cdot 2 \cdot 1 \cdot 3 \cdot 2 \cdot 1} = 4200 \text{ ways.}$$

(Notice the large number of ways this task can be accomplished!)

Example 2.12 In the setting of Example 2.11, suppose three employees of a certain ethnic group all get assigned to job I. What is the probability of this happening under a random assignment of employees to jobs?

Solution

We have seen in Example 2.11 that there are 4200 ways of assigning the ten workers to the three jobs. The event of interest assigns three specified employees to job I. It remains to determine how many ways the other seven employees can be assigned to jobs II and III, which is

$$\frac{7!}{4!\,3!} = \frac{7(6)(5)}{3(2)(1)} = 35.$$

Thus, the chance of assigning three specific workers all to job I is

$$\frac{35}{4200} = \frac{1}{120}$$

which is very small, indeed!

EXERCISES

2.10 An experiment consists of observing two vehicles in succession move through the intersection of two streets.

(a) List the possible outcomes, assuming each vehicle can go straight, turn right, or turn left.

(b) Assuming the outcomes to be equally likely, find the probability that at least one vehicle turns left. (Would this assumption always be reasonable?)

(c) Assuming the outcomes to be equally likely, find the probability that at most one vehicle makes a turn.

2.11 A commercial building is designed with two entrances, say I and II. Two customers arrive and enter the building.

(a) List the elements of a sample space for this observational experiment.

(b) If all elements in (a) are equally likely, find the probability that both customers use door I; that both customers use the same door.

2.12 A corporation has two construction contracts that are to be assigned to one or more of three firms bidding for these contracts. (One firm could receive both contracts.)

(a) List the possible outcomes for the assignment of contracts to the firms.

(b) If all outcomes are equally likely, find the probability that both contracts go to the same firm.

(c) Under the assumptions of (b), find the probability that one specific firm, say firm I, gets at least one contract.

2.13 Among five portable generators produced by an assembly line in one day, two are defective. If two generators are selected for sale, find the probability that both will be nondefective. (Assume the two selected for sale are chosen so that every possible sample of size two has the same probability of being selected.)

2.14 Seven applicants have applied for two jobs. How many ways can the jobs be filled if

(a) the first person chosen receives a higher salary than the second?

(b) there are no differences between the jobs?

2.15 A package of six light bulbs contain two defective bulbs. If three bulbs are selected for use, find the probability that none is defective.

2.16 How many four-digit serial numbers can be formed if no digit is to be repeated within any one number? (The first digit may be a zero.)

2.17 A fleet of eight taxis is to be divided among three airports, A, B, and C, with two going to A, five to B, and one to C.

(a) In how many ways can this be done?

(b) What is the probability that the cab driven by Jones ends up at airport C?

2.18 Show that $\binom{n}{r} = \binom{n-1}{r-1} + \binom{n-1}{r}$, $1 \le r \le n$.

2.19 Five employees of a firm are ranked from 1 to 5 in their abilities to program a computer. Three of these employees are selected to fill equivalent programming jobs. If all possible choices of three (out of the five) are equally likely, find the probability that

(a) the employee ranked number 1 is selected.

(b) the highest-ranked employee among those selected has rank 2 or lower.

(c) the employees ranked 4 and 5 are selected.

2.20 For a certain style of new automobile the colors blue, white, black, and green are in equal demand. Three successive orders are placed for automobiles of this style. Find the probability that

(a) one blue, one white, and one green are ordered.

(b) two blues are ordered.

(c) at least one black is ordered.

(d) exactly two of the orders are for the same color.

2.21 A firm is placing three orders for supplies among five different distributors. Each order is randomly assigned to one of the distributors; and a distributor can receive multiple orders. Find the probability that

(a) all orders go to different distributors.

(b) all orders to to the same distributor.

(c) exactly two of the three orders go to one particular distributor.

2.22 An assembly operation for a computer circuit board consists of four operations that can be performed in any order.

(a) In how many ways can the assembly operation be performed?

(b) One of the operations involves soldering wire to a microchip. If all possible assembly orderings are equally likely, what is the probability that the soldering comes first or second?

2.23 Nine impact wrenches are to be divided evenly among three assembly lines.

(a) In how many ways can this be done?

(b) Two of the wrenches are used and seven are new. What is the probability that a particular line (line A) gets both used wrenches?

2.5

CONDITIONAL PROBABILITY AND INDEPENDENCE

Sometimes a sample space for an experiment can be narrowed by extra information available on events in question. For instance, a sample space appropriate for measuring weights of all *people* could be narrowed considerably if it is known that only *adults* are to be investigated in the study. This extra information is referred to as *conditional information.* Let's look at a specific example in more detail.

Of 100 students completing an introductory statistics course, 20 were business majors. Ten students received As in the course, and three of these were business majors. These facts are easily displayed on a Venn diagram, such as the one that follows, where A represents those students receiving As and B represents business majors.

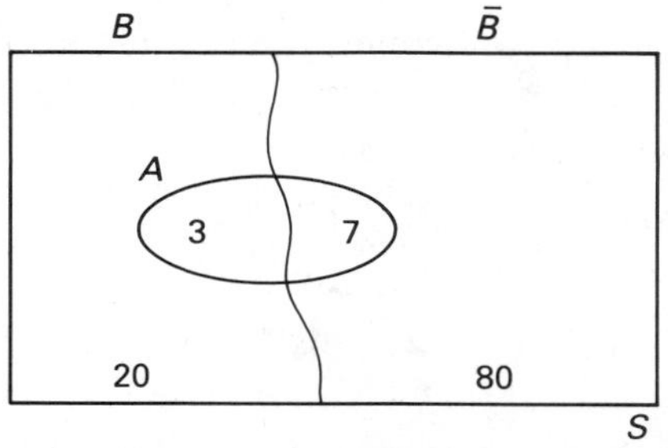

Now, for a randomly selected student from this class, $P(A) = .1$ and $P(B) = .2$. But, suppose we know that the randomly selected student is a business major. Then, we may want to know the probability that the student received an A, *given* that he or she is a business major. Among the twenty business majors, three received As. Thus, $P(A \text{ given } B)$, written $P(A \mid B)$, is 3/20.

From the Venn diagram we see that the conditional (or given) information reduces the effective sample space to just the twenty business majors. Among them, three received As. Note that

$$P(A \mid B) = \frac{3}{20} = \frac{P(AB)}{P(B)} = \frac{3/100}{20/100}.$$

This fact motivates Definition 2.5.

Definition 2.5

If A and B are any two events, then the **conditional probability** of A given B, denoted by $P(A \mid B)$, is

$$P(A \mid B) = \frac{P(AB)}{P(B)}$$

provided $P(B) \neq 0$.

Example 2.13 From five motors, of which one is defective, two motors are to be selected at random for use on a particular day. Find the probability that the second motor selected is not defective, given that the first was not defective.

Solution

Let N_i denote that the ith motor selected is not defective. We want $P(N_2 \mid N_1)$. From Definition 2.5,

$$P(N_2 \mid N_1) = \frac{P(N_1 N_2)}{P(N_1)}.$$

Looking at the twenty possible outcomes given in Figure 2.7, we can see that event N_1 contains sixteen of these outcomes, and $N_1 N_2$ contains twelve. Thus, since the twenty outcomes are equally likely,

$$P(N_2 \mid N_1) = \frac{P(N_1 N_2)}{P(N_1)} = \frac{12/20}{16/20} = \frac{12}{16} = \frac{3}{4}.$$

Does this answer seem intuitively reasonable?

Conditional probabilities satisfy the three axioms of probability (Definition 2.4), as can be easily shown. First, because $AB \subset B$, then $P(AB) \leq P(B)$. Also, $P(AB) \geq 0$ and $P(B) \geq 0$, so that

$$0 \leq P(A \mid B) = \frac{P(AB)}{P(B)} \leq 1.$$

Second,

$$P(S \mid B) = \frac{P(SB)}{P(B)} = \frac{P(B)}{P(B)} = 1.$$

Third, if $A_1, A_2, \ldots,$ are mutually exclusive events, then so are

$A_1B, A_2B, \ldots,$ and

$$P\left(\bigcup_{i=1}^{\infty} A_i \mid B\right) = \frac{P\left[\left(\bigcup_{i=1}^{\infty} A_i\right)B\right]}{P(B)}$$

$$= \frac{P\left[\bigcup_{i=1}^{\infty} (A_iB)\right]}{P(B)} = \frac{\sum_{i=1}^{\infty} P(A_iB)}{P(B)}$$

$$= \sum_{i=1}^{\infty} \frac{P(A_iB)}{P(B)} = \sum_{i=1}^{\infty} P(A_i \mid B).$$

Conditional probability plays a key role in many practical applications of probability. In these applications, important conditional probabilities are often affected drastically by seemingly small changes in the basic information from which the probabilities are derived. The following discussion of a medical application of probability illustrates the point.

A screening test indicates the presence or absence of a disease, and such tests are often used by physicians to detect diseases. Virtually all screening tests, however, have errors associated with their use. On thinking about the possible errors for a moment, it is clear that two different kinds of errors are possible; the test could show a person to have the disease when he or she does not (false positive) or fail to show that a person has the disease when it is present (false negative). Measures of these two types of errors are conditional probabilities called *sensitivity* and *specificity*.

The following diagram will help in defining and interpreting these measures. The + indicates presence of the disease under study and the − indicates absence of the disease. The true diagnosis may never be known, but often can be determined by more intensive follow-up tests.

		True Diagnosis +	True Diagnosis −	
Test Result	+	a	b	$a + b$
	−	c	d	
		$a + c$	$b + d$	$a + b + c + d = n$

In this scenario, n people are tested and $a + b$ are shown by the test to have the disease. Of these, a actually have the disease and b do not have the disease (false positives). Of the $c + d$ who test negative, c actually have the disease (false negatives). Using these labels,

$$\text{sensitivity} = \frac{a}{a + c},$$

the conditional probability of a positive test given that the person has the disease, and

$$\text{specificity} = \frac{d}{b + d},$$

the conditional probability of a negative test given that the person does not have the disease.

Obviously, a good test should have sensitivity and specificity both close to one. If sensitivity is close to one, then c (the number of false negatives) must be small. If specificity is close to one, then b (the number of false positives) must be small. Even when sensitivity and specificity are both close to one, a screening test can produce misleading results if not carefully applied. To see this, we look at one other important measure, the *predictive value* of a test given by

$$\text{predictive value} = \frac{a}{a + b}.$$

The predictive value is the conditional probability of the person actually having the disease given a positive test result. Clearly, a good test should have a high predictive value, but this is not always possible even for highly sensitive and specific tests. The reason that all three measures cannot always be close to one simultaneously lies in the fact that predictive value is affected by the prevalence rate of the disease (that is, the proportion of the population under study that actually has the disease). We illustrate with three numerical situations given as I, II, and III on the diagram.

True Diagnosis

			+	−	
I	Test Result	+	90	10	100
		−	10	90	
			100	100	200

			+	−	
II	Test Result	+	90	100	190
		−	10	900	
			100	1000	1100

			+	−	
III	Test Result	+	90	1000	1090
		−	10	9000	
			100	10,000	10,100

Among the 200 people under study in I, 100 have the disease (a prevalence rate of 50 %). The test has sensitivity and specificity each equal to .90, and the predictive value is $90/100 = .90$. This is a good situation; the test is doing well.

In II, the prevalence rate changes to 100/1100, or 9 %. Even though the sensitivity and specificity are still .90, the predictive value has dropped to $90/190 = .47$. In III, the prevalence rate is 100/10,000 or about 1 % and the predictive value has dropped farther to .08. Thus, only 8 % of those tested positive actually have the disease, even though the test has high sensitivity and specificity. What does this imply about using screening tests on large populations in which the prevalence rate for the disease being studied is low? An assessment of the answer to this question involves a careful look at conditional probabilities.

If the extra information in knowing that an event B has occurred does not change the probability of A—that is, if $P(A \mid B) = P(A)$—then events A and B are said to be independent. Because

$$P(A \mid B) = \frac{P(AB)}{P(B)},$$

the condition $P(A \mid B) = P(A)$ is equivalent to

$$\frac{P(AB)}{P(B)} = P(A)$$

or

$$P(AB) = P(A)P(B).$$

Definition 2.6

Two events A and B are said to be **independent** if

$$P(A \mid B) = P(A)$$

or

$$P(B \mid A) = P(B).$$

This is equivalent to stating that

$$P(AB) = P(A)P(B).$$

Example 2.14 Suppose that a foreman must select one worker for a special job, from a pool of four available workers, numbered 1, 2, 3, and 4. He selects the worker by mixing the four names and randomly selecting one. Let A denote the event that worker 1 or 2 is selected, B the event that worker 1 or 3 is selected, and C the event that worker 1 is selected. Are A and B independent? Are A and C independent?

Solution

Because the name is selected at random, a reasonable assumption for the probabilistic model is to assign a probability of 1/4 to each individual worker. Then $P(A) = 1/2$, $P(B) = 1/2$, and $P(C) = 1/4$. Because the intersection AB contains only worker 1, $P(AB) = 1/4$. Now $P(AB) = 1/4 = P(A)P(B)$, so A and B are independent. Since AC also contains only worker 1, $P(AC) = 1/4$. But $P(AC) = 1/4 \neq P(A)P(C)$, so A and C *are not* independent. A and C are said to be *dependent* because the occurrence of C changes the probability that A occurs.

2.6

RULES OF PROBABILITY

We now establish four rules that will help in calculating probabilities of events. First recall that the complement $\bar{A}$ of an event A is the set of all outcomes in a sample space S that are not in A. Thus $\bar{A}$ and A are mutually exclusive and their union is S. That is,

$$\bar{A} \cup A = S.$$

It follows that

$$P(\bar{A} \cup A) = P(\bar{A}) + P(A) = P(S) = 1$$

or

$$P(\bar{A}) = 1 - P(A).$$

Thus, we have established the following theorem.

Theorem 2.5 If $\bar{A}$ is the complement of an event A is a sample space S, then $P(\bar{A}) = 1 - P(A)$.

Example 2.15 A quality control inspector has ten assembly lines from which to choose products for testing. Each morning of a five-day week, she randomly selects one of the lines to work on for the day. Find the probability that a line is chosen more than once during the week.

Solution

It is easier, here, to think in terms of complements, and first find the probability that no line is chosen more than once. If no line is repeated, five different lines must be chosen on successive days, which can be done in

$$P_5^{10} = \frac{10!}{5!} = 10(9)(8)(7)(6) \text{ ways.}$$

The total number of possible outcomes for the selection of five lines without restriction is $(10)^5$, by an extension of the multiplication rule. Thus,

$$P(\text{no line is chosen more than once}) = \frac{P_5^{10}}{(10)^5} = \frac{10(9)(8)(7)(6)}{(10)^5} = .30$$

and

$$P(\text{a line is chosen more than once}) = 1.0 - .30 = .70.$$

Axiom 3 applies to $P(A \cup B)$ if A and B are disjoint. But, what happens when A and B are not disjoint? Theorem 2.6 gives the answer.

Theorem 2.6 If A and B are any two events, then

$$P(A \cup B) = P(A) + P(B) - P(AB).$$

If A and B are mutually exclusive, then

$$P(A \cup B) = P(A) + P(B).$$

Proof

From a Venn diagram for the union of A and B (see Figure 2.1) it is easy to see that

$$A \cup B = A\bar{B} \cup \bar{A}B \cup AB$$

and that the three events on the right-hand side of the equality are mutually exclusive. Hence,

$$P(A \cup B) = P(A\bar{B}) + P(\bar{A}B) + P(AB).$$

Now,

$$A = A\bar{B} \cup AB$$

and

$$B = \bar{A}B \cup AB$$

so that

$$P(A) = P(A\bar{B}) + P(AB)$$

and

$$P(B) = P(\bar{A}B) + P(AB).$$

It follows that

$$P(A\bar{B}) = P(A) - P(AB)$$

and

$$P(\bar{A}B) = P(B) - P(AB).$$

Substituting into the first equation for $P(A \cup B)$, we have

$$\begin{aligned} P(A \cup B) &= P(A) - P(AB) + P(B) - P(AB) + P(AB) \\ &= P(A) + P(B) - P(AB), \end{aligned}$$

and the proof is complete. ■

The formula for the probability of the union of k events, $A_1, A_2, \ldots, A_k$, is derived in similar fashion, and is given by

$$\begin{aligned} P(A_1 \cup A_2 \cup \cdots \cup A_k) = \sum_{i=1}^{k} P(A_i) - \sum_{i<j}\sum (A_iA_j) \\ + \sum_{i<j<l}\sum\sum (A_iA_jA_l) - \cdots + \cdots \\ - (-1)^k P(A_1A_2 \cdots A_k). \end{aligned}$$

The next rule is actually just a rearrangement of the definition of conditional probability, for the case in which a conditional probability may be known and we want to find the probability of an intersection.

Theorem 2.7 If A and B are any two events, then

$$\begin{aligned} P(AB) &= P(A)P(B \mid A) \\ &= P(B)P(A \mid B). \end{aligned}$$

If A and B are independent, then

$$P(AB) = P(A)P(B).$$

We illustrate the use of these three theorems in the following examples.

Example 2.16 Records indicate that for the parts coming out of a hydraulic repair shop at an airplane rework facility, 20% will have a shaft defect, 10% will have a bushing defect, and 75% will be defect free. For an item chosen at random from this output, find the probability of

A: the item has at least one type of defect.

B: the item has only a shaft defect.

Solution

The percentages given imply that 5% of the items have both a shaft and a bushing defect. Let D_1 denote the event that an item has a shaft defect, and D_2 the event that it has a bushing defect. Then,

$$A = D_1 \cup D_2$$

and

$$\begin{aligned} P(A) &= P(D_1 \cup D_2) \\ &= P(D_1) + P(D_2) - P(D_1 D_2) \\ &= .20 + .10 - .05 \\ &= .25. \end{aligned}$$

Another possible solution is to observe that the complement of A is the event that an item has no defects. Thus,

$$\begin{aligned} P(A) &= 1 - P(\bar{A}) \\ &= 1 - .75 = .25. \end{aligned}$$

To find $P(B)$, note that the event D_1 that the item has a shaft defect is the union of the events that it has *only* a shaft defect (B) and the event that it has both defects $(D_1 D_2)$. That is,

$$D_1 = B \cup D_1 D_2$$

where B and $D_1 D_2$ are mutually exclusive. Therefore,

$$P(D_1) = P(B) + P(D_1 D_2)$$

or

$$\begin{aligned} P(B) &= P(D_1) - P(D_1 D_2) \\ &= .20 - .05 = .15. \end{aligned}$$

You should sketch these events on a Venn diagram and verify the results just derived.

Example 2.17 A section of an electrical circuit has two relays in parallel, as shown in Figure 2.8. The relays operate independently, and when a switch is thrown, each will close properly with probability only .8. If both relays are open, find the probability that current will flow from s to t when the switch is thrown.

FIGURE 2.8 Two relays in parallel.

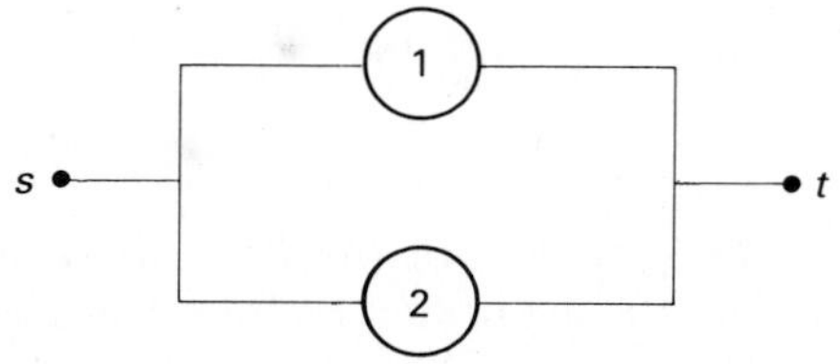

Solution

Let O denote an open relay and C a closed relay. The four outcomes for this experiment are shown by

$$\begin{array}{lll} & \text{Relay} & \text{Relay} \\ & 1 & 2 \\ E_1 = \{(& O, & O)\} \\ E_2 = \{(& O, & C)\} \\ E_3 = \{(& C, & O)\} \\ E_4 = \{(& C, & C)\}. \end{array}$$

Since the relays operate independently, we can find the probabilities for these outcomes as follows:

$$P(E_1) = P(O)P(O) = (.2)(.2) = .04$$
$$P(E_2) = P(O)P(C) = (.2)(.8) = .16$$
$$P(E_3) = P(C)P(O) = (.8)(.2) = .16$$
$$P(E_4) = P(C)P(C) = (.8)(.8) = .64$$

if A denotes the event that current will flow from s to t, then

$$A = E_2 \cup E_3 \cup E_4$$

or

$$\bar{A} = E_1$$

(At least one of the relays must close for current to flow.) Thus,

$$\begin{aligned} P(A) &= 1 - P(\bar{A}) \\ &= 1 - P(E_1) \\ &= 1 - .04 = .96 \end{aligned}$$

which is the same as $P(E_2) + P(E_3) + P(E_4)$.

Example 2.18 A monkey is to be taught to recognize colors by tossing one red, one black, and one white ball into boxes of the respective colors, one ball to a box. If the monkey has not learned the colors, and merely tosses one ball in each box at random, find the probability of

(a) No color matches.
(b) Exactly one color match.

Solution

This problem can be solved by listing outcomes because only three balls are involved, but a more general method of solution will be illustrated. Define the following events:

A_1: color match in the red box

A_2: color match in the black box

A_3: color match in the white box.

There are $3! = 6$ equally likely ways of randomly tossing the balls into the boxes with one ball in each box. Also, there are only $2! = 2$ ways of tossing the balls into the boxes if one particular box is required to have a color match. Hence,

$$P(A_1) = P(A_2) = P(A_3) = \frac{2}{6} = \frac{1}{3}.$$

Similarly, it follows that

$$\begin{aligned} P(A_1A_2) &= P(A_1A_3) = P(A_2A_3) \\ &= P(A_1A_2A_3) = \frac{1}{6}. \end{aligned}$$

(a) Note that

$$\begin{aligned} P(\text{no color matches}) &= 1 - P(\text{at least one color match}) \\ &= 1 - P(A_1 \cup A_2 \cup A_3) \\ &= 1 - [P(A_1) + P(A_2) + P(A_3) - P(A_1A_2) \\ &\qquad - P(A_1A_3) - P(A_2A_3) + P(A_1A_2A_3)] \\ &= 1 - \left[3\left(\frac{1}{3}\right) - 3\left(\frac{1}{6}\right) + \left(\frac{1}{6}\right)\right] \\ &= \frac{1}{3}. \end{aligned}$$

(b) We leave it to the reader to show that

$$\begin{aligned} P(\text{exactly one match}) &= P(A_1) + P(A_2) + P(A_3) \\ &\quad - 2[P(A_1A_2) + P(A_1A_3) + P(A_2A_3)] \\ &\quad + 3P(A_1A_2A_3) \\ &= 3\left(\frac{1}{3}\right) - 2(3)\left(\frac{1}{6}\right) + 3\left(\frac{1}{6}\right) = \frac{1}{2}. \end{aligned}$$

The fourth rule we present in this section is based on the notion of a partition of a sample space. Events $B_1, B_2, \ldots, B_k$ are said to partition a sample space S if

1. $B_iB_j = \emptyset$ for any pair i and j ($\emptyset$ denotes the null set)
2. $B_1 \cup B_2 \cdots \cup B_k = S$.

For example, the set of tires in an auto assembly warehouse may be partitioned according to suppliers or employees of a firm may be partitioned according to level of education. We illustrate a partition for the case $k = 2$ in Figure 2.9.

The key idea involving a partition is to observe that an event A (see Figure 2.9) can be written as the union of mutually exclusive events AB_1 and AB_2. That is,

$$A = AB_1 \cup AB_2$$

and thus,

$$P(A) = P(AB_1) + P(AB_2).$$

If conditional probabilities $P(A \mid B_1)$ and $P(A \mid B_2)$ are known, then $P(A)$ can be found by writing

$$P(A) = P(B_1)P(A \mid B_1) + P(B_2)P(A \mid B_2).$$

In problems dealing with partitions, it is frequently of interest to find

FIGURE 2.9 A partition of S into B_1 and B_2.

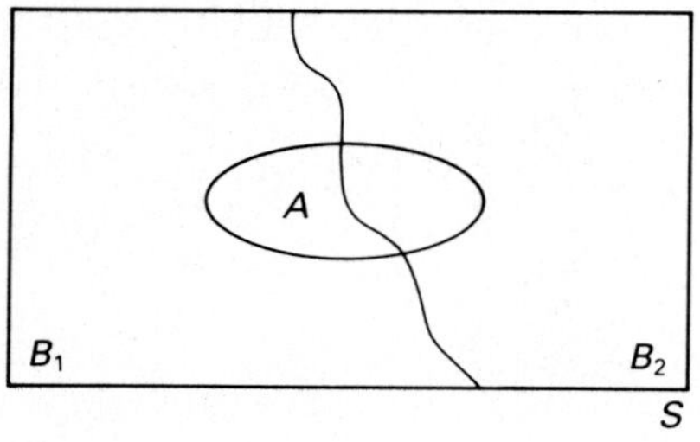

probabilities of the form $P(B_1 \mid A)$, which can be written

$$P(B_1 \mid A) = \frac{P(B_1A)}{P(A)}$$

$$= \frac{P(B_1)P(A \mid B_1)}{P(B_1)P(A \mid B_1) + P(B_2)P(A \mid B_2)}.$$

This result is a special case of Bayes's Rule.

Theorem 2.8 If $B_1, B_2, \ldots, B_k$ forms a partition of S, and A is any event in S, then

$$P(B_j \mid A) = \frac{P(B_j)P(A \mid B_j)}{\sum_{i=1}^{k} P(B_i)P(A \mid B_i)}.$$

The proof is an extension of the result shown earlier for $k = 2$.

Example 2.19 A company buys tires from two suppliers, 1 and 2. Supplier 1 has a record of delivering tires containing 10 % defectives, whereas supplier 2 has a defective rate of only 5 %. Suppose 40 % of the current supply came from supplier 1. If a tire is taken from this supply and observed to be defective, find the probability that it came from supplier 1.

Solution

Let B_i denote the event that a tire comes from supplier i, $i = 1, 2$, and note that B_1 and B_2 form a partition of the sample space for the experiment of selecting one tire. Let A denote the event that the selected tire is defective. Then,

$$P(B_1 \mid A) = \frac{P(B_1)P(A \mid B_1)}{P(B_1)P(A \mid B_1) + P(B_2)P(A \mid B_2)}$$

$$= \frac{.40(.10)}{.40(.10) + (.60)(.05)}$$

$$= \frac{.04}{.04 + .03} = \frac{4}{7}.$$

Supplier 1 has a greater probability of being the party supplying the defective tire than does supplier 2.

In practice, approximations for probabilities of events often come from real data. The data may not provide exact probabilities, but the approximations frequently are good enough to provide a basis for clearer insights into the problem at hand. We illustrate the use of real data in probability calculations with Examples 2.20 and 2.21.

Example 2.20 The data in Table 2.1 is a summary of employment in the United States in 1982.

TABLE 2.1 **Employment in the United States, 1982 (figures in millions)**

Civilian noninstitutional population	170
Civilian labor force	110
Employed	99
Unemployed	11
Not in the labor force	62

Source: U.S. Department of Labour, Bureau of Labor Statistics.

If an arbitrarily selected U.S. resident were asked, in 1982, to fill out a questionnaire on employment, find the probability that the resident would have been

(a) in the labor force.
(b) employed.
(c) employed and in the labor force.
(d) employed given that he or she was known to be in the labor force.
(e) either not in the labor force or unemployed.

Solution

Let L denote the event that the resident is in the labor force and E that he or she is employed.

(a) Because 172 million people make up the population under study and 110 million are in the labor force,

$$P(L) = \frac{110}{172}.$$

(b) Similarly, 99 million are employed; thus,

$$P(E) = \frac{99}{172}.$$

(c) The employed persons are a subset of those in the labor force; in other

$$P(B_j \mid A) = \frac{P(B_j)(P(A \mid B_j))}{\sum P(B_i)P(A \mid B_i)}$$

words, $EL = E$. Hence,

$$P(EL) = P(E) = \frac{99}{172}.$$

(d) Among the 110 million people known to be in the labor force, 99 million are employed. Therefore,

$$P(E \mid L) = \frac{99}{110}.$$

Note that this can also be found by using Definition 2.5, as follows:

$$P(E \mid L) = \frac{P(EL)}{P(L)} = \frac{99/172}{110/172} = \frac{99}{110}.$$

(e) The event that the resident is not in the labor force $\bar{L}$ is mutually exclusive from the event that he or she is unemployed. Therefore,

$$\begin{aligned} P(\bar{L} \cup \bar{E}) &= P(\bar{L}) + P(\bar{E}) \\ &= \frac{62}{172} + \frac{11}{172} = \frac{73}{172}. \end{aligned}$$

The next example uses a data set that is a little more complicated in structure.

Example 2.21 Table 2.2 presents information on fires reported in 1978. All figures in the table are percents, and the main body of the table shows percentages according to cause. For example, the 22 in the upper left corner shows that 22 % of fires in the family homes were caused by heating.

TABLE 2.2 **Percent of fires in residences by cause, 1978**

Cause of fire	Family homes	Apartments	Mobile homes	Hotels/ motels	Other	All locations
Heating	22	6	22	8	45	19
Cooking	15	24	13	7	0	16
Incendiary substance	10	15	7	16	8	11
Smoking	7	18	6	36	19	10
Electrical	8	5	15	7	28	8
Other	38	32	37	26	0	36
All causes	73	20	3	2	2	

Source: National Fire Incident Reporting Service.

If a residential fire is called into a fire station, find the probability that it is

(a) a fire caused by heating.
(b) a fire in a family home.
(c) a fire caused by heating given that this was in an apartment.
(d) an apartment fire caused by heating
(e) a fire in a family home given that it was caused by heating.

Solution

(a) Over all locations 19 % of fires are caused by heating (see the right-hand column of Table 2.2). Thus,

$$P(\text{heating fire}) = .19.$$

(b) Over all causes 73 % of the fires are in family homes (see the bottom row of Table 2.2). Thus,

$$P(\text{family home fire}) = .73.$$

(c) The third column from the left deals only with apartment fires. Thus, 6 % of all apartment fires are caused by heating. Note that this is a conditional probability and can be written

$$P(\text{heating fire} \mid \text{apartment fire}) = .06.$$

(d) In part (c), we know the fire was in an apartment, and the probability in question was conditional on that information. Here we must find the probability of an intersection between apartment fire (say, event A) and heating fire (say, event H). We know $P(H \mid A)$ directly from the table, so

$$\begin{aligned} P(AH) &= P(A)P(H \mid A) \\ &= (.20)(.06) = .012. \end{aligned}$$

In other words, only 1.2 % of all reported fires are caused by heating in apartments.

(e) The figures in the body of Table 2.2 give probabilities of causes, given the location of the fire. We are now asked to find the probability of a location given the cause. This is exactly the situation to which Bayes's Rule (Theorem 2.8) applies. The locations form a partition of the set of all fires into five different groups. We then have, with obvious shortcuts in wording,

$$P(\text{family} \mid \text{heating}) = \frac{P(\text{family})P(\text{heating} \mid \text{family})}{P(\text{heating})}.$$

Now $P(\text{heating}) = .19$ from Table 2.2. However, it might be informative to show how this figure is derived from the other columns of the table. We have

$$\begin{aligned} P(\text{heating}) &= P(\text{family})P(\text{heating} \mid \text{family}) \\ &\quad + P(\text{apartment})P(\text{heating} \mid \text{apartment}) \\ &\quad + P(\text{mobile})P(\text{heating} \mid \text{mobile}) \\ &\quad + P(\text{hotel})P(\text{heating} \mid \text{hotel}) \\ &\quad + P(\text{other})P(\text{heating} \mid \text{other}) \\ &= (.73)(.22) + (.20)(.06) + (.03)(.22) + (.02)(.08) \\ &\quad + (.02)(.45) \\ &= .19. \end{aligned}$$

Then,

$$P(\text{family} \mid \text{heating}) = \frac{(.73)(.22)}{.19} = .85.$$

In other words, among heating fires 85 % are found in family homes.

EXERCISES

2.24 Vehicles coming into an intersection can turn left or right or go straight ahead. Two vehicles enter an intersection in succession. Find the probability that at least one of the two turns left given that at least one of the two vehicles turns. What assumptions have you made?

2.25 A purchasing office is to assign a contract for computer paper and a contract for microcomputer disks to any one of three firms bidding for these contracts. (Any one firm could receive both contracts.) Find the probability that

(a) firm I receives a contract given that both contracts do not go to the same firm.

(b) firm I receives both contracts.

(c) firm I receives the contract for paper given that it does not receive the contract for disks.

What assumptions have you made?

2.26 The data at the top of page 46 give the number of accidental deaths overall, and for three specific causes, for the United States in 1984. You are told that a certain person recently died in an accident. Approximate the probability that

(a) it was a motor vehicle accident.

(b) it was a motor vehicle accident if you know the person to be male.

(c) it was a motor vehicle accident if you know the person to be between 15 and 24 years of age.

(d) it was a fall if you know the person to be over age 75.

(e) the person was male.

	All types	Motor vehicle	Falls	Drowning
All ages	92,911	46,263	11,937	5,388
Under 5	3,652	1,132	114	638
5–14	4,198	2,263	68	532
15–24	19,801	14,738	399	1,353
25–44	25,498	15,036	963	1,549
45–64	15,273	6,954	1,624	763
65–74	8,424	3,020	1,702	281
75 and over	16,065	3,114	7,067	272
Male	64,053	32,949	6,210	4,420
Female	28,858	13,314	5,727	968

Source: The World Almanac and Book of Facts, 1989 edition, copyright © Newspaper Enterprise Association, Inc. 1988, New York, NY 10166. Used by permission.

2.27 The data below show the distribution of arrival times at work by mode of travel for workers in the central business district of a large city. The figures are percentages. (The columns should add to 100, but some do not because of rounding.)

Arrival time	Transit	Drove alone	Shared ride with family member	Car pool	All
Before 7:15	17	16	16	19	18
7:15–7:45	35	30	30	42	34
7:45–8:15	32	31	43	35	33
8:15–8:45	10	14	8	2	10
After 8:45	5	11	3	2	6

Source: C. Hendrickson, and E. Plank, "The Flexibility of Departure Times for Work Trips," *Transportation Research* 18A no. 1 (1984): 25–36. Used by permission.

If a randomly selected worker is asked about his or her travel to work, find the probability that the worker

(a) arrives before 7:15 given that the worker drives alone.
(b) arrives at or after 7:15 give that the worker drives alone.
(c) arrives before 8:15 given that the worker rides in a car pool.

Can you find the probability that the worker drives alone, using only these data?
No - they give you percentages instead of numbers

2.28 Table 2.2 shows percentages of fires by cause for certain residential locations. If a fire is reported to be residential, find the probability that it is

(a) caused by smoking.
(b) in a mobile home.
(c) caused by smoking given that it was in a mobile home.
(d) in a mobile home given that it was caused by smoking.

2.29 An incoming lot of silicon wafers is to be inspected for defectives by an engineer in a microchip manufacturing plant. In a tray containing twenty wafers, assume four are defective. Two wafers are to be randomly selected for inspection. Find the probability that

(a) neither is defective.
(b) at least one of the two is defective.
(c) neither is defective given that at least one is not defective.

2.30 In the setting of Exercise 2.29, answer the same three questions if only two among the twenty wafers are assumed to be defective.

2.31 Show that, for any event A and B,

(a) $P(AB) \geq P(A) + P(B) - 1$.
(b) the probability that exactly one of the events occurs is $P(A) + P(B) - 2P(AB)$.

2.32 Resistors are produced by a certain firm and marketed as 10-ohm resistors. However, the actual ohms of resistance produced by the resistors may vary. It is observed that 5% of the values are below 9.5 ohms and 10% are above 10.5 ohms. If two of these resistors, randomly selected, are used in a system, find the probability that

(a) both have actual values between 9.5 and 10.5 ohms.
(b) at least one has an actual value in excess of 10.5 ohms.

2.33 Consider the following segment of an electric circuit with three relays. Current will flow from a to b if there is at least one closed path when the relays are switched to "closed." However, the relays may malfunction. Suppose they close properly only with probability .9 when the switch is thrown, and suppose they operate independently of one another. Let A denote the event that current will flow from a to b when the relays are switched to "closed."

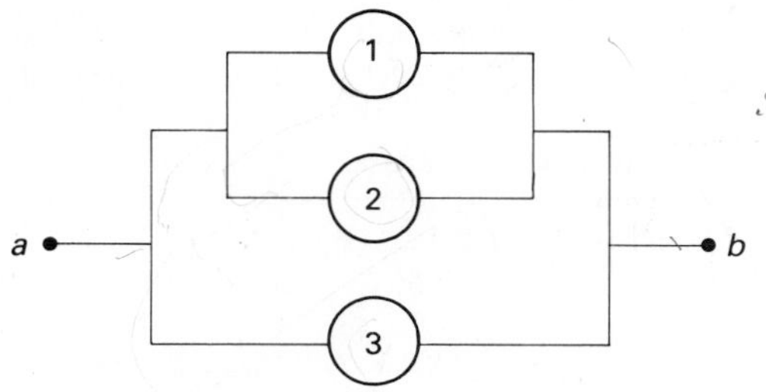

(a) Find $P(A)$.
(b) Find the probability that relay 1 is closed properly, given that current is known to be flowing from a to b.

2.34 With relays operating as in Exercise 2.33, compare the probability of current flowing from a to b in the series system that follows:

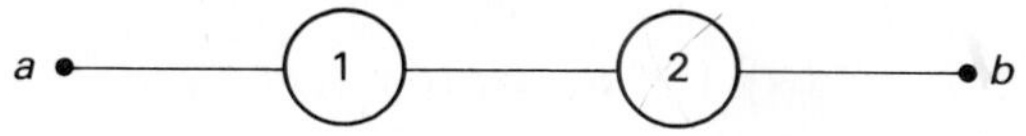

with the probability of flow in the parallel system as follows:

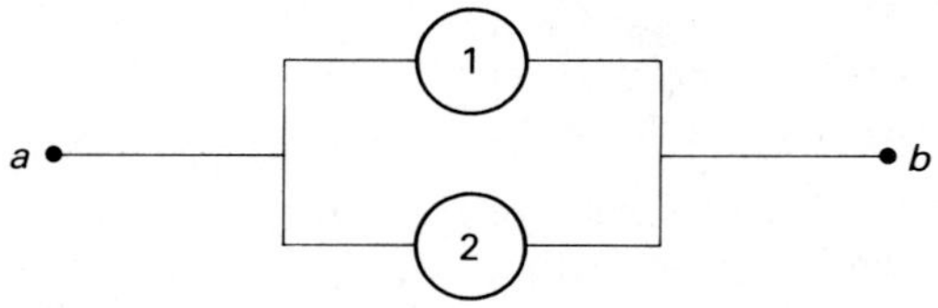

2.35 Electric motors coming off two assembly lines are pooled for storage in a common stockroom, and the room contains an equal number of motors from each line. Motors from that room are periodically sampled and tested. It is known that 10 % of the motors from Line I are defective and 15 % of the motors from Line II are defective. If a motor is randomly selected from the stockroom and found to be defective, find the probability that it came from Line I.

2.36 Two methods, A and B, are available for teaching a certain industrial skill. The failure rate is 20 % for A and 10 % for B. However, B is more expensive and, hence, used only 30 % of the time (A is used the other 70 %). A worker is taught the skill by one of the methods, but fails to learn it correctly. What is the probability that he was taught by method A?

2.37 A diagnostic test for a certain disease is said to be 90 % accurate in that, if a person has the disease, the test will detect it with probability .9. Also, if a person does not have the disease, the test will report that he or she doesn't have it with probability .9. Only 1 % of the population has the disease in question. If the diagnostic test reports that a person chosen at random from the population has the disease, what is the conditional probability that the person, in fact, has the disease? Are you surprised by the size of the answer? Would you call this diagnostic test reliable?

2.38 A proficiency examination for a certain skill was given to 100 employees of a firm. Forty of the employees were men. Sixty of the employees passed the examination, in that they scored above a preset level for satisfactory performance. The breakdown among men and women passing the test was as follows.

	Male (M)	Female (F)
Pass (P)	24	36
Fail ($\bar{P}$)	16	24

100

Suppose an employee is selected randomly from the 100 who took the examination.

(a) Find the probability that the employee passed given that he was a man.

(b) Find the probability that the employee was a man given a passing grade.

(c) Are events P and M independent?

(d) Are events P and F independent?

2.39 By using Venn diagrams or similar arguments, show that, for events A, B, and C,

$$P(A \cup B \cup C) = P(A) + P(B) + P(C) - P(AB) - P(AC) - P(BC) + P(ABC).$$

2.40 By using the definition of conditional probability, show that

$$P(ABC) = P(A)P(B \mid A)P(C \mid AB).$$

2.41 There are twenty-three students in a classroom. What is the probability that at least two of them have the same birthday (day and month)? Assume that the year has 365 days. State your assumptions.

SUPPLEMENTARY EXERCISES

2.42 A coin is tossed four times and the outcome is recorded for each toss.

(a) List the outcomes for the experiment.

(b) Let A be the event that the experiment yields three heads. List the outcomes in A.

(c) Make a reasonable assignment of probabilities to the outcomes and find $P(A)$.

2.43 A hydraulic rework shop in a factory turned out seven rebuilt pumps today. Suppose three pumps are still defective. Two of the seven are selected for thorough testing and then classified as defective or not defective.

(a) List the outcomes for this experiment.

(b) Let A be the event that the selection includes no defectives. List the outcomes in A.

(c) Assign probabilities to the outcomes and find $P(A)$.

2.44 The national maximum speed limit (NMSL) of 55 miles per hour was placed in force in the United States in early 1974. The data below show the percent of vehicles found to travel at various speeds for three types of highways in 1973 (before the NMSL), 1974 (the year the NMSL was put in force), and 1975.

	Rural interstate						Rural primary						Rural secondary					
Vehicle Speed (mph)	**Car**			**Truck**			**Car**			**Truck**			**Car**			**Truck**		
	73	**74**	**75**	**73**	**74**	**75**	**73**	**74**	**75**	**73**	**74**	**75**	**73**	**74**	**75**	**73**	**74**	**75**
30–35	0	0	0	0	0	0	0	1	0	1	1	1	4	4	2	6	7	5
35–40	0	0	0	1	1	0	3	2	2	5	5	3	6	9	6	9	11	7
40–45	0	1	1	2	2	2	5	7	5	9	10	7	11	14	10	15	16	13
45–50	2	7	5	5	11	8	13	18	14	20	21	19	19	25	21	22	25	23
50–55	5	24	23	15	29	29	16	29	29	21	30	30	19	23	26	19	21	26
55–60	13	37	41	27	36	40	22	27	32	22	23	27	19	16	22	17	14	19
60–65	21	21	22	25	15	16	19	11	12	14	7	10	11	6	9	7	4	5
65–70	29	7	6	18	5	4	13	3	5	6	2	2	7	2	3	4	2	1
70–75	19	2	2	5	1	1	6	2	1	1	1	1	3	1	1	0	0	1
75–80	7	1	0	2	0	0	1	0	0	0	0	0	1	0	0	1	0	0
80–85	4	0	0	0	0	0	0	0	0	0	0	0	0	0	0	0	0	0

Source: D. B. Kamerud, "The 55 MPH Speed Limit: Costs, Benefits, and Implied Trade-Offs," *Transportation Research* 17A, no. 1 (1983): 51–64.

(a) Find the probability that a randomly observed car on a rural interstate was traveling less than 55 mph in 1973; less than 55 mph in 1974; less than 55 mph in 1975.

(b) Answer the questions in (a) for a randomly observed truck.

(c) Answer the questions in (a) for a randomly observed car on a rural secondary road.

2.45 An experiment consists of tossing a pair of dice.

(a) Use the combinatorial theorems to determine the number of outcomes in the sample space, S.

(b) Find the probability that the sum of the numbers appearing on the dice is equal to 7.

2.46 Show that $\binom{3}{0} + \binom{3}{1} + \binom{3}{2} + \binom{3}{3} = 2^3$. Note that in general $\sum_{i=0}^{n} \binom{n}{i} = 2^n$.

2.47 Of the persons arriving at a small airport, 60 % fly on major airlines, 30 % fly on privately owned airplanes, and 10 % fly on commercially owned airplanes not belonging to an airline. Of the persons arriving on major airlines, 50 % are traveling for business reasons, whereas this figure is 60 % for those arriving on private planes and 90 % for those arriving on other commercially owned planes. For a person selected randomly from a group of arrivals, find the probability that

(a) the person is traveling on business.

(b) the person is traveling on business and on a private airplane.

(c) the person is traveling on business given that he or she arrived on a commercial airliner.

(d) the person arrived on a private plane given that he or she is traveling on business.

2.48 In how many ways can a committee of three be selected from ten people?

2.49 How many different telephone numbers can be formed from a seven-digit number if the first digit cannot be zero?

2.50 A personnel director for a corporation has hired ten new engineers. If three (distinctly different) positions are open at a particular plant, in how many ways can he fill the position?

2.51 An experimenter wishes to investigate the effect of three variables—pressure, temperature, and type of catalyst—on the yield in a refining process. If the experimenter intends to use three settings each for temperature and pressure and two types of catalysts, how many experimental runs will have to be conducted to run all possible combinations of pressure, temperature, and type of catalyst?

2.52 An inspector must perform eight tests on a randomly selected keyboard coming off an assembly line. The sequence in which the tests are conducted is important because the time lost between tests will vary. If an efficiency expert were to study all possible sequences to find the one that required the minimum length of time, how many sequences would be included in his study?

2.53 (a) Two cards are drawn from a fifty two-card deck. What is the probability that the draw will yield an ace and a face card?
(b) Five cards are drawn from a fifty two-card deck. What is the probability that all five cards will be spades? Of the same suit?

2.54 The quarterback on a certain football team completes 60% of his passes. If he tries three passes, assumed independent, in a given quarter,

(a) what is the probability that he will complete all three?
(b) at least one?
(c) at least two?

2.55 Two men each tossing a balanced coin obtain a "match" if both coins are heads or if both coins are tails. If the process is repeated three times,

(a) What is the probability of three matches?
(b) What is the probability that all six tosses (three for each man) result in tails?

Coin tossing provides a model for many practical experiments. Suppose that the "coin tosses" represented the answers given by two students for three specific true–false questions on an examination. If the two students gave three matches for answers, would the low probability acquired in part (a) suggest collusion?

2.56 Refer to Exercise 2.55. What is the probability that the pair of coins are tossed four times before a match occurs; that is, a match occurs for the first time on the fourth toss?

2.57 Suppose that the probability of exposure to the flu during an epidemic is .6. Experience has shown that a serum is 80% successful in preventing an inoculated person from acquiring the flu, if exposed. A person not inoculated faces a probability of .90 of acquiring the flu if exposed. Two persons, one inoculated and one not, are capable of performing a highly specialized task in a business. Assume that they are not at the same location, are not in contact with the same people, and cannot expose each other. What is the probability that at least one will get the flu?

2.58 Two gamblers bet \$1 each on the successive tosses of a coin. Each has a bank of \$6.

(a) What is the probability that they break even after six tosses of the coin?
(b) What is the probability that one player, say Jones, wins all the money on the tenth toss of the coin?

2.59 Suppose that the streets of a city are laid out in a grid with streets running north–south and east–west. Consider the following scheme for patrolling an area of sixteen blocks by sixteen blocks. A patrolman commences walking at the intersection in the center of the area. At the corner of each block, he randomly elects to go north, south, east, or west.

(a) What is the probability that he will reach the boundary of his patrol area by the time he walks the first eight blocks?
(b) What is the probability that he will return to the starting point after walking exactly four blocks?

2.60 Consider two mutually exclusive events, A and B, such that $P(A) > 0$ and $P(B) > 0$. Are A and B independent? Give a proof for your answer.

2.61 An accident victim will die unless in the next 10 minutes he receives an amount of type A Rh-positive blood, which can be supplied by a single donor. It requires 2 minutes to type a prospective donor's blood and 2 minutes to complete the transfer of blood. A large number of untyped donors are available, 40 % of which have type A Rh-positive blood. What is the probability that the accident victim will be saved if only one blood-typing kit is available?

2.62 An assembler of electric fans uses motors from two sources. Company A supplies 90 % of the motors and company B supplies the other 10 %. Suppose it is known that 5 % of the motors supplied by company A are defective and 3 % of the motors supplied by company B are defective. An assembled fan is found to have a defective motor. What is the probability that this motor was supplied by company B?

2.63 Show that, for three events, A, B, and C,

$$P[(A \cup B) \mid C] = P(A \mid C) + P(B \mid C) - P[(A \cap B \mid C)].$$

2.64 If A and B are independent events, show that A and $\bar{B}$ also are independent.

2.65 Three events, A, B, and C, are said to be independent if

$$P(AB) = P(A)P(B),$$
$$P(AC) = P(A)P(C),$$
$$P(BC) = P(B)P(C),$$

and

$$P(ABC) = P(A)P(B)P(C).$$

Suppose that a balanced coin is independently tossed two times. Define the following events:

A: head appears on the first toss
B: head appears on the second toss
C: both tosses yield the same outcome.

Are A, B, and C independent?

2.66 A line from a to b has midpoint c. A point is chosen at random on the line and marked x (the point x being chosen at random implies that x is equally likely to fall in any subinterval of fixed length l). Find the probability that the line segments ax, bx, and ac can be joined to form a triangle.

2.67 Eight tires of different brands are ranked from 1 to 8 (best to worst) according to mileage performance. If four of these tires are chosen at random by a customer, find the probability that the best tire among those selected by the customer is actually ranked third among the original eight.

2.68 Suppose that n indistinguishable balls are to be arranged in N distinguishable boxes so that each distinguishable arrangement is equally likely. If $n \geq N$, show that the probability that no box will be empty is given by

$$\frac{\binom{n-1}{N-1}}{\binom{N+n-1}{N-1}}.$$

2.69 Relays in a section of an electrical circuit operate independently, and each one closes properly with probability .9 when a switch is thrown. The following two designs, each involving four relays, are presented for a section of a new circuit. Which has the higher probability of current flowing from a to b when the switch is thrown?

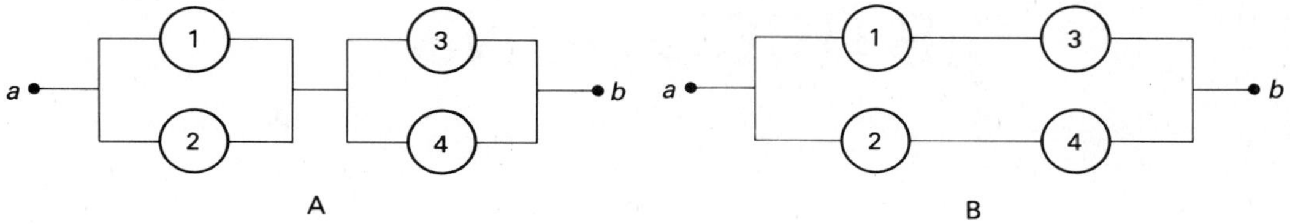

2.70 A blood test for hepatitis has the following accuracy:

	Test result	
Patient	+	−
With hepatitis	.90	.10
Without hepatitis	.01	.99

The disease rate in the general population is 1 in 10,000.

(a) What is the probability that a person who gets a positive blood test result actually has hepatitis?

(b) A patient is sent for a blood test because he has lost his appetite and has jaundice. The physician knows that this type of patient has probability 1/2 of having hepatitis. If this patient gets a positive result on his blood test, what is the probability that he has hepatitis?

CHAPTER 3

DISCRETE PROBABILITY DISTRIBUTIONS

3.1

RANDOM VARIABLES AND THEIR PROBABILITY DISTRIBUTIONS

Most of the experiments we encounter generate outcomes that can be interpreted in terms of real numbers, such as heights of children, number of voters favoring a certain candidate, tensile strength of wires, and numbers of accidents at specified intersections. These numerical outcomes, with values that can change from experiment to experiment, are called *random variables.* We will look at an illustrative example of a random variable before we give a more formal definition.

A section of an electrical circuit has two relays, numbered 1 and 2, operating in parallel. The current will flow when a switch is thrown if either one or both of the relays close. The probability of a relay closing properly is .8, which is the same for each relay. The relays operate independently, we assume. Let E_i denote the event that relay i closes properly when the switch is thrown. Then, $P(E_i) = .8$.

When the switch is thrown, a numerical outcome of some interest to the operator of this system is X, the number of relays that close properly. Now, X can take on only three possible values, because the number of relays that close must be 0, 1 or 2. We can find the probabilities associated with these values of X by relating them to the underlying events, E_i. Thus, we have

$$\begin{aligned} P(X = 0) &= P(\bar{E}_1\bar{E}_2) \\ &= P(\bar{E}_1)P(\bar{E}_2) \\ &= (.2)(.2) = .04 \end{aligned}$$

because $X = 0$ means neither relay closes and the relays operate independently. Similarly,

$$\begin{aligned} P(X = 1) &= P(E_1\bar{E}_2 \cup \bar{E}_1E_2) \\ &= P(E_1\bar{E}_2) + P(\bar{E}_1E_2) \\ &= P(E_1)P(\bar{E}_2) + P(\bar{E}_1)P(E_2) \\ &= (.8)(.2) + (.2)(.8) \\ &= .32 \end{aligned}$$

and

$$\begin{aligned} P(X = 2) &= P(E_1E_2) \\ &= P(E_1)P(E_2) \\ &= (.8)(.8) = .64. \end{aligned}$$

The values of X, along with their probabilities, are more useful for keeping track of the operation of this system than the underlying events, E_i, for the *number* of properly closing relays is the key to whether the system will work. The current will flow if X is at least one, and this event has probability

$$\begin{aligned} P(X \geq 1) &= P(X = 1 \text{ or } X = 2) \\ &= P(X = 1) + P(X = 2) \\ &= .32 + .64 \\ &= .96. \end{aligned}$$

Note that we have mapped the outcomes of an experiment into a set of three meaningful real numbers, and attached a probability to each. Such situations provide the motivation for Definitions 3.1 and 3.2.

Definition 3.1

A **random variable** is a real-valued function whose domain is a sample space.

Random variables will be denoted by upper-case letters such as X, Y, and Z.

The actual numerical values that a random variable can assume will be denoted by lower-case letters, such as x, y, and z. We can then talk about the "probability that X takes on the value x," $P(X = x)$, denoted by $p(x)$.

In the relay example, the random variable, X, has only three possible values and it is a relatively simple matter to assign probabilities to these values. Such a random variable is called *discrete*.

Definition 3.2

A random variable X is said to be **discrete** if it can take on only a finite number, or a countable infinity, of possible values x.

In this case,

1. $P(X = x) = p(x) \geq 0$.
2. $\sum_x P(X = x) = 1$, where the sum is over all possible values x.

The function $p(x)$ is called the *probability function* of X.

The probability function is sometimes called the *probability mass function* of X, to denote the idea that a mass of probability is piled up at discrete points.

It is often convenient to list the probabilities for a discrete random variable on a table. With X defined as the number of closed relays in the problem just discussed, the table is as follows:

x	$p(x)$
0	.04
1	.32
2	.64
Total	1.00

This listing is one way of representing the *probability distribution* of X.

Note that the probability function $p(x)$ satisfies two properties:

1. $0 \leq p(x) \leq 1$ for any x.
2. $\sum_x p(x) = 1$, where the sum is over all possible values of x.

Functional forms for the probability function $p(x)$ of commonly occurring random variables will be given in later sections. We now illustrate another method for arriving at a tabular presentation of a discrete probability distribution.

Example 3.1 The output of circuit boards from two assembly lines set up to produce identical boards is mixed into one storage tray. As inspectors examine the boards, it is difficult to determine whether a board comes from line A or line B. A probabilistic assessment of this question is often helpful. Suppose a storage tray contains ten circuit boards, of which six came from line A and four from line B. An inspector selects two of these identically appearing boards for inspection. He is interested in X, the number of inspected boards from line A. Find the probability distribution for X.

Solution

The experiment consists of two selections, each of which can result in two outcomes. Let A_i denote the event that the ith board comes from line A and B_i the event that it comes from line B. Then the probability of selecting two boards from line A ($X = 2$) is

$$P(A_1 A_2) = P(A \text{ on 1st})P(A \text{ on 2nd} \mid A \text{ on 1st}).$$

The multiplicative law of probability is used, and the probability on the second selection depends upon what happened on the first selection. There are other possibilities for outcomes that will result in other values of X. These outcomes are conveniently listed on the tree in Figure 3.1. The probabilities for the various selections are given on the branches of the tree.

FIGURE 3.1 Outcomes for Example 3.1.

1st Selection	2nd Selection	Probability	Value of X
$\frac{6}{10}$ A	$\frac{5}{9}$ A	$\frac{30}{90}$	2
	$\frac{4}{9}$ B	$\frac{24}{90}$	1
$\frac{4}{10}$ B	$\frac{6}{9}$ A	$\frac{24}{90}$	1
	$\frac{3}{9}$ B	$\frac{12}{90}$	0

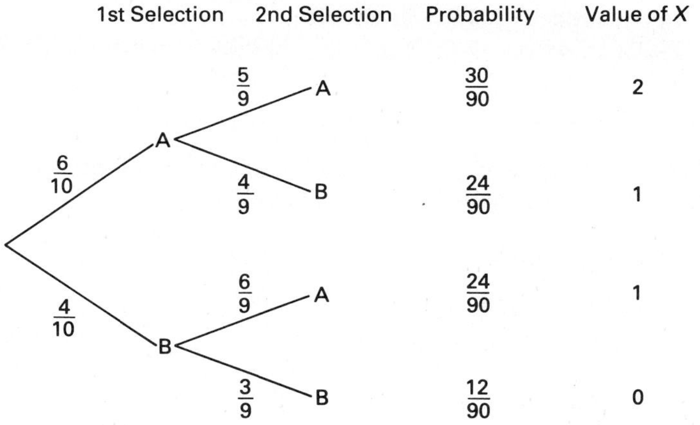

It is easily seen that X has three possible outcomes, with probabilities as follows:

x	$p(x)$
0	$\frac{12}{90}$
1	$\frac{48}{90}$
2	$\frac{30}{90}$
Total	1.00

The reader should envision this concept extended to more selections from trays of various structure.

We sometimes study the behavior of random variables by looking at their *cumulative* probabilities. That is, for any random variable X we may look at $P(X \leq b)$ for any real number b. This is the cumulative probability for X evaluated at b. Thus, we can define a function $F(b)$ as

$$F(b) = P(X \leq b).$$

Definition 3.3

The **distribution function** $F(b)$ for a random variable X is defined as

$$F(b) = P(X \leq b).$$

If X is discrete,

$$F(b) = \sum_{x=-\infty}^{b} p(x)$$

where $p(x)$ is the probability function.
The distribution function is often called the *cumulative distribution function* (c.d.f.).

The random variable X, denoting the number of relays closing properly (defined at the beginning of this section), has probability distribution given by

$$P(X = 0) = .04$$
$$P(X = 1) = .32$$
$$P(X = 2) = .64.$$

Note that

$$P(X \leq 1.5) = P(X \leq 1.9) = P(X \leq 1) = .36$$

The distribution function for this random variable then has the form

$$F(b) = \begin{cases} 0 & b < 0 \\ .04 & 0 \leq b < 1 \\ .36 & 1 \leq b < 2 \\ 1.00 & 2 \leq b \end{cases}.$$

This function is graphed in Figure 3.2.

FIGURE 3.2 A distribution function for a discrete random variable.

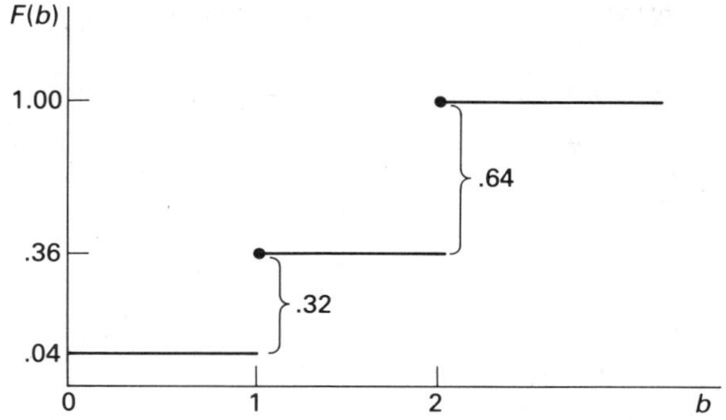

EXERCISES

3.1 Among ten applicants for an open position, six are women and four are men. Suppose three applicants are randomly selected from the applicant pool for final interviews. Find the probability distribution for X, the number of women applicants among the final three.

3.2 The median annual income for household heads in a certain city is \$18,000. Four such household heads are randomly selected for an opinion poll.

(a) Find the probability distribution of X, the number, out of the four, that have annual incomes below \$18,000.

(b) Would you say that it is unusual to see all four below \$18,000 in this type of poll? (What is the probability of this event?)

3.3 Wade Boggs of the Boston Red Sox hit .363 in 1987. (He got a hit on 36.3% of his official times at bat.) In a typical game, he was up to bat three official times. Find the probability distribution for X, the number of hits in a typical game. What assumptions are involved in the answer? Are the assumptions reasonable? Is it unusual for a good hitter to go 0 for 3 in one game?

3.4 A commercial building has two entrances, numbered I and II. Three people enter the building at 9:00 A.M. Let X denote the number that select entrance I. Assuming the people choose entrances independently and at random, find the probability distribution for X. Were any additional assumptions necessary for your answer?

3.5 Table 2.2 on page 43 gives information on causes of residential fires. Suppose four independent residential fires are reported in one day, and let X denote the number, out of the four, that are in family homes.

(a) Find the probability distribution for X in tabular form.

(b) Find the probability that at least one of the four fires is in a family home.

3.6 It was observed that 40% of the vehicles crossing a certain toll bridge are commmercial trucks. Four vehicles will cross the bridge in the next minute. Find the probability distribution for X, the number of commercial trucks among the four, if the vehicle types are independent of one another.

3.7 Of the people entering a blood bank to donate blood, 1 in 3 have type O^+ blood, and 1 in 15 have type O^- blood. For the next three people entering the blood bank, let X denote the number with O^+ blood and Y the number with O^- blood. Assuming independence among the people, with respect to blood type, find the probability distributions for X and Y. Also find the probability distribution for $X + Y$, the number of people with type O blood.

3.8 Daily sales records for a computer manufacturing firm show that it will sell 0, 1, or 2 mainframe computer systems with probabilities as listed:

Number of sales	0	1	2
Probability	.7	.2	.1

(a) Find the probability distribution for X, the number of sales in a two-day period, assuming that sales are independent from day to day.

(b) Find the probability that at least one sale is made in the two-day period.

3.9 Four microchips are to be placed in a computer. Two of the four chips are randomly selected for inspection before assembly of the computer. Let X denote the number of defective chips found among the two inspected. Find the probability distribution for X if

(a) two of the microchips were defective.

(b) one of the four microchips was defective.

(c) none of the microchips was defective.

3.10 When turned on, each of the three switches in the following diagram work properly with probability .9. If a switch is working properly, current can flow through it when it is turned on. Find the probability distribution for Y, the number of closed paths from a to b, when all three switches are turned on.

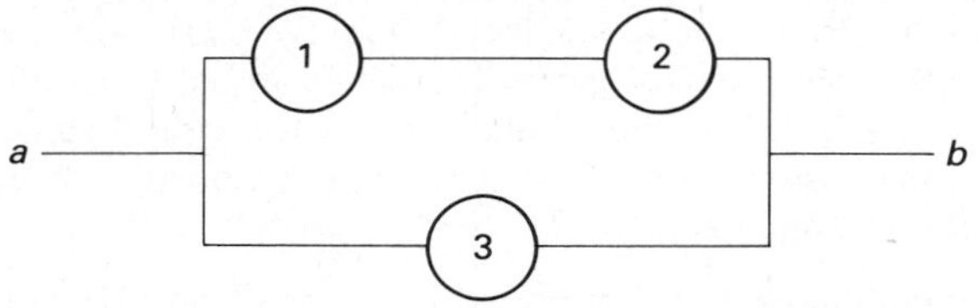

3.2

EXPECTED VALUES OF RANDOM VARIABLES

Because a probability can be thought of as the long-run relative frequency of occurrence for an event, a probability distribution can be interpreted as showing the long-run relative frequency of occurrence for numberical outcomes associated with an experiment. Suppose, for example, that you and a friend are matching balanced coins. Each of you tosses a coin. If the upper

FIGURE 3.3 Relative frequency of winnings.

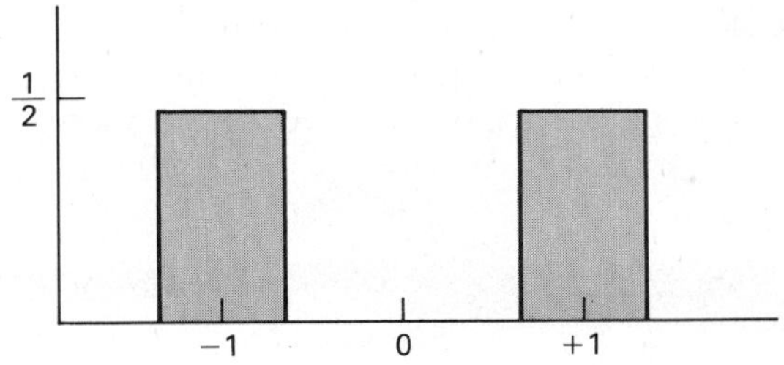

faces match, you win \$1.00; if they do not match, you lose \$1.00 (your friend wins \$1.00). The probability of a match is .5 and, in the long run, you will win about half of the time. Thus, a relative frequency distribution of your winnings should look like Figure 3.3. Note that the negative sign indicates a loss to you.

On the average, how much will you win per game over the long run? If Figure 3.3 is a correct display of your winnings, you win −1 half of the time and +1 half of the time for an average of

$$(-1)\left(\frac{1}{2}\right) + (1)\left(-\frac{1}{2}\right) = 0.$$

This average is sometimes called your expected winnings per game, or the *expected value* of your winnings. (An expected value of 0 indicates that this is a fair game.) The general definition of expected value is given in Definition 3.4.

> **Definition 3.4**
>
> The **expected value** of a discrete random variable X having probability distribution $p(x)$ is givey by[1]
>
> $$E(X) = \sum_x xp(x).$$
>
> (The sum is over all values of x for which $p(x) > 0$.)
>
> We sometimes use the notation
>
> $$E(X) = \mu.$$

Now payday has arrived and you and your friend up the stakes to \$10 per game of matching coins. You now win −10 or +10 with equal probability; your expected winnings per game is

$$(-10)\left(\frac{1}{2}\right) + (10)\left(\frac{1}{2}\right) = 0,$$

[1] We assume absolute convergence when the range of X is countable; we talk about an expectation only when it is assumed to exist.

and the game is still fair. The new stakes can be thought of as a function of the old in the sense that if X represents your winnings per game when you were playing for \$1.00, then $10X$ represents your winnings per game when playing for \$10.00. Such functions of random variables arise often, and the extension of the definition of expected value to cover these cases is given in Theorem 3.1.

Theorem 3.1 If X is a discrete random variable with probability distribution $p(x)$ and if $g(x)$ is any real valued function of X, then

$$E[g(X)] = \sum_x g(x)p(x).$$

Proof

The proof of Theorem 3.1 comes directly from the definition of expected value. Suppose the possible values of X are labeled $x_1, x_2, x_3, \ldots, x_i \cdots$. Now, $g(X)$ is also a random variable taking on values $g(x_i)$, some of which may be identical. Therefore,

$$P[g(X) = y_i] = \sum_i p(x_i) = q(y_i)$$

where the sum extends over all values of x_i that are mapped into y_i by the function $g(x)$. We will denote this sum by $q(y_i)$. Beginning with the result we want to verify, we have

$$E[g(X)] = \sum_{j=1}^{\infty} g(x_j)p(x_j)$$

$$= \sum_{j=1}^{\infty} \left[\sum_i g(x_i)p(x_i)\right]$$

(where the inner sum is over all indices so that $g(x_i) = y_i$)

$$= \sum_{j=1}^{\infty} y_j \sum_i p(x_i)$$

$$= \sum_{j=1}^{\infty} y_j q(y_i)$$

which is the $E[g(X)]$ by Definition 3.4. ∎

You and your friend decided to complicate the payoff picture in the coin-matching game by allowing you to win \$1 if the match is tails and \$2 if the match is heads. You still lose \$1 if the coins do not match. Quickly you see

FIGURE 3.4 Relative frequency of winnings.

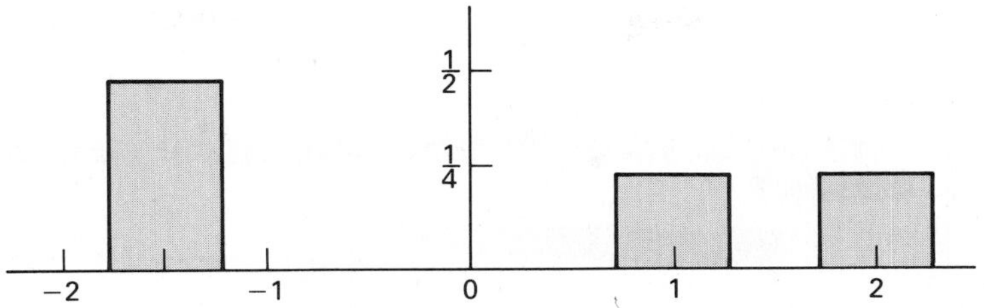

that this is not a fair game, because your expected winnings are

$$(-1)\left(\frac{1}{2}\right) + (1)\left(\frac{1}{4}\right) + (2)\left(\frac{1}{4}\right) = .25.$$

You compensate for this by paying your friend \$1.50 if the coins do not match. Then, your expected winnings per game are

$$(-1.5)\left(\frac{1}{2}\right) + (1)\left(\frac{1}{4}\right) + (2)\left(\frac{1}{4}\right) = 0,$$

and the game is now fair. What is the difference between this game and the original one in which payoffs were \$1? The difference certainly cannot be explained by the expected value, since both games are fair. You can win more but also lose more with the new payoffs, and the difference between the two games can be partially explained in terms of the variability of your winnings across many games. This increased variability can be seen in Figure 3.4, the relative frequency for your winnings in the new game, which spreads out more than Figure 3.3. Formally, variation is often measured by the *variance* and a related quantity called the *standard deviation*.

Definition 3.5

The **variance** of a random variable X with expected value μ is given by

$$V(X) = E(X - \mu)^2.$$

We sometimes use the notation

$$E(X - \mu)^2 = \sigma^2.$$

The smallest value σ^2 can assume is zero, and that would occur if all the probability was at a single point (that is, X takes on a constant value with probability one). The variance will become larger as the points with positive probability spread out more.

Observe that the variance squares the units in which we are measuring. A measure of variation that maintains the original units is the standard deviation.

> ***Definition 3.6***
>
> The **standard deviation** of a random variable X is the square root of the variance, given by
>
> $$\sigma = \sqrt{\sigma^2} = \sqrt{E(X - \mu)^2}.$$

For the game of Figure 3.3, the variance of your winnings is (with $\mu = 0$)

$$\begin{aligned}\sigma^2 &= E(X - \mu)^2 \\ &= (-1)^2\left(\frac{1}{2}\right) + (1)^2\left(\frac{1}{2}\right) = 1.\end{aligned}$$

It follows that $\sigma = 1$, as well. For the game of Figure 3.4, the variance of your winnings is

$$\begin{aligned}\sigma^2 &= (-1.5)^2\left(\frac{1}{2}\right) + (1)^2\left(\frac{1}{4}\right) + (1)^2\left(\frac{1}{4}\right) \\ &= 2.375\end{aligned}$$

and

$$\sigma = 1.54.$$

Which game would you rather play?

The standard deviation can be thought of as the size of a "typical" deviation between an observed outcome and the expected value. For Figure 3.3 each outcome (-1 or $+1$) deviates precisely one standard deviation from the expected value. For Figure 3.3 the positive values average 1.5 units from the expected value of 0 (as does the negative value), which is approximately one standard deviation.

The mean (expected value) and variance often are used to summarize the information in a large set of data. Table 3.1 gives cancer mortality rates for white males in 67 counties of Florida for the 1970s. These data can be displayed in a histogram as shown in Figure 3.5. To produce this figure, the interval from 155 (a little smaller than the smallest data value) to 265 (a little larger than the largest data value) is divided into 11 convenient intervals of 10 units each. The graph shows the midpoints of these intervals. The height of a bar represents the fraction (relative frequency) of data values falling into that interval.

TABLE 3.1 **Cancer mortality rates (per 100,000) for white males in Florida (1970s)**

156.5	159.7	167.0	167.8	170.8	171.0	171.4	175.0	177.4	178.4
179.3	179.4	179.8	181.3	181.4	181.4	185.1	185.2	185.7	185.8
186.3	189.4	190.3	190.4	191.2	191.7	191.8	192.4	192.9	193.1
193.3	194.3	195.0	195.0	195.1	195.4	197.2	200.0	200.1	200.2
201.0	202.7	203.4	203.9	204.4	207.8	208.0	209.2	210.2	213.5
213.8	214.1	215.2	219.4	219.7	222.0	224.2	224.3	230.4	233.4
234.1	235.1	235.2	237.1	240.4	241.7	256.8			

Source: U.S. Environmental Protection Agency and National Cancer Institute, "U.S. Cancer Mortality Rates and Trends 1950–1979," Vol. 1 (Washington, D.C.: U.S. Government Printing Office, 1988).

If a county is chosen at random (so that each county has a probability of selection equal to 1/67), the chance that its mortality rate, X, exceeds 225 is 9/67, the sum of the relative frequencies in the last three bars on the right. The mean value of X is given by

$$\mu = \sum xp(x)$$

$$= \sum x\left(\frac{1}{67}\right) = 199.34,$$

which is simply the average of the 67 measurements. Similarly,

$$\sigma = \sqrt{\sum (x - \mu)^2 p(x)}$$

$$= \sqrt{\sum (x - \mu)^2 \left(\frac{1}{67}\right)} = 21.56.$$

FIGURE 3.5 Relative frequency histogram of data in Table 3.1.

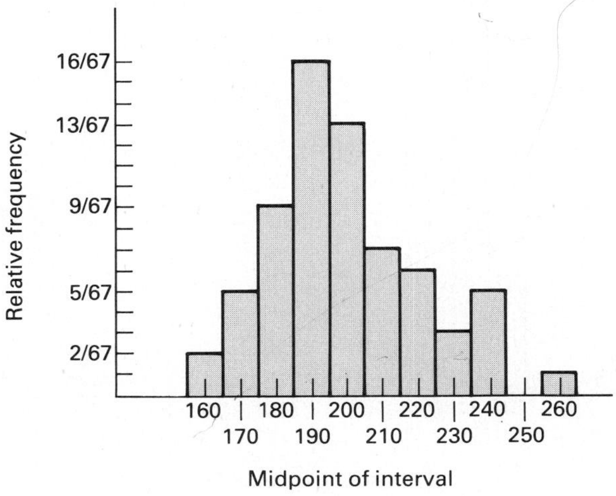

TABLE 3.2 **Cancer mortality rates (per 100,000) for white males in Nebraska (1970s)**

37.2	96.1	114.4	116.9	121.1	129.6	135.1	136.6	144.4	146.1
146.7	147.3	147.5	148.1	148.2	149.6	150.4	150.6	151.8	152.7
153.6	153.9	154.6	156.1	156.6	157.2	157.9	158.5	159.8	161.1
161.3	161.9	162.4	162.6	162.6	163.0	163.2	163.9	164.9	165.8
166.8	167.1	169.0	171.4	171.5	172.3	174.1	175.6	176.1	177.4
178.9	179.3	179.5	179.7	179.8	180.2	181.9	182.8	183.1	183.6
184.8	184.9	188.0	188.2	188.3	189.5	189.7	191.4	191.6	191.6
192.3	192.5	192.8	194.4	194.6	197.1	197.6	197.6	198.7	200.7
203.2	203.5	211.1	222.0	223.3	230.2	240.0	387.1	398.3	440.7
577.9	622.5	692.1							

Source: U.S. Environmental Protection Agency and National Cancer Institute, "U.S. Cancer Mortality Rates and Trends 1950–1979," Vol. 1 (Washington, D.C.: U.S. Government Printing Office, 1988).

An interpretation of these numbers is that a "typical" county in Florida had a mortality rate of about 200 ± 22 for the 1970s.

If we look at the same measurement for another state, the picture changes. Table 3.2 shows cancer mortality rates in 93 counties of Nebraska for the 1970s, and Figure 3.6 gives a histogram for these data. For a randomly selected county mortality rate from Nebraska, $\mu = 192.47$ and $\sigma = 95.16$. Note that Figures 3.5 and 3.6 have approximately the same mean, but differ considerably in standard deviation, a measure of variability (or spread) in the

FIGURE 3.6 Relative frequency histogram of data in Table 3.2.

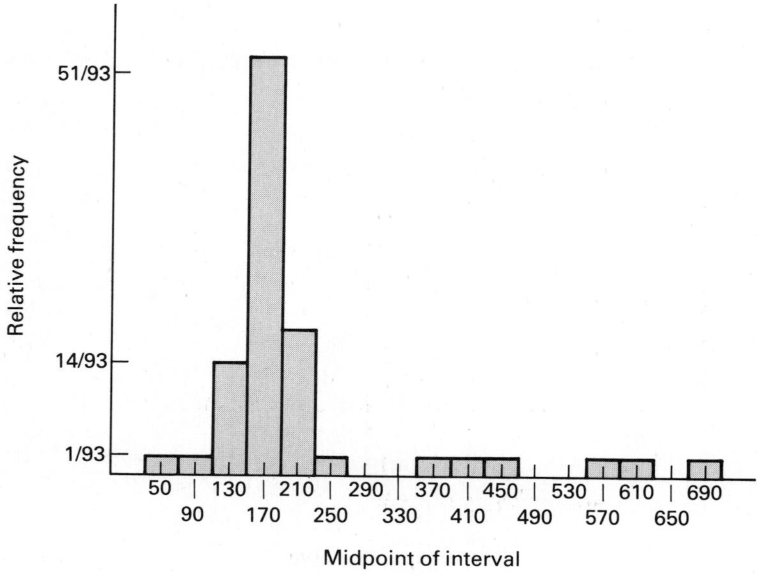

data. It would not be surprising to see a mortality rate around 300 in Nebraska, but this never seem to occur in Florida (at least, it did not occur in the 1970s). We now provide other examples and extensions of these basic results.

Example 3.2 The manager of a stockroom in a factory knows from his study of records that the daily demand (number of times used) for a certain tool has the following probability distribution:

Demand	0	1	2
Probability	.1	.5	.4

(That is, 50 % of the daily records show that the tool was used one time.) If X denotes the daily demand, find $E(X)$ and $V(X)$.

Solution

From Definition 3.4, we see that

$$E(X) = \sum_x xp(x)$$
$$= 0(.1) + 1(.5) + 2(.4) = 1.3$$

that is, the tool is used an average of 1.3 times per day.

From Definition 3.5, we see that

$$V(X) = E(X - \mu)^2$$
$$= \sum_x (x - \mu)^2 p(x)$$
$$= (0 - 1.3)^2(.1) + (1 - 1.3)^2(.5) + (2 - 1.3)^2(.4)$$
$$= (1.69)(.1) + (.09)(.5) + (.49)(.4)$$
$$= .410.$$

Our work in manipulating expected values is greatly facilitated by making use of the two results of Theorem 3.2. Often, $g(X)$ is a linear function and, in this case, the calculations of expected value and variance are especially simple.

Theorem 3.2 For any random variable X and constants a and b,

1. $E(aX + b) = aE(X) + b$.
2. $V(aX + b) = a^2V(X)$.

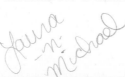

We sketch a proof of this theorem for a discrete random variable X having probability distribution given by $p(x)$. By Theorem 3.1,

$$\begin{aligned}
E(aX + b) &= \sum_x (ax + b)p(x) \\
&= \sum_x [(ax)p(x) + bp(x)] \\
&= \sum_x axp(x) + \sum_x bp(x) \\
&= a\sum_x xp(x) + b\sum_x p(x) \\
&= aE(X) + b.
\end{aligned}$$

Note that $\sum_x p(x)$ must equal unity. Also, by Definition 3.5,

$$\begin{aligned}
V(aX + b) &= E[(aX + b) - E(aX + b)]^2 \\
&= E[aX + b - (aE(X) + b)]^2 \\
&= E[aX - aE(X)]^2 \\
&= E[a^2(X - E(X))^2] \\
&= a^2E[X - E(X)]^2 \\
&= a^2V(X).
\end{aligned}$$

We illustrate the use of these results in the following example.

Example 3.3 In Example 3.2, suppose it costs the factory \$10 each time the tool is used. Find the mean and variance of the daily costs for use of this tool.

Solution

Recall that the X of Example 3.2 is the daily demand. The daily cost of using this tool is the $10X$. We have, by Theorem 3.2,

$$\begin{aligned}
E(10X) = 10E(X) &= 10(1.3) \\
&= 13.
\end{aligned}$$

The factory should budget \$13 per day to cover the cost of using the tool.

Also, by Theorem 3.2,

$$\begin{aligned}
V(10X) = (10)^2V(X) &= 100(.410) \\
&= 41.
\end{aligned}$$

We will make use of this value in a later example.

Theorem 3.2 leads us to a more efficient computational formula for a variance, as given in Theorem 3.3.

Theorem 3.3 If X is a random variable with a mean of μ, then

$$V(X) = E(X^2) - \mu^2.$$

Proof

The proof is as follows. Starting with the definition of variance, we have

$$\begin{aligned} V(X) &= E(X - \mu)^2 = E(X^2 - 2X\mu + \mu^2) \\ &= E(X^2) - E(2X\mu) + E(\mu^2) \\ &= E(X^2) - 2\mu E(X) + \mu^2 \\ &= E(X^2) - 2\mu^2 + \mu^2 \\ &= E(X^2) - \mu^2. \end{aligned}$$

Example 3.4 Use the result of Theorem 3.3 to compute the variance of X as given in Example 3.2.

Solution

In Example 3.2, X has probability distribution given by

x	0	1	2
$p(x)$	.1	.5	.4

and we saw that $E(X) = 1.3$. Now,

$$\begin{aligned} E(X^2) &= \sum_x x^2 p(x) \\ &= (0)^2(.1) + (1)^2(.5) + (2)^2(.4) \\ &= 0 + .5 + 1.6 \\ &= 2.1. \end{aligned}$$

By Theorem 3.3,

$$\begin{aligned} V(X) &= E(X^2) - \mu^2 \\ &= 2.1 - (1.3)^2 = .41. \end{aligned}$$

We have computed means and variances for a number of probability distributions and argued that these two quantities give us some useful information on the center and spread of the probability mass. Now suppose we know only the mean and variance for a probability distribution. Can we say anything specific about probabilities for certain intervals about the mean? The answer is yes, and two useful results on the relationship between mean, standard deviation, and relative frequency will now be disussed.

The first result applies only to mound-shaped, reasonably symmetric distributions such as the one pictured in Figure 3.5. In such cases, approximately 68 % of the observations will lie within one standard deviation of the mean, and approximately 95 % of the observations will lie within two standard deviations of the mean. For the data on which Figure 3.5 is based, $\mu = 199.34$ and $\sigma = 21.56$. The interval

$$(\mu - \sigma, \mu + \sigma)$$

or

$$(177.78, 220.70)$$

contains $46/67 = .686$ or approximately 68 % of the observations. The interval

$$(\mu - 2\sigma, \mu + 2\sigma)$$

or

$$(156.22, 242.46)$$

contains $66/67 = .985$ or a little more than 95 % of the observations.

Contrasting Figure 3.5 to Figure 3.6, we see that Figure 3.6 is not nearly as symmetric, having a long tail to the right. For the data of Figure 3.6, $\mu = 192.47$ and $\sigma = 95.16$. The interval $(\mu - \sigma, \mu + \sigma)$ contains about 91 % of the measurements and the interval $(\mu - 2\sigma, \mu + 2\sigma)$ contains about 93 %. So the 68 %–95 % rule does not work well in this case. If we have only a mean and standard deviation for a probability distribution and know nothing about the shape of the distribution, we must be careful in applying the 68 %–95 % rule. It may provide a poor approximation.

The second result is a theorem relating μ, σ and relative frequencies (probabilities) that gives very rough approximations but works in all cases.

Theorem 3.4 Tchebysheff's Theorem:
Let X be a random variable with a mean of μ and a variance of σ^2. Then for any positive k

$$P(|X - \mu| < k\sigma) \geq 1 - \frac{1}{k^2}.$$

The proof of Theorem 3.4 begins with the definition of $V(X)$ and then makes substitutions in the sum defining this quantity. Now,

$$\begin{aligned} V(X) = \sigma^2 &= \sum_{-\infty}^{\infty} (x - \mu)^2 p(x) \\ &= \sum_{-\infty}^{\mu-k\sigma} (x - \mu)^2 p(x) + \sum_{\mu-k\sigma}^{\mu+k\sigma} (x - \mu)^2 p(x) \\ &\quad + \sum_{\mu+k\sigma}^{\infty} (x - \mu)^2 p(x). \end{aligned}$$

(The first sum stops at the largest value of x smaller than $\mu - k\sigma$ and the third sum begins at the smallest value of x larger than $\mu + k\sigma$; the middle sum collects the remaining terms.) Observe that the middle sum is never negative, and for both of the outside sums,

$$(x - \mu)^2 \geq k^2\sigma^2.$$

Eliminating the middle sum and substituting for $(x - \mu)^2$ in the other two yields

$$\sigma^2 \geq \sum_{-\infty}^{\mu-k\sigma} k^2\sigma^2 p(x) + \sum_{\mu+k\sigma}^{\infty} k^2\sigma^2 p(x)$$

or

$$\sigma^2 \geq k^2\sigma^2\left[\sum_{-\infty}^{\mu-k\sigma} p(x) + \sum_{\mu+k\sigma}^{\infty} p(x)\right]$$

or

$$\sigma^2 \geq k^2\sigma^2 P(|X - \mu| \geq k\sigma).$$

It follows that

$$P(|X - \mu| \geq k\sigma) \leq \frac{1}{k^2}$$

or

$$P(|X - \mu| < k\sigma) \geq 1 - \frac{1}{k^2}.$$

The inequality in the statement of the theorem is equivalent to

$$P(\mu - k\sigma < X < \mu + k\sigma) \geq 1 - \frac{1}{k^2}.$$

To interpret this result, let $k = 2$, for example. Then the interval $\mu - 2\sigma$ to $\mu + 2\sigma$ must contain at least $1 - 1/k^2 = 1 - 1/4 = 3/4$ of the probability mass for the random variable. We give more specific illustrations in the following two examples.

Example 3.5 The daily production of electric motors at a certain factory averaged 120 with a standard deviation of 10.

(a) What can be said about the fraction of days that will have a production level between 100 and 140?

(b) Find the shortest interval certain to contain at least 90 % of the daily production levels.

Solution

(a) The interval 100 to 140 is $\mu - 2\sigma$ to $\mu + 2\sigma$, with $\mu = 120$ and $\sigma = 10$. Thus, $k = 2$,

$$1 - \frac{1}{k^2} = 1 - \frac{1}{4} = \frac{3}{4}$$

and at least 75 % of the days will have total production in this interval. (This could be closer to 95 % if the daily production figures show a mound-shaped, symmetric relative frequency distribution.)

(b) To find k, we must set $(1 - 1/k^2)$ equal to .9 and solve for k. That is,

$$1 - \frac{1}{k^2} = .9$$

$$\frac{1}{k^2} = .1$$

$$k^2 = 10$$

or

$$k = \sqrt{10} = 3.16.$$

The interval

$$\mu - 3.16\sigma \text{ to } \mu + 3.16\sigma$$

or

$$120 - 31.6(10) \text{ to } 120 + 3.16(10)$$

or

$$88.4 \text{ to } 151.6$$

will then contain at least 90 % of the daily production levels.

Example 3.6 The daily cost for use of a certain tool has a mean of \$13 and a variance of 41 (see Example 3.3). How often will this cost exceed \$30?

Solution

First, we must find the distance between the mean and 30, in terms of the standard deviation of the distribution of costs. We have

$$\frac{30 - \mu}{\sqrt{\sigma^2}} = \frac{30 - 13}{\sqrt{41}} = \frac{17}{6.4} = 2.66$$

or 30 is 2.66 standard deviations above the mean. Letting $k = 2.66$ in Theorem 3.4, we have that the interval

$$\mu - 2.66\sigma \text{ to } \mu + 2.66\sigma$$

or

$$13 - 2.66(6.4) \text{ to } 13 + 2.66(6.4)$$

or

$$-4 \text{ to } 30$$

must contain at least

$$1 - \frac{1}{k^2} = 1 - \frac{1}{(2.66)^2} = 1 - .14 = .86$$

of the probability. Since the daily cost cannot be negative, at most .14 of the probability mass can exceed 30. Thus the cost cannot exceed \$30 more than 14 % of the time.

EXERCISES

3.11 You are to pay \$1.00 to play a game consisting of drawing one ticket at random from a box of numbered tickets. You win the amount (in dollars) of the number on the ticket you draw. Two boxes are available with numbered tickets as shown below:

I [0, 1, 2]   II [0, 0, 0, 1, 4]

(a) Find the expected value and variance of your *net* gain per play with box I.
(b) Repeat point (a) for box II.
(c) Given that you have decided to play, which box would you choose and why?

3.12 College students in Florida are required to pass a college level academic skills test (CLAST) before completing their college work. The score for 60 students on the mathematics portion of this test are shown below:

225	264	272	272	279	282	282	287	287	287	289
292	292	295	295	295	295	300	300	303	303	303
306	306	306	310	313	313	313	313	313	313	317
317	317	320	320	320	324	324	324	324	324	324
324	324	324	329	334	334	334	334	346	346	353
363	363	375	375	375						

(a) If one student's score is selected randomly to represent these 60, find μ and σ.
(b) Do you think the 68 %–95 % rule will work well to interpret these and other similar sets of scores?

3.13 Daily sales records for a computer manufacturing firm show that it will sell 0, 1, or 2 mainframe computer systems with probabilities as listed:

Number of Sales	0	1	2
Probability	.7	.2	.1

Find the expected value, variance, and standard deviation for daily sales.

3.14 Approximately 10 % of the glass bottles coming off a production line have serious defects in the glass. If two bottles are randomly selected for inspection, find the expected value and variance of the number of inspected bottles with serious defects.

3.15 Two construction contracts are to be randomly assigned to one or more of three firms, I, II, and III. A firm may receive more than one contract. If each contract has a potential profit of $90,000, find the expected potential profit for firm I. Also, find the expected potential profit for firms I and II together.

3.16 If two balanced coins are tossed, what is the expected value and variance of the number of heads observed?

3.17 The number of breakdowns for a university computer system is closely monitored by the director of the computing center, since it is critical to the efficient operation of the center. The number averages 4 per week, with a standard deviation of .8 per week.

(a) Find an interval that must include at least 90 % of the weekly figures on number of breakdowns.

(b) The center director promises that the number of breakdowns will rarely exceed 8 in a one week period. Is the director safe in this claim? Why?

3.18 Keeping an adequate supply of spare parts on hand is an important function of the parts department of a large electronics firm. The monthly demand for microcomputer printer boards was studied for some months and found to average 28 with a standard deviation of 4. How many printer boards should be stocked at the beginning of each month to ensure that the demand will exceed the supply with probability less than .10?

3.19 An important feature of golf cart batteries is the number of minutes they will perform before needing to be recharged. A certain manufacturer advertises batteries that will run, under a 75 amp discharge test, for an average of 100 minutes, with a standard deviation of 5 minutes.

(a) Find an interval that must contain at least 90 % of the performance periods for batteries of this type.

(b) Would you expect many batteries to die out in less than 80 minutes? Why?

3.20 Costs of equipment maintenance are an important part of a firm's budget. Each visit by a field representative to check out a malfunction in a word processing system costs $40. The word processing system is expected to malfunction approximately 5 times per month, and the standard deviation of the number of malfunctions is 2.

(a) Find the expected value and standard deviation of the monthly cost of visits by the field representative.

(b) How much should the firm budget per month so the costs of these visits will be covered at least 75 % of the time?

At this point, it may seem like every problem has its own unique probability distribution, and we must start from the basics to construct such a distribution each time a new problem comes up. Fortunately, that is not the case. Certain basic probability distributions can be developed as models for a large number of practical problems. In the remainder of this chapter, we consider some fundamental discrete distributions, looking at the theoretical assumptions that underlie these distributions as well as the means, variances, and applications of the distributions.

3.3 THE BERNOULLI DISTRIBUTION

Consider the inspection of a single item taken from an assembly line. Suppose a 0 is recorded if the item is not defective, and a 1 is recorded if the item is defective. If X is the random variable denoting the condition of the inspected item, then $X = 0$ with probability $(1 - p)$ and $X = 1$ with probability p, where p denotes the probability of observing a defective item. The probability distribution of X is then given by

$$p(x) = p^x(1 - p)^{1-x} \qquad x = 0, 1$$

where $p(x)$ denotes the probability that $X = x$. Such a random variable is said to have a *Bernoulli distribution,* or to represent the outcome of a single Bernoulli trial. Any random variable denoting the presence or absence of a certain condition in an observed phenomenon will possess a distribution of this type. Frequently, one outcome is termed *success* and the other *failure.*

Suppose that we repeatedly observe items of this type, and record a value of X for each item observed. What is the average of X we would expect to see? By Definition 3.4 the expected value of X is given by

$$\begin{aligned} E(X) &= \sum_x xp(x) \\ &= 0p(0) + 1p(1) \\ &= 0(1 - p) + 1(p) = p. \end{aligned}$$

Thus, if 10% of the items are defective, we expect to observe an *average* of .1 defective per item inspected. (In other words, we would expect to see one defective item for every ten inspected.)

For the Bernoulli random variable X, the variance (see Theorem 3.3) is

$$\begin{aligned} V(X) &= E(X^2) - [E(X)]^2 \\ &= \sum_x x^2p(x) - p^2 \\ &= 0(1 - p) + 1(p) - p^2 \\ &= p - p^2 = p(1 - p). \end{aligned}$$

The Bernoulli random variable will be used as a building block to form other probability distributions, such as the binomial distribution of Section 3.4. The properties of the Bernoulli distribution are summarized here.

The Bernoulli Distribution

$$p(x) = p^x(1-p)^{1-x} \qquad x = 0, 1 \quad \text{for} \quad 0 \le p \le 1$$
$$E(X) = p \qquad V(X) = p(1-p)$$

3.4 THE BINOMIAL DISTRIBUTION

Instead of inspecting a single item, as we do with the Bernoulli random variable, suppose we now independently inspect n items and record values for $X_1, X_2, \ldots, X_n$, where $X_i = 1$ if the ith inspected item is defective and $X_i = 0$ otherwise. We, in fact, have observed a sequence of n independent Bernoulli random variables. One especially interesting function of $X_1, \ldots, X_n$ is the sum

$$Y = \sum_{i=1}^{n} X_i$$

which denotes the number of defectives among the n sampled items.

We can easily find the probability distribution for Y under the assumption that $P(X_i = 1) = p$, where p remains constant over all trials. For simplicity let's look at the specific case of $n = 3$. The random variable Y can then take on four possible values, 0, 1, 2, and 3. For Y to be 0 all three X_i's must be 0. Thus,

$$\begin{aligned} P(Y = 0) &= P(X_1 = 0, X_2 = 0, X_3 = 0) \\ &= P(X_1 = 0)P(X_2 = 0)P(X_3 = 0) \\ &= (1-p)^3. \end{aligned}$$

Now if $Y = 1$, then exactly one value of X_i is 1 and the other two are 0. Thus,

$$\begin{aligned} P(Y = 1) &= P[(X_1 = 1, X_2 = 0, X_3 = 0) \cup (X_1 = 0, X_2 = 1, X_3 = 0) \\ &\quad \cup (X_1 = 0, X_2 = 0, X_3 = 1)] \\ &= P(X_1 = 1, X_2 = 0, X_3 = 0) + P(X_1 = 0, X_2 = 1, X_3 = 0) \\ &\quad + P(X_1 = 0, X_2 = 0, X_3 = 1) \end{aligned}$$

(because the three possibilities are mutually exclusive)

$$\begin{aligned}
&= P(X_1 = 1)P(X_2 = 0)P(X_3 = 0) \\
&\quad + P(X_1 = 0)P(X_2 = 1)P(X_3 = 0) \\
&\quad + P(X_1 = 0)P(X_2 = 0)P(X_3 = 1) \\
&= p(1 - p)^2 + p(1 - p)^2 + p(1 - p)^2 \\
&= 3p(1 - p)^2.
\end{aligned}$$

For $Y = 2$ two values of X_i must be 1 and one must be 0, which can also occur in three mutually exclusive ways. Hence,

$$\begin{aligned}
P(Y = 2) &= P(X_1 = 1, X_2 = 1, X_3 = 0) \\
&\quad + P(X_1 = 1, X_2 = 0, X_3 = 1) \\
&\quad + P(X_1 = 0, X_2 = 1, X_3 = 1) \\
&= 3p^2(1 - p).
\end{aligned}$$

The event $Y = 3$ can occur only if all values of X_i are 1, so

$$\begin{aligned}
P(Y = 3) &= P(X_1 = 1, X_2 = 1, X_3 = 1) \\
&= P(X_1 = 1)P(X_2 = 1)P(X_3 = 1) \\
&= p^3.
\end{aligned}$$

Notice that the coefficient in each of the expressions for $P(Y = y)$ is the number of ways of selecting y positions, in the sequence, in which to place 1s. Because there are three positions in the sequence, this number amounts to $\binom{3}{y}$. Thus, we can write

$$P(Y = y) = \binom{3}{y} p^y (1 - p)^{3-y} \qquad y = 0, 1, 2, 3, \text{ when } n = 3.$$

For general n, the probability that Y takes on a specific value, say y, is given by the term $p^y(1 - p)^{n-y}$ multiplied by the number of possible outcomes resulting in exactly y defectives being observed. This number is the number of ways of selecting y positions for defectives in the n possible positions of the sequence, and it is given by

$$\binom{n}{y} = \frac{n!}{y!\,(n - y)!}$$

where $n! = n(n - 1) \cdots 1$ and $0! = 1$. Thus, in general,

$$P(Y = y) = p(y) = \binom{n}{y} p^y (1 - p)^{n-y} \qquad y = 0, 1, \ldots, n.$$

The probability distribution just given is referred to as the *binomial distribution.*

To summarize, a random variable Y possesses a binomial distribution if

1. the experiment consists of a fixed number n of identical trials;
2. each trial can result in one of only two possible outcomes, called *success* or *failure*;
3. the probability of success, p, is constant from trial to trial;
4. the trials are independent;
5. Y is defined to be the number of successes among the n trials.

Many experimental situations result in a random variable that can be adequately modeled by the binomial distribution. In addition to the number of defectives in a sample of n items, examples may include the number of employees favoring a certain retirement policy out of n employees interviewed, the number of pistons in an eight-cylinder engine that are misfiring, and the number of electronic systems sold this week out of the n that are manufactured. A general formula for $p(x)$ identifies a family of distributions indexed by certain constants called *parameters*. Thus n and p are the parameters of the binomial distribution.

Example 3.7 Suppose that a large lot of fuses contains 10% defectives. If four fuses are randomly sampled from the lot, find the probability that exactly one fuse is defective. Find the probability that at least one fuse, in the sample of four, is defective.

Solution

We assume that the four trials are independent and that the probability of observing a defective is the same (.1) for each trial. This would be approximately true if the lot indeed were *large*. (If the lot contained only a few fuses, removal of one fuse would substantially change the probability of observing a defective on the second draw.) Thus the binomial distribution provides a reasonable model for this experiment, and we have, with Y denoting the number of defectives,

$$p(1) = \binom{4}{1}(.1)^1(.9)^3 = .2916.$$

To find $P(Y \geq 1)$, observe that

$$\begin{aligned} P(Y \geq 1) &= 1 - P(Y = 0) = 1 - p(0) \\ &= 1 - \binom{4}{0}(.1)^0(.9)^4 \\ &= 1 - (.9)^4 \\ &= .3439. \end{aligned}$$

Discrete distributions, like the binomial, can arise in situations where the underlying problem involves a continuous random variable that is not discrete. The following example provides an illustration.

Example 3.8 In a study of life lengths for a certain type of battery, it was found that the probability of a battery life X exceeding 4 hours is .135. If three such batteries are in use in independently operating systems, find the probability that only one of the batteries lasts 4 hours or more.

Solution

Letting Y denote the number of batteries lasting 4 hours or more, we can reasonably assume Y to have a binomial distribution with $p = .135$. Hence,

$$P(Y = 1) = p(1) = \binom{3}{1}(.135)^1(.865)^2$$
$$= .303.$$

There are numerous ways of finding $E(Y)$ and $V(Y)$ for a binomially distributed random variable Y. We might use the basic definition and compute

$$E(Y) = \sum_x yp(y)$$
$$= \sum_{y=0}^{n} y\binom{n}{y}p^y(1 - p)^{n-y}$$

but direct evaluation of this expression is a little tricky. Another approach is to make use of the results on linear functions of random variables, to be presented in Chapter 5. We will show in Chapter 5 that, because the binomial Y arose as a sum of independent Bernoulli random variables, $X_1, \ldots, X_n$

$$E(Y) = E\left[\sum_{i=1}^{n} X_i\right] = \sum_{i=1}^{n} E(X_i)$$
$$= \sum_{i=1}^{n} p = np$$

and

$$V(Y) = \sum_{i=1}^{n} V(X_i) = \sum_{i=1}^{n} p(1 - p) = np(1 - p).$$

Example 3.9 Referring to Example 3.7, page 78, suppose that the four fuses sampled from the lot were shipped to a customer before being tested, on a guarantee basis. Assume that the cost of making the shipment good is given by $C = 3Y^2$, where Y denotes the number of defectives in the shipment of four. Find the expected repair cost.

Solution

We know that

$$E(C) = E(3Y^2) = 3E(Y^2)$$

and it now remains to find $E(Y^2)$. From Theorem 3.3,

$$V(Y) = E(Y - \mu)^2 = E(Y^2) - \mu^2.$$

Since $V(Y) = np(1 - p)$ and $\mu = E(Y) = np$, we see that

$$\begin{aligned} E(Y^2) &= V(Y) + \mu^2 \\ &= np(1 - p) + (np)^2. \end{aligned}$$

For Example 3.7, $p = .1$ and $n = 4$, and hence,

$$\begin{aligned} E(C) &= 3E(Y^2) = 3[np(1 - p) + (np)^2] \\ &= 3[(4)(.1)(.9) + (4)^2(.1)^2] \\ &= 1.56. \end{aligned}$$

If the costs were originally in dollars, we could expect to pay an average of \$1.56 in repair costs for each shipment of four fuses.

Table 2 of the Appendix gives cumulative binomial probabilities for selected values of n and p. The entries in the table are values of

$$\sum_{y=0}^{a} p(y) = \sum_{y=0}^{a} \binom{n}{y} p^y (1 - p)^{n-y}, \qquad a = 0, 1, \ldots, n - 1.$$

The following example illustrates the use of Table 2.

Example 3.10 An industrial firm supplies ten manufacturing plants with a certain chemical. The probability that any one firm calls in an order on a given day is .2, and this is the same for all ten plants. Find the probability that, on the given day, the number of plants calling in orders is

(a) at most 3.
(b) at least 3.
(c) exactly 3.

Solution

Let Y denote the number of plants calling in orders on the day in question. If the plants order independently, then Y can be modeled to have a binomial distribution with $p = .2$.

(a) We then have

$$\begin{aligned} P(Y \le 3) &= \sum_{y=0}^{3} p(y) \\ &= \sum_{y=0}^{3} \binom{10}{y} (.2)^y (.8)^{10-y} \\ &= .879 \end{aligned}$$

from Table 2(b), column $p = .2$ and row $a = 3$.

(b) Note that

$$\begin{aligned} P(Y \ge 3) &= 1 - P(Y \le 2) \\ &= 1 - \sum_{y=0}^{2} \binom{10}{y} (.2)^y (.8)^{10-y} \\ &= 1 - .678 = .322. \end{aligned}$$

(c) Observe that

$$\begin{aligned} P(Y = 3) &= P(Y \le 3) - P(Y \le 2) \\ &= .879 - .678 = .201 \end{aligned}$$

from results just established.

The examples used to this point have specified n and p to calculate probabilities or expected values. Sometimes, however, it is necessary to choose n to achieve a specified probability. Example 3.11 illustrates the point.

Example 3.11 The guidance system for a rocket operates correctly with probability p when called upon. Independent but identical backup systems are installed in the rocket so that the probability that at least one system will operate correctly when called upon is no less than .99. Let n denote the number of guidance systems in the rocket. How large must n be to achieve the specified probability of at least one guidance system operating if

(a) $p = .9$?

(b) $p = .8$?

Solution

Let Y denote the number of correctly operating systems. If the systems are identical and independent, Y has a binomial distribution. Thus,

$$\begin{aligned} P(Y \ge 1) &= 1 - P(Y = 0) \\ &= 1 - \binom{n}{0} p^0 (1-p)^n \\ &= 1 - (1-p)^n. \end{aligned}$$

The conditions specify that n must be such that $P(Y \geq 1) = .99$ or more.

(a) When $p = .9$,

$$P(Y \geq 1) = 1 - (1 - .9)^n \geq .99$$

results in

$$1 - (.1)^n \geq .99$$

or

$$(.1)^n \leq 1 - .99 = .01$$

so $n = 2$. That is, installing two guidance systems will satisfy the specifications.

(b) When $p = .8$,

$$P(Y \geq 1) = 1 - (1 - .8)^n \geq .99$$

results in

$$(.2)^n \leq .01.$$

Now $(.2)^2 = .04$, and $(.2)^3 = .008$, and we must go to $n = 3$ systems so that

$$P(Y \geq 1) = 1 - (.2)^2 = .992 > .99.$$

Note: We cannot achieve the .99 probability exactly because Y can assume only integer values.

We now move on to a discussion of other discrete random variables, but the binomial distribution, summarized here, will be used frequently throughout the text.

The Binomial Distribution

$$p(y) = \binom{n}{y} p^y (1 - p)^{n-y} \qquad y = 0, 1, \ldots, n \text{ for } 0 \leq p \leq 1$$

$$E(Y) = np \qquad V(Y) = np(1 - p)$$

EXERCISES

3.21 Let X denote a random variable having a binomial distribution with $p = .2$ and $n = 4$. Find

(a) $P(X = 2)$
(b) $P(X \geq 2)$
(c) $P(X \leq 2)$
(d) $E(X)$
(e) $V(X)$.

3.22 Let X denote a random variable having a binomial distribution with $p = .4$ and $n = 20$. Use Table 2 of the Appendix to evaluate

(a) $P(X \leq 6)$ (b) $P(X \geq 12)$
(c) $P(X = 8)$.

3.23 A machine that fills boxes of cereal underfills a certain proportion, p. If twenty-five boxes are randomly selected from the output of this machine, find the probability that no more than two are underfilled when

(a) $p = .1$. (b) $p = .2$.

3.24 In testing the lethal concentration of a chemical found in polluted water, it is found that a certain concentration will kill 20 % of the fish that are subjected to it for 24 hours. If 20 fish are placed in a tank containing this concentration of chemical, find the probability that, after 24 hours,

(a) exactly 14 survive (b) at least 10 survive.
(c) at most 16 survive.

3.25 Referring to Exercise 3.24,

(a) find the number expected to survive, out of 20.
(b) find the variance of the number of survivors, out of 20.

3.26 Among persons donating blood to a clinic 80 % have Rh^+ blood (that is, the Rhesus factor is present in their blood). Five people donate blood at the clinic on a particular day.

(a) Find the probability that at least 1 of the 5 does not have the Rh factor.
(b) Find the probability that at most 4 of the 5 have Rh^+ blood.

3.27 The clinic needs 5 Rh^+ donors on a certain day. How many people must donate blood to have the probability of at least 5 Rh^+ donors over .90?

3.28 The *U.S. Statistical Abstract* reports that the median family income in the United States for 1985 was \$27,735. Among four randomly selected families, find the probability that

(a) all four had incomes above \$27,735 in 1985.
(b) one of the four had an income below \$27,735 in 1985.

3.29 According to B. E. Sullivan (1984, p. 119), 55 % of United States corporations say that one of the most important factors in locating a corporate headquarters is the "quality of life" for the employees. If five firms are contacted by the governor of Florida concerning possible relocation to that state, find the probability that at least three say that quality of life is an important factor in their decision. What assumptions have you made in finding this answer?

3.30 Goranson and Hall (1980, pp. 279–280) explain that the probability of detecting a crack in an airplane wing is the product of p_1, the probability of inspecting a plane with a wing crack; p_2, the probability of inspecting the detail in which the crack is located; and p_3, the probability of detecting the damage.

(a) What assumptions justify the multiplication of these probabilities?

(b) Suppose $p_1 = .9$, $p_2 = .8$, and $p_3 = .5$ for a certain fleet of planes. If three planes are inspected from this fleet, find the probability that a wing crack will be detected in at least one of them.

3.31 A missile protection system consists of n radar sets operating independently, each with probability .9 of detecting an aircraft entering a specified zone. (All radar sets cover the same zone.) If an airplane enters the zone, find the probability that it will be detected if

(a) $n = 2$. (b) $n = 4$.

3.32 Refer to Exercise 3.31. How large must n be to have a .99 probability of detecting an aircraft entering the zone?

3.33 A complex electronic system is built with a certain number of backup components in its subsystems. One subsystem has four identical components, each with probability of .2 of failing in less than 1000 hours. The subsystem will operate if any two or more of the four components are operating. Assuming that the components operate independently, find the probability that

(a) exactly two of the four components last longer than 1000 hours.

(b) the subsystem operates longer than 1000 hours.

3.34 An oil exploration firm is to drill ten wells, with each well having probability .1 of successfully producing oil. It costs the firm \$10,000 to drill each well. A successful well will bring in oil worth \$500,000.

(a) Find the firm's expected gain from the ten wells.

(b) Find the standard deviation of the firm's gain.

3.35 A firm sells four items randomly selected from a large lot known to contain 10% defectives. Let Y denote the number of defectives among the four sold. The purchaser of the item will return the defectives for repair, and the repair cost is given by

$$C = 3Y^2 + Y + 2.$$

Find the expected repair cost.

3.36 From a large lot of new tires, n are to be sampled by a potential buyer, and the number of defectives, X, is to be observed. If at least one defective is observed in the sample of n, the entire lot is to be rejected by the potential buyer. Find n so that the probability of detecting at least one defective is approximately .90 if

(a) 10% of the lot is defective.

(b) 5% of the lot is defective.

3.37 Ten motors are packaged for sale in a certain warehouse. The motors sell for \$100 each, but a double-your-money-back guarantee is in effect for any defectives the purchaser might receive. Find the expected net gain for the seller if the probability of any one motor being defective is .08. (Assume the quality of any one motor is independent of that of the others.)

3.5

THE GEOMETRIC DISTRIBUTION

Suppose a series of test firings of a rocket engine can be represented by a sequence of independent Bernoulli random variables with $X_i = 1$ if the ith trial results in a successful firing and $X_i = 0$ otherwise. Assume that the probability of a successful firing is constant for the trials, and let this probability be denoted by p. For this problem we might be interested in the number of the trial on which the first successful firing occurs. If Y denotes the number of the trial on which the first success occurs, then

$$\begin{aligned} P(Y = y) = p(y) &= P(X_1 = 0, X_2 = 0, \ldots, X_{y-1} = 0, X_y = 1) \\ &= P(X_1 = 0)P(X_2 = 0) \cdots P(X_{y-1} = 0)P(X_y = 1) \\ &= (1 - p)^{y-1}p \qquad y = 1, 2, \ldots \end{aligned}$$

because of the independence of the trials. This formula is referred to as the *geometric probability distribution.* Note that this random variable can take on a countably infinite number of possible values.

In addition to the rocket-firing example just given, here are other situations that result in a random variable whose probability can be modeled by a geometric distribution: the number of customers contacted before the first sale is made, the number of years a dam is in service before it overflows, and the number of automobiles going through a radar check before the first speeder is detected. The following example illustrates the use of the geometric distribution.

Example 3.12 A recruiting firm finds that 30% of the applicants for a certain industrial job have advanced training in computer programming. Applicants are interviewed sequentially and selected at random from the pool. Find the probability that the first applicant having advanced training in programming is found on the fifth interview.

Solution

The probability of finding a suitably trained applicant will remain relatively constant from trial to trial if the pool of applicants is reasonably large. It then makes sense to define Y as the number of the trial on which the first applicant having advanced training in programming is found, and model Y as having a geometric distribution. Thus,

$$P(Y = 5) = p(5) = (.7)^4(.3) = .072.$$

From the basic definition

$$\begin{aligned} E(Y) &= \sum_y yp(y) = \sum_{y=1}^{\infty} yp(1-p)^{y-1} \\ &= p\sum_{y=1}^{\infty} y(1-p)^{y-1} \\ &= p[1 + 2(1-p) + 3(1-p)^2 + \cdots]. \end{aligned}$$

The infinite series can be split up into a triangular array of series as follows:

$$\begin{aligned} E(Y) = p[1 &+ (1-p) + (1-p)^2 + \cdots \\ &+ (1-p) + (1-p)^2 + \cdots \\ &+ (1-p)^2 + \cdots \\ &+ \cdots]. \end{aligned}$$

Each line on the right side is an infinite, decreasing geometric progression with common ratio $(1 - p)$. Thus the first line inside the bracket sums to $1/p$, the second to $(1 - p)/p$, the third to $(1 - p)^2/p$, and so on.[2] On accumulating these totals, we then have

$$\begin{aligned} E(Y) &= p\left[\frac{1}{p} + \frac{1-p}{p} + \frac{(1-p)^2}{p} + \cdots\right] \\ &= 1 + (1-p) + (1-p)^2 + \cdots \\ &= \frac{1}{1-(1-p)} = \frac{1}{p}. \end{aligned}$$

This answer for $E(Y)$ should seem intuitively realistic. For example, if 10% of a certain lot of items are defective, and if an inspector looks at randomly selected items one at a time, then he should expect to wait until the tenth trial to see the first defective.

The variance of the geometric distribution will be derived in Chapter 5. The result, however, is

$$V(Y) = \frac{1-p}{p^2}.$$

Example 3.13 Referring to Example 3.12, pages 85–86, let Y denote the number of the trial on which the first applicant having advanced training in computer programming is found. Suppose that the first applicant with advanced training is offered the position, and the applicant accepts. If each interview costs \$30.00, find the expected value and variance of the total cost of interviewing until the job is filled. Within what interval would this cost be expected to fall?

[2] Recall that $a + ax + ax^2 + ax^3 + \cdots = a/(1 - x)$, if $|x| < 1$.

Solution

Because Y is the number of the trial on which the interviewing process ends, the total cost of interviewing is $C = 30Y$. Now,

$$E(C) = 30E(Y) = 30\left(\frac{1}{p}\right)$$

$$= 30\left(\frac{1}{.3}\right) = 100$$

and

$$V(C) = (30)^2 V(Y) = \frac{900(1-p)}{p^2}$$

$$= \frac{900(.7)}{(.3)^2}$$

$$= 7000.$$

[handwritten margin notes:]
$$1 - \frac{1}{k^2} = .75$$
$$\frac{1}{k^2} = .25$$
$$k = 2$$

The standard deviation of C is then $\sqrt{V(C)} = \sqrt{7000} = 83.67$. Tchebysheff's Theorem (see Section 3.2) says that C will lie within two standard deviations of its mean at least 75 % of the time. Thus, it is quite likely that C will be between

$$100 - 2(83.67) \text{ and } 100 + 2(83.67)$$

or

$$-67.34 \text{ and } 267.34.$$

Since the lower bound is negative, that end of the interval is meaningless. However, we can still say that it is quite likely that the total cost of such an interviewing process will be less than $267.34.

The Geometric Distribution

$$p(y) = p(1-p)^{y-1} \qquad y = 1, 2, \ldots \text{ for } 0 < p < 1$$

$$E(Y) = \frac{1}{p} \qquad\qquad V(Y) = \frac{1-p}{p^2}$$

3.6

THE NEGATIVE BINOMIAL DISTRIBUTION

In Section 3.5, we saw that the geometric distribution models the probabilistic behavior of the number of the trial on which the *first success* occurs in a sequence of independent Bernoulli trials. But what if we were interested in the number of the trial for the second success, or third success, or, in general, the rth success? The distribution governing the probabilistic behavior in these cases is called the *negative binomial distribution.*

Let Y denote the number of the trial on which the rth success occurs in a sequence of independent Bernoulli trials with p denoting the common probability of success. We can derive the distribution of Y from known facts. Now,

$$\begin{aligned} P(Y = y) &= P[(\text{1st } (y-1) \text{ trials contain } (r-1) \text{ successes} \\ &\qquad \text{and } y\text{th trial is a success}] \\ &= P[(\text{1st } (y-1) \text{ trials contain } (r-1) \text{ successes}] \\ &\qquad \times P[y\text{th trial is a success}]. \end{aligned}$$

Because the trials are independent, the joint probability can be written as a product of probabilities. The first probability statement is identical to that resulting in a binomial model, and hence,

$$\begin{aligned} P(Y = y) = p(y) &= \binom{y-1}{r-1} p^{r-1}(1-p)^{y-r} \times p \\ &= \binom{y-1}{r-1} p^{r}(1-p)^{y-r} \qquad y = r, r+1, \ldots . \end{aligned}$$

Example 3.14 As in Example 3.12, 30% of the applicants for a certain position have advanced training in computer programming. Suppose that three jobs requiring advanced programming training are open. Find the probability that the third qualified applicant is found on the fifth interview, if the applicants are interviewed sequentially and at random.

Solution

Again we assume independent trials with .3 being the probability of finding a qualified candidate on any one trial. Let Y denote the number of the trial on which the third qualified candidate is found. Then Y can reasonably be assumed to have a negative binomial distribution, so

$$\begin{aligned} P(Y = 5) = p(5) &= \binom{4}{2}(.3)^3(.7)^2 \\ &= 6(.3)^3(.7)^2 \\ &= .079. \end{aligned}$$

The expected value, or mean, and the variance for the negative binomially distributed Y are found easily by analogy with the geometric distribution. Recall that Y denotes the number of the trial on which the rth success occurs.

Let W_1 denote the number of the trial on which the first success occurs; W_2 the number of trials between the first success and the second success, including the trial of the second success; W_3 the number of trials between the second success and the third success; and so forth. The results of the trials are then as diagramed next (F standing for failure and S for success).

$$\underbrace{FF\cdots FS}_{W_1}\underbrace{F\cdots FS}_{W_2}\underbrace{F\cdots FS}_{W_3}.$$

It is easy to observe that $Y = \sum_{i=1}^{r} W_i$, where the W_i's are independent and each has a geometric distribution. Thus, by results to be derived in Chapter 5,

$$E(Y) = \sum_{i=1}^{r} E(W_i) = \sum_{i=1}^{r} \left(\frac{1}{p}\right) = \frac{r}{p}$$

and

$$V(Y) = \sum_{i=1}^{r} V(W_i) = \sum_{i=1}^{r} \frac{1-p}{p^2} = \frac{r(1-p)}{p^2}.$$

Example 3.15 A large stockpile of used pumps contains 20% that are unusable and need repair. A repairman is sent to the stockpile with three repair kits. He selects pumps at random and tests them one at a time. If a pump works, he goes on to the next one. If a pump doesn't work, he uses one of his repair kits on it. Suppose that it takes 10 minutes to test a pump if it works, and 30 minutes to test and repair a pump that does not work. Find the expected value and variance of the total time it takes the repairman to use up his three kits.

Solution

Letting Y denote the number of the trial on which the third defective pump is found, we see that Y has a negative binomial distribution with $p = .2$. The total time T taken to use up the three repair kits is

$$T = 10(Y - 3) + 3(30) = 10Y + 3(20).$$

(Each test takes 10 minutes, but the repairs take 20 extra minutes.)

It follows that

$$\begin{aligned} E(T) &= 10E(Y) + 3(20) \\ &= 10\left(\frac{3}{.2}\right) + 3(20) \\ &= 150 + 60 = 210 \end{aligned}$$

and

$$\begin{aligned} V(T) &= (10)^2 V(Y) \\ &= (10)^2\left[\frac{3(.8)}{(.2)^2}\right] \\ &= 100[60] \\ &= 6000. \end{aligned}$$

Thus, the total time to use up the kits has an expected value of 210 minutes, with a standard deviation of $\sqrt{6000} = 77.46$ minutes.

The negative binomial distribution is used to model a wide variety of phenomena, from number of defects per square yard in fabrics to numbers of individuals in an insect population after many generations.

The Negative Binomial Distribution

$$p(y) = \binom{y-1}{r-1} p^r (1-p)^{y-r} \qquad y = r, r+1, \ldots \text{ for } 0 < p < 1$$

$$E(Y) = \frac{r}{p} \qquad V(Y) = \frac{r(1-p)}{p^2}$$

EXERCISES

3.38 Let Y denote a random variable having a geometric distribution, with probability of success on any trial denoted by p.

(a) Find $P(Y \geq 2)$ if $p = .1$.

(b) Find $P(Y > 4 \mid Y > 2)$ for general p. Compare with the unconditional probability $P(Y > 2)$.

3.39 Let Y denote a negative binomial random variable with $p = .4$. Find $P(Y \geq 4)$ if

(a) $r = 2$. (b) $r = 4$.

3.40 Suppose that 10% of the engines manufactured on a certain assembly line are defective. If engines are randomly selected one at a time and tested, find the probability that the first nondefective engine is found on the second trial.

3.41 Referring to Exercise 3.40, find the probability that the third nondefective engine is found

(a) on the fifth trial. (b) on or before the fifth trial.

3.42 Referring to Exercise 3.40, given that the first two engines are defective, find the probability that at least two more engines must be tested before the first nondefective is found.

3.43 Referring to Exercise 3.40, find the mean and variance of the number of the trial on which

(a) the first nondefective engine is found.

(b) the third nondefective engine is found.

3.44 The employees of a firm that manufactures insulation are being tested for indications of asbestos in their lungs. The firm is requested to send three employees who have positive indications of asbestos on to a medical center for further testing. If 40 % of the employees have positive indications of asbestos in their lungs, find the probability that ten employees must be tested to find three positives.

3.45 Referring to Exercise 3.44, if each test costs \$20, find the expected value and variance of the total cost of conducting the tests to locate three positives. Do you think it is highly likely that the cost of completing these tests would exceed \$350?

3.46 If one-third of the persons donating blood at a clinic have O^+ blood, find the probability that the

(a) first O^+ donor is the fourth donor of the day.

(b) second O^+ donor is the fourth donor of the day.

3.47 A geological study indicates that an exploratory oil well drilled in a certain region should strike oil with probability .2. Find the probability that the

(a) first strike of oil comes on the third well drilled.

(b) third strike of oil comes on the fifth well drilled.

What assumptions are necessary for your answers to be correct?

3.48 In the setting of Exercise 3.47, suppose a company wants to set up three producing wells. Find the expected value and variance of the number of wells that must be drilled to find the three successful ones.

3.49 A large lot of tires contains 10 % defectives. Four are to be chosen to place on a car.

(a) Find the probability that six tires must be selected from the lot to get four good ones.

(b) Find the expected value and variance of the number of selections that must be made to get four good tires.

3.50 The telephone lines coming into an airline reservation office are all occupied about 60 % of the time.

(a) If you are calling this office, what is the probability that you complete your call on the first try? second try? third try?

(b) If you and a friend both must complete separate calls to this reservation office, what is the probability that it takes a total of four tries for the two of you?

3.51 An appliance comes in two colors, white and brown, which are in equal demand. A certain dealer in these appliances has three of each color in stock, although this is not known to the customers. Customers arrive and independently order these appliances. Find the probability that

(a) the third white is ordered by the fifth customer.
(b) the third brown is ordered by the fifth customer.
(c) all of the whites are ordered before any of the browns.
(d) all of the whites are ordered before all of the browns.

3.7

THE POISSON DISTRIBUTION

A number of probability distributions come about through limiting arguments applied to other distributions. One such very useful distribution is called the *Poisson distribution.*

Consider the development of a probabilistic model for the number of accidents occurring at a particular highway intersection in a period of one week. We can think of the time interval as being split up into n subintervals such that

$$P(\text{one accident in a subinterval}) = p$$
$$P(\text{no accidents in a subinterval}) = 1 - p.$$

Note that we are assuming the same value of p holds for all subintervals, and the probability of more than one accident in any one subinterval is zero. If the occurrence of accidents can be regarded as independent from subinterval to subinterval, then the total number of accidents in the time period (which equals the total number of subintervals containing one accident) will have a binomial distribution.

Although there is no unique way to choose the subintervals and we therefore know neither n nor p, it seems reasonable to assume that as n increases, p should decrease. Thus, we want to look at the limit of the binomial probability distribution as $n \to \infty$ and $p \to 0$. To get something interesting, we take the limit under the restriction that the mean, np in the binomial case, remains constant at a value we will call λ.

Now, with $np = \lambda$ or $p = \lambda/n$, we have

$$\begin{aligned}
&\lim_{n\to\infty} \binom{n}{y}\left(\frac{\lambda}{n}\right)^y\left(1 - \frac{\lambda}{n}\right)^{n-y} \\
&= \lim_{n\to\infty} \frac{\lambda^y}{y!}\left(1 - \frac{\lambda}{n}\right)^n \frac{n(n-1)\cdots(n-y+1)}{n^y}\left(1 - \frac{\lambda}{n}\right)^{-y} \\
&= \frac{\lambda^y}{y!}\lim_{n\to\infty}\left(1 - \frac{\lambda}{n}\right)^n\left(1 - \frac{\lambda}{n}\right)^{-y}\left(1 - \frac{1}{n}\right)\left(1 - \frac{2}{n}\right)\cdots\left(1 - \frac{y-1}{n}\right).
\end{aligned}$$

Noting that

$$\lim_{n\to\infty}\left(1-\frac{\lambda}{n}\right)^n = e^{-\lambda}$$

and that all other terms involving n tend to unity, we have the limiting distribution

$$p(y) = \frac{\lambda^y}{y!}e^{-\lambda} \qquad y = 0, 1, 2, \ldots .$$

Recall that λ denotes the mean number of occurrences in one time period (a week for the example under consideration), and hence if t nonoverlapping time periods were considered, the mean would be λt.

This distribution, called the *Poisson distribution with parameter* λ, can be used to model counts in areas or volumes, as well as in time. For example, we may use this distribution to model the number of flaws in a square yard of textile, the number of bacteria colonies in a cubic centimeter of water, or the number of times a machine fails in the course of a workday. We illustrate the use of the Poisson distribution in the following example.

Example 3.16 For a certain manufacturing industry the number of industrial accidents averages three per week. Find the probability that no accidents will occur in a given week.

Solution

If accidents tend to occur independently of one another, and if they occur at a constant rate over time, the Poisson model provides an adequate representation of the probabilities. Thus,

$$p(0) = \frac{3^0}{0!}e^{-3} = e^{-3} = .05.$$

Table 3 of the Appendix gives values for cumulative Poisson probabilities of the form

$$\sum_{y=0}^{a} e^{-\lambda}\frac{\lambda^y}{y!}, \qquad a = 0, 1, 2, \ldots$$

for selected values of λ.

The following example illustrates the use of Table 3.

Example 3.17 Refer to Example 3.16 and let Y denote the number of accidents in the given week. Find $P(Y \le 4)$, $P(Y \ge 4)$, and $P(Y = 4)$.

Solution

From Table 3,

$$P(Y \le 4) = \sum_{y=0}^{4} \frac{(3)^y}{y!} e^{-3} = .815.$$

Also,

$$\begin{aligned} P(Y \ge 4) &= 1 - P(Y \le 3) \\ &= 1 - .647 = .353 \end{aligned}$$

and

$$\begin{aligned} P(Y = 4) &= P(Y \le 4) - P(Y \le 3) \\ &= .815 - .647 = .168. \end{aligned}$$

We can intuitively determine what the mean and variance of a Poisson distribution should be by recalling the mean and variance of a binomial distribution and the relationship between the two distributions. A binomial distribution has mean np and variance $np(1 - p) = np - (np)p$. Now if n gets large and p remains at $np = \lambda$, the variance $np - (np)p = \lambda - \lambda p$ should tend toward λ. In fact, the Poisson distribution does have both mean and variance equal to λ.

The mean of the Poisson distribution is easily derived formally if one remembers a simple Taylor series expansion of e^x; namely,

$$e^x = 1 + x + \frac{x^2}{2!} + \frac{x^3}{3!} + \cdots.$$

Then,

$$\begin{aligned} E(Y) &= \sum_y yp(y) = \sum_{y=0}^{\infty} y \frac{\lambda^y}{y!} e^{-\lambda} = \sum_{y=1}^{\infty} y \frac{\lambda^y}{y!} e^{-\lambda} \\ &= \lambda e^{-\lambda} \sum_{y=1}^{\infty} \frac{\lambda^{y-1}}{(y-1)!} \\ &= \lambda e^{-\lambda}\left(1 + \lambda + \frac{\lambda^2}{2!} + \frac{\lambda^3}{3!} + \cdots\right) \\ &= \lambda e^{-\lambda} e^{\lambda} = \lambda. \end{aligned}$$

The formal derivation of the fact that

$$V(Y) = \lambda$$

is left as an exercise for the interested reader.

Example 3.18 The manager of an industrial plant is planning to buy a new machine of either type A or type B. For each day's operation the number of repairs, X, that machine A requires is a Poisson random variable with mean $.10t$, where t denotes the time (in hours) of daily operation. The number of daily repairs, Y, for machine B is Poisson with mean $.12t$. The daily cost of operating A is $C_A(t) = 10t + 30X^2$; for B it is $C_B(t) = 8t + 30Y^2$. Assume that the repairs take negligible time and each night the machines are to be cleaned, so they operate like new machines at the start of each day. Which machine minimizes the expected daily cost if a day consists of

(a) 10 hours?
(b) 20 hours?

Solution

The expected cost for A is

$$\begin{aligned} E[C_A(t)] &= 10t + 30E(X^2) \\ &= 10t + 30[V(X) + (E(X))^2] \\ &= 10t + 30[.10t + .01t^2] \\ &= 13t + .3t^2. \end{aligned}$$

Similarly,

$$\begin{aligned} E[C_B(t)] &= 8t + 30E(Y^2) \\ &= 8t + 30[.12t + .0144t^2] \\ &= 11.6t + .432t^2. \end{aligned}$$

For part (a),

$$E[C_A(10)] = 13(10) + .3(10)^2 = 160$$

and

$$E[C_B(10)] = 11.6(10) + .432(10)^2 = 159.2$$

which results in the choice of machine B.

For part (b),

$$E[C_A(20)] = 380$$

and

$$E[C_B(20)] = 404.8$$

which results in the choice of machine A. In conclusion, B is more economical for short time periods because of its smaller hourly operating cost. However, for long time periods, A is more economical because it tends to be repaired less frequently.

The Poisson Distribution

$$p(y) = \frac{\lambda^y}{y!}e^{-\lambda} \qquad y = 0, 1, 2, \ldots$$

$$E(Y) = \lambda \qquad V(Y) = \lambda$$

The Poisson distribution will appear again in Section 3.11.

EXERCISES

3.52 Let Y denote a random variable having a Poisson distribution with mean $\lambda = 2$. Find
(a) $P(Y = 4)$
(b) $P(Y \geq 4)$
(c) $P(Y < 4)$
(d) $P(Y \geq 4 \mid Y \geq 2)$.

3.53 The number of telephone calls coming into the central switchboard of an office building averages four per minute.
(a) Find the probability that no calls will arrive in a given 1-minute period.
(b) Find the probability that at least two calls will arrive in a given 1-minute period.
(c) Find the probability that at least two calls will arrive in a given 2-minute period.

3.54 The quality of computer disks is measured by sending the disks through a certifier that counts the number of missing pulses. A certain brand of computer disks has averaged .1 missing pulse per disk.
(a) Find the probability that the next inspected disk will have no missing pulse.
(b) Find the probability that the next inspected disk will have more than one missing pulse.
(c) Find the probability that neither of the next two inspected disks will contain any missing pulse.

3.55 The national maximum speed limit (NMSL) of 55 miles per hour was placed in force in the United States in early 1974. The benefits of this law have been studied by D. B. Kamerud who reports that the fatality rate for interstate highways with the NMSL in 1975 is approximately 16 per 10^9 vehicle miles.
(a) Find the probability of at most 15 fatalities occurring in 10^9 vehicle miles.
(b) Find the probability of at least 20 fatalities occurring in 10^9 vehicle miles.
(Assume that fatalities per vehicle mile follows a Poisson distribution.)

3.56 In the article cited in Exercise 3.55, the projected fatality rate for 1975 if the NMSL had not been in effect would have been 25 per 10^9 vehicle miles. Under these conditions,
(a) find the probability of at most 15 fatalities occurring in 10^9 vehicle miles.
(b) find the probability of at least 20 fatalities occurring in 10^9 vehicle miles.
(c) compare the answers in parts (a) and (b) to those in Exercise 3.55.

3.57 In a timesharing computer system, the number of teleport inquiries averages .2 per millisecond and follows a Poisson distribution.

(a) Find the probability that no inquiries are made during the next millisecond.

(b) Find the probability that no inquiries are made during the next 3 milliseconds.

3.58 Rebuilt ignition systems leave an aircraft rework facility at the rate of three per hour, on the average. The assembly line needs four ignition systems in the next hour. What is the probability that they will be available?

3.59 Customer arrivals at a checkout counter in a department store have a Poisson distribution with an average of eight per hour. For a given hour find the probability that

(a) exactly eight customers arrive.

(b) no more than three customers arrive.

(c) at least two customers arrive.

3.60 Referring to Exercise 3.59, if it takes approximately 10 minutes to service each customer, find the mean and variance of the total service time connected to the customer arrivals for 1 hour. (Assume that an unlimited number of servers are available, so that no customer has to wait for service.) Is it highly likely that total service time would exceed 200 minutes?

3.61 Referring to Exercise 3.59, find the probability that exactly two customers arrive in the 2-hour period of time

(a) between 2:00 P.M. and 4:00 P.M. (one continuous 2-hour period).

(b) between 1:00 P.M. and 2:00 P.M. and between 3:00 P.M. and 4:00 P.M. (two separate 1-hour periods for a total of 2 hours).

3.62 The number of imperfections in the weave of a certain textile has a Poisson distribution with a mean of four per square yard.

(a) Find the probability that a 1-square-yard sample will contain at least one imperfection.

(b) Find the probability that a 3-square-yard sample will contain at least one imperfection.

3.63 Referring to Exercise 3.62, the cost of repairing the imperfections in the weave is $10 per imperfection. Find the mean and standard deviation of the repair costs for an 8-square-yard bolt of the textile in question.

3.64 The number of bacteria colonies of a certain type in samples of polluted water has a Poisson distribution with a mean of two per cubic centimeter.

(a) If four 1-cubic-centimeter samples are independently selected from this water, find the probability that at least one sample will contain one or more bacteria colonies.

(b) How many 1-cubic-centimeter samples should be selected to have a probability of approximately .95 of seeing at least one bacteria colony?

3.65 Let Y have a Poisson distribution with mean λ. Find $E[Y(Y - 1)]$ and use the result to show that $V(Y) = \lambda$.

3.66 A food manufacturer uses an extruder (a machine that produces bite-size foods like cookies and many snack foods) that produces revenue for the firm at the rate of $200 per hour when in operation. However, the extruder breaks down on the average of two times for every 10 hours of operation. If Y denotes the number of breakdowns during the time of operation, the revenue generated by the machine is given by

$$R = 200t - 50Y^2$$

where t denotes hours of operation. The extruder is shut down for routine maintenance on a regular schedule, and operates like a new machine after this maintenance. Find the optimal maintenance interval, t_0, to maximize the expected revenue between shutdowns.

3.67 The number of cars entering a parking lot is a random variable having a Poisson distribution with a mean of four per hour. The lot holds only twelve cars.

(a) Find the probability that the lot fills up in the first hour. (Assume all cars stay in the lot longer than 1 hour.)

(b) Find the probability that fewer than twelve cars arrive during an 8-hour day.

3.8

THE HYPERGEOMETRIC DISTRIBUTION

The distributions already discussed in this chapter have as their basic building block a series of *independent* Bernoulli trials. The examples, such as sampling from large lots, depict situations in which the trials of the experiment generate, for all practical purposes, independent outcomes. Suppose that we have a relatively small lot consisting of N items, of which k are defective. If two items are sampled sequentially, then the outcome for the second draw is very much influenced by what happened on the first draw, provided that the first item drawn remains out of the lot. A new distribution must be developed to handle this situation involving *dependent* trials.

In general suppose a lot consists of N items, of which k are of one type (called *successes*) and $N - k$ are of another type (called *failures*). Suppose that n items are sampled randomly and sequentially from the lot, with none of the sampled items being replaced. (This is called *sampling without replacement.*) Let $X_i = 1$ if the ith draw results in a success and $X_i = 0$ otherwise, $i = 1, \ldots, n$, and let Y denote the total number of successes among the n sampled items. To develop the probability distribution for Y, let us start by looking at a special case for $Y = y$. One way for y successes to occur is to have

$$X_1 = 1, X_2 = 1, \ldots, X_y = 1, X_{y+1} = 0, \ldots, X_n = 0.$$

We know that

$$P(X_1 = 1, X_2 = 2) = P(X_1 = 1)P(X_2 = 1 \mid X_1 = 1)$$

and this result can be extended to give

$$\begin{aligned}P(X_1 = 1, X_2 = 1, \ldots, X_y = 1, X_{y+1} = 0, \ldots, X_n = 0)\\ = P(X_1 = 1)P(X_2 = 1 \mid X_1 = 1)P(X_3 = 1 \mid X_2 = 1, X_1 = 1) \cdots\\ P(X_n = 0 \mid X_{n-1} = 0, \ldots, X_{y+1} = 0, X_y = 1, \ldots, X_1 = 1).\end{aligned}$$

Now,

$$P(X_1 = 1) = \frac{k}{N}$$

if the item is randomly selected, and similarly,

$$P(X_2 = 1 \mid X_1 = 1) = \frac{k-1}{N-1}$$

because, at this point, one of the k successes has been removed. Using this idea repeatedly, we see that

$$\begin{aligned}P(X_1 = 1, \ldots, X_y = 1, X_{y+1} = 0, \ldots, X_n = 0)\\ = \left(\frac{k}{N}\right)\left(\frac{k-1}{N-1}\right) \cdots \left(\frac{k-y+1}{N-y+1}\right)\\ \times \left(\frac{N-k}{N-y}\right) \cdots \left(\frac{N-k-n+y+1}{N-n+1}\right)\end{aligned}$$

provided $y \le k$. A more compact way to write the above expression is to employ factorials, arriving at the formula

$$\frac{\dfrac{k!}{(k-y)!} \times \dfrac{(N-k)!}{(N-k-n+y)!}}{\dfrac{N!}{(N-n)!}}.$$

(The reader can check the equivalence of the two expressions.)

Any specified arrangement of y successes and $(n - y)$ failures will have the same probability as the one just derived for all successes followed by all failures; the terms merely will be rearranged. Thus to find $P(Y = y)$, we need only to count how many of these arrangements are possible. Just as in the binomial case, the number of such arrangements is $\binom{n}{y}$. Hence, we have

$$\begin{aligned}P(Y = y) &= \binom{n}{y} \frac{\dfrac{k!}{(k-y)!} \times \dfrac{(N-k)!}{(N-k-n+y)!}}{\dfrac{N!}{(N-n)!}}\\ &= \frac{\binom{k}{y}\binom{N-k}{n-y}}{\binom{N}{n}}.\end{aligned}$$

Of course, $0 \le y \le k \le N$, $0 \le y \le n \le N$. This formula is referred to as the *hypergeometric probability distribution.* Note that it arises from a situation quite similar to the binomial, except that the trials are *dependent.*

Experiments that result in a random variable possessing a hypergeometric distribution usually involve counting the number of successes in a sample taken from a small lot. Examples could include counting the number of men that show up on a committee of five randomly selected from among twenty employees or counting the number of Brand A alarm systems sold in three sales from a warehouse containing two Brand A and four Brand B systems.

Example 3.19 A personnel director selects two employees for a certain job from a group of six employees, of which one is female and five are male. Find the probability that the woman is selected for one of the jobs.

Solution

If the selections are made at random, and if Y denotes the number of women selected, then the hypergeometric distribution would provide a good model for the behavior of Y. Hence,

$$P(Y = 1) = p(1) = \frac{\binom{1}{1}\binom{5}{1}}{\binom{6}{2}} = \frac{1 \times 5}{15} = \frac{1}{3}.$$

Here $N = 6$, $k = 1$, $n = 2$, and $y = 1$.

It might be instructive to see this calculation from basic principles, letting $X_i = 1$ if the ith draw results in the woman, and $X_i = 0$ otherwise. Then,

$$\begin{aligned} P(Y = 1) &= P(X_1 = 1, X_2 = 0) + P(X_1 = 0, X_2 = 1) \\ &= P(X_1 = 1)P(X_2 = 0 \mid X_1 = 1) \\ &\quad + P(X_1 = 0)P(X_2 = 1 \mid X_1 = 0) \\ &= \left(\frac{1}{6}\right)\left(\frac{5}{5}\right) + \left(\frac{5}{6}\right)\left(\frac{1}{5}\right) = \frac{1}{3}. \end{aligned}$$

The techniques needed to derive the mean and variance of the hypergeometric distribution will be given in Chapter 5. The results are

$$E(Y) = n\left(\frac{k}{N}\right)$$

$$V(Y) = n\left(\frac{k}{N}\right)\left(1 - \frac{k}{N}\right)\left(\frac{N - n}{N - 1}\right).$$

Observe that, because the probability of selecting a success on one draw is k/N, the mean of the hypergeometric distribution has the same form as the mean of the binomial distribution. Also the variance of the hypergeometric is like the variance of the binomial, multiplied by $(N - n)/(N - 1)$, a correction factor for dependent samples.

Example 3.20 In an assembly-line production of industrial robots, gear box assemblies can be installed in 1 minute each if holes have been properly drilled in the boxes, and in 10 minutes each if holes must be redrilled. Twenty gear boxes are in stock, and it is assumed that two will have improperly drilled holes. Five gear boxes must be selected from the twenty available for installation in the next five robots in line.

(a) Find the probability that all five gear boxes will fit properly.

(b) Find the expected value, variance, and standard deviation of the time it takes to install these five gear boxes.

Solution

(a) In this problem, $N = 20$; and the number of nonconforming boxes is assumed to be $k = 2$, according to the manufacturer's usual standards. Let Y' denote the number of nonconforming boxes (number with improperly drilled holes) in the sample of 5. Then,

$$P(Y = 0) = \frac{\binom{2}{0}\binom{18}{5}}{\binom{20}{5}}$$

$$= \frac{(1)(8568)}{15{,}504} = .55.$$

(b) The total time, T, taken to install the boxes (in minutes) is

$$T = 10Y + (5 - Y)$$
$$= 9Y + 5$$

since each of Y nonconforming boxes takes 10 minutes to install, and the others take only 1 minute. To find $E(T)$ and $V(T)$, we first need $E(Y)$ and $V(Y)$.

$$E(Y) = n\left(\frac{k}{N}\right) = 5\left(\frac{2}{20}\right) = .5$$

and

$$V(Y) = n\left(\frac{k}{N}\right)\left(1 - \frac{k}{N}\right)\left(\frac{N-n}{N-1}\right)$$
$$= 5(.1)(1 - .1)\left(\frac{20-5}{20-1}\right)$$
$$= .355.$$

It follows that

$$E(T) = 9E(Y) + 5$$
$$= 9(.5) + 5 = 9.5$$

and

$$V(T) = (9)^2 V(Y)$$
$$= 81(.355) = 28.755.$$

Thus, installation time should average 9.5 minutes, with a standard deviation of $\sqrt{28.755} = 5.4$ minutes.

The Hypergeometric Distribution

$$p(y) = \frac{\binom{k}{y}\binom{N-k}{n-y}}{\binom{N}{n}} \qquad y = 0, 1, \ldots, k \text{ with } \binom{b}{a} = 0 \text{ if } a > b$$

$$E(Y) = n\left(\frac{k}{N}\right) \qquad V(Y) = n\left(\frac{k}{N}\right)\left(1 - \frac{k}{N}\right)\left(\frac{N-n}{N-1}\right)$$

EXERCISES

3.68 From a box containing four white and three red balls, two balls are selected at random, without replacement. Find the probability that
- (a) exactly one white ball is selected.
- (b) at least one white ball is selected.
- (c) two white balls are selected, given that at least one white ball is selected.
- (d) the second ball drawn is white.

3.69 A warehouse contains ten printing machines, four of which are defective. A company randomly selects five of the machines for purchase. What is the probability that all five of the machines are not defective?

3.70 Referring to Exercise 3.69, the company purchasing the machines returns the defective ones for repair. If it costs $50 to repair each machine, find the mean and variance of the total repair cost. In what interval would you expect the repair costs on these five machines to lie? (Use Tchebysheff's Theorem.)

3.71 A corporation has a pool of six firms, four of which are local, from which they can purchase certain supplies. If three firms are randomly selected without replacement, find the probability that

(a) at least one selected firm is not local.

(b) all three selected firms are local.

3.72 A foreman has ten employees from whom he must select four to perform a certain undesirable task. Among the ten employees three belong to a minority ethnic group. The foreman selected all three minority employees (plus one other) to perform the undesirable task. The minority group then protested to the union steward that they were discriminated against by the foreman. The foreman claimed that the selection was completely at random. What do you think?

3.73 Specifications call for a type of thermistor to test out at between 9000 and 10,000 ohms at 25°C. From ten thermistors available three are to be selected for use. Let Y denote the number among the three that do not conform to specifications. Find the probability distribution for Y (tabular form) if

(a) the ten contain two thermistors not conforming to specifications.

(b) the ten contain four thermistors not conforming to specifications.

3.74 Used photocopying machines are returned to the supplier, cleaned, and then sent back out on lease agreements. Major repairs are not made and, as a result, some customers receive malfunctioning machines. Among eight used photocopiers in supply today, three are malfunctioning. A customer wants to lease four of these machines immediately. Hence four machines are quickly selected and sent out, with no further checking. Find the probability that the customer receives

(a) no malfunctioning machines.

(b) at least one malfunctioning machine.

(c) three malfunctioning machines.

3.75 An eight-cylinder automobile engine has two misfiring spark plugs. If all four plugs are removed from one side of the engine, what is the probability that the two misfiring ones are among them?

3.76 The "worst-case" requirements are defined in the design objectives for a brand of computer terminal. A quick preliminary test indicates that four out of a lot of ten such terminals failed the worst-case requirements. Five of the ten are randomly selected for further testing. Let Y denote the number, among the five, that failed the preliminary test. Find

(a) $P(Y \geq 1)$ (b) $P(Y \geq 3)$

(c) $P(Y \geq 4)$ (d) $P(Y \geq 5)$.

3.77 An auditor checking the accounting practices of a firm samples three accounts from an accounts receivable list of eight. Find the probability that the auditor sees at least one past due account if there are

(a) two such accounts among the eight.
(b) four such accounts among the eight.
(c) seven such accounts among the eight.

3.78 A group of six software packages available to solve a linear programming problem have been ranked from 1 to 6 (best to worst). An engineering firm selects two of these packages for purchase without looking at the ratings. Let Y denote the number of packages purchased by the firm that are ranked from 3 to 6. Show the probability distribution for Y in tabular form.

3.79 Lot acceptance sampling procedures for an electronics manufacturing firm call for sampling n items from a lot of N items, and accepting the lot if $Y \le c$, where Y is the number of nonconforming items in the sample. For an incoming lot of twenty printer covers, five are to be sampled. Find the probability of accepting the lot if $c = 1$ and the actual number of nonconforming covers in the lot is

(a) 0. (b) 1. (c) 2. (d) 3. (e) 4.

3.80 In the setting and terminology of Exercise 3.79, answer the same questions if $c = 2$.

3.81 Two assembly lines (I and II) have the same rate of defectives in their production of voltage regulators. Five regulators are sampled from each line and tested. Among the total of ten tested regulators, four were defective. Find the probability that exactly two of the defectives came from line I.

3.9

THE MOMENT-GENERATING FUNCTION

We saw in earlier sections that if $g(Y)$ is a function of a random variable Y with probability distribution given by $p(y)$, then

$$E[g(Y)] = \sum_y g(y)p(y).$$

A special function with many theoretical uses in probability theory is the expected value of e^{tY}, for a random variable Y, and this expected value is called the *moment-generating function* (mgf). We denote mgfs by $M(t)$, and thus,

$$M(t) = E(e^{tY}).$$

The expected values of powers of a random variable are often called *moments*. Thus $E(Y)$ is the first moment and $E(Y^2)$ the second moment of Y. One use for the moment-generating function is that, in fact, it does generate moments of Y. When $M(t)$ exists, it is differentiable in a neighborhood of the origin $t = 0$, and the derivatives may be taken inside the expectation. Thus,

$$M^{(1)}(t) = \frac{dM(t)}{dt} = \frac{d}{dt}E[e^{tY}]$$

$$= E\left[\frac{d}{dt}e^{tY}\right] = E[Ye^{tY}].$$

Now, if we set $t = 0$, we have

$$M^{(1)}(0) = E(Y).$$

Going on to the second derivative,

$$M^{(2)}(t) = E(Y^2e^{tY})$$

and

$$M^{(2)}(0) = E(Y^2).$$

In general,

$$M^{(k)}(0) = E(Y^k).$$

It often is easier to evaluate $M(t)$ and its derivatives than to find the moments of the random variable directly. Other theoretical uses of the mgf will be seen in later chapters.

Example 3.21 Evaluate the moment-generating function for the geometric distribution and use it to find the mean and variance of this distribution.

Solution

For the geometric variable Y, we have

$$M(t) = E(e^{tY}) = \sum_{y=1}^{\infty} e^{ty}p(1-p)^{y-1}$$

$$= pe^t \sum_{y=1}^{\infty} (1-p)^{y-1}(e^t)^{y-1}$$

$$= pe^t \sum_{y=1}^{\infty} [(1-p)e^t]^{y-1}$$

$$= pe^t\{1 + [(1-p)e^t] + [(1-p)e^t]^2 + \cdots$$

$$= pe^t\left[\frac{1}{1-(1-p)e^t}\right]$$

because the series is geometric with common ratio $(1 - p)e^t$.

To evaluate the mean, we have

$$M^{(1)}(t) = \frac{[1 - (1 - p)e^t]pe^t - pe^t[-(1 - p)e^t]}{[1 - (1 - p)e^t]^2}$$

$$= \frac{pe^t}{[1 - (1 - p)e^t]^2}$$

and

$$M^{(1)}(0) = \frac{p}{[1 - (1 - p)]^2} = \frac{1}{p}.$$

To evaluate the variance, we first need

$$E(Y^2) = M^{(2)}(0).$$

Now,

$$M^{(2)}(t) = \frac{[1 - (1 - p)e^t]^2 pe^t - pe^t\{2[1 - (1 - p)e^t](-1)(1 - p)e^t\}}{[1 - (1 - p)e^t]^4}$$

and

$$M^{(2)}(0) = \frac{p^3 + 2p^2(1 - p)}{p^4} = \frac{p + 2(1 - p)}{p^2}.$$

Hence,

$$V(Y) = E(Y^2) - [E(Y)]^2$$

$$= \frac{p + 2(1 - p)}{p^2} - \frac{1}{p^2} = \frac{p + 2(1 - p)}{p^2} = \frac{1 - p}{p^2}.$$

Moment-generating functions have some very important properties that make them extremely useful in finding expected values and determining the probability distributions of random variables. These properties will be discussed in detail in Chapters 4 and 5, but one such property is given in Exercise 3.85.

3.10

THE PROBABILITY-GENERATING FUNCTION

In an important class of discrete random variables, Y takes integral values, $Y = 0, 1, 2, 3, \ldots,$ and consequently represents a count. The binomial, geometric, hypergeometric, and Poisson random variables all fall in this class. The following examples present practical situations that result in integral-valued random variables. One, involving the theory of queues (waiting lines), is concerned with the number of persons (or objects) awaiting service at a particular point in time. Knowledge of the behavior of this random variable is important in designing manufacturing plants where production consists of a sequence of operations each of which requires a different length of time to complete. An insufficient number of service stations for a particular production operation can result in a bottleneck, the formation of a queue of products waiting to be serviced, which would slow down the manufacturing operation. Queuing theory also is important in determining the number of checkout counters needed for a supermarket and in the design of hospitals and clinics.

Integer-valued random variables are very important in studies of population growth, too. For example, epidemiologists are interested in the growth of bacterial populations and, also, in the growth of the number of persons afflicted by a particular disease. The number of elements in each of these populations will be an integral-valued random variable.

A mathematical device that is very useful in finding the probability distributions and other properties of integral-valued random variables is the probability-generating function, $P(t)$, defined by

$$P(t) = E(t^Y).$$

If Y is an integer-valued random variable with

$$P(Y = i) = p_i, \qquad i = 0, 1, 2, \ldots,$$

then

$$P(t) = E(t^Y) = p_0 + p_1 t + p_2 t^2 + \cdots.$$

The reason for calling $P(t)$ a probability-generating function is clear when we compare $P(t)$ with the moment-generating function, $M(t)$. Particularly, the coefficient of t^i in $P(t)$ is the probability p_i. If we know $P(t)$ and can expand it into a series, we can determine $p(y)$ as the coefficient of t^y. Repeated differentiation of $P(t)$ yields *factorial moments* for the random variable Y.

Definition 3.7

The kth **factorial moment** for a random variable Y is defined to be

$$\mu_{[k]} = E[Y(Y-1)(Y-2)\cdots(Y-k+1)]$$

where k is a positive integer.

When a probability generating function exists, it can be differentiated in a neighborhood of $t = 1$. Thus, with

$$P(t) = E(t^Y)$$

we have

$$P^{(1)}(t) = \frac{dP(t)}{dt} = E\left[\frac{dt^Y}{dt}\right]$$
$$= E(Yt^{Y-1})$$

and

$$P^{(1)}(1) = E(Y).$$

Similarly,

$$P^{(2)}(t) = E[Y(Y - 1)t^{Y-2}]$$

and

$$P^{(2)}(1) = E[Y(Y - 1)] = \mu_{[2]}.$$

In general,

$$P^{(k)}(t) = E[Y(Y - 1)\cdots(Y - k + 1)t^{y-k}]$$

and

$$P^{(k)}(t) = E[Y(Y - 1)\cdots(Y - k + 1)t^{y-k}]$$
$$= \mu_{[k]}.$$

Example 3.22 Find the probability-generating function for the geometric random variable, and use it to find the mean.

Solution

Note that $p_0 = 0$ because Y cannot assume this value. Then,

$$P(t) = E(t^Y) = \sum_{y=1}^{\infty} t^y q^{y-1} p = \sum_{y=1}^{\infty} \frac{p}{q}(qt)^y$$
$$= \frac{p}{q}[qt + (qt)^2 + (qt)^3 + \cdots].$$

The terms of the series are those of an infinite geometric progression. We can let $t \le 1$, so that $qt < 1$. Then,

$$P(t) = \frac{p}{q}\left\{\frac{qt}{1 - qt}\right\} = \frac{pt}{1 - qt}.$$

Now,

$$P^{(1)}(t) = \frac{d}{dt}\left\{\frac{pt}{1 - qt}\right\} = \frac{(1 - qt)p - (pt)(-q)}{(1 - qt)^2}.$$

Setting $t = 1$,

$$P^{(1)}(1) = \frac{p^2 + pq}{p^2} = \frac{p(p + q)}{p^2} = \frac{1}{p},$$

which is the mean of a geometric random variable.

Since we already have the moment-generating function to assist in finding the moment of a random variable, we might ask of what value is $P(t)$. The answer is that it may be exceedingly difficult to find $M(t)$ but easy to find $P(t)$. Or, $P(t)$ may be easier to work with in a particular setting. Thus $P(t)$ simply provides an additional tool for finding the moments of a random variable. It may or may not be useful in a given situation.

Finding the moments of a random variable is not the major use of the probability-generating function. Its primary application is in deriving the probability function (and hence the probability distribution) for other related integral-valued random variables. For these applications, see Feller (1968), Parzen (1964), and Section 6.6.

EXERCISES

3.82 Find the moment-generating function for the Bernoulli random variable.

3.83 Show that the moment-generating function for the binomial random variable is given by

$$M(t) = [pe^t + (1 - p)]^n.$$

Use this result to derive the mean and variance for the binomial distribution.

3.84 Show that the moment-generating function for the Poisson random variable with mean λ is given by

$$M(t) = e^{\lambda(e^t - 1)}.$$

Use this result to derive the mean and variance for the Poisson distribution.

3.85 If X is a random variable with moment-generating function $M(t)$, and Y is a function of X given by $Y = aX + b$, show that the moment-generating function for Y is $e^{tb}M(at)$.

3.86 Use the result of Exercise 3.85 to show that

$$E(Y) = aE(X) + b$$

and

$$V(Y) = a^2V(X).$$

3.87 Find the probability-generating function for a binomial random variable of n trials, with probability of success p. Use this function to find the mean and variance of the binomial.

3.11

THE POISSON PROCESS

In this section, we introduce the application of probability theory to modeling random phenomena that change over time, the so-called stochastic processes. Thus, a time parameter is introduced and the random variable, $Y(t)$, is regarded as a function of time. (Later we shall see that the concept of *time* can be generalized to include space.) Examples are endless, but $Y(t)$ could represent the size of a biological population at time t, the cost of operating a complex industrial system for t time units, the distance displaced by a particle in time t, or the number of customers waiting to be served at a checkout counter t time units after its opening.

Although we do not intend to produce an exhaustive discussion of stochastic processes, a few elementary concepts are in order. A stochastic process $Y(t)$ is said to have *independent* increments if, for any set of time points $t_0 < t_1 < \cdots < t_n$, the random variables $[Y(t_i) - Y(t_{i-1})]$ and $[Y(t_j) - Y(t_{j-1})]$ are independent for $i \neq j$. (See Section 5.3 for a discussion of independent random variables.) The process is said to have *stationary* independent increments if, in addition, the random variable $[Y(t_2 + h) - Y(t_1 + h)]$ and $[Y(t_2) - Y(t_1)]$ have identical distributions for any $h > 0$. We begin our discussion of specific processes by considering an elementary counting process with wide applicability, the Poisson process.

Let $Y_{(t)}$ denote the number of occurrences of some event in the time interval $(0, t)$, $t > 0$. This could be the number of accidents at a particular intersection, the number of times a computer breaks down, or any similar count. We derived the Poisson distribution in Section 3.7 by looking at a limiting form of the binomial distribution, but now we derive the Poisson process by working directly from a set of axioms.

The Poisson process is defined as satisfying the following axioms:

Axiom 1: $P[Y(0) = 0] = 1$.
Axiom 2: $Y(t)$ has stationary independent increments. (The number of occurrences in two nonoverlapping time intervals of the same length are independent and have the same probability distribution.)
Axiom 3: $P\{[Y(t + h) - Y(t)] = 1\} = \lambda h + o(h)$, where $h > 0$, $o(h)$ is a generic notation for a term such that $o(h)/h \to 0$ as $h \to 0$, and λ is a constant.
Axiom 4: $P\{[Y(t + h) - Y(t)] > 1\} = o(h)$.

Axiom 3 says that the chance of one occurrence in a small interval of time, h, is roughly proportional to h, and Axiom 4 states that the chance of more than one such occurrence tends to reach zero faster than h itself does. These two axioms imply that

$$P\{[Y(t + h) - Y(t)] = 0\} = 1 - \lambda h - o(h).$$

The probability distribution now will be derived from the axioms. Consider the interval $(0, t + h)$ partitioned into the two disjoint pieces $(0, t)$ and $(t, t + h)$. Then we can write

$$P[Y(t + h) = k] = \sum_{j=0}^{k} P\{Y(t) = j, [Y(t + h) - Y(t)] = k - j\}$$

because the events inside the right-hand probability are mutually exclusive. Because independent increments are assumed, we have

$$P[Y(t + h) = k] = \sum_{j=0}^{k} P[Y(t) = j]P\{[Y(t + h) - Y(t)] = k - j\}.$$

By stationarity and Axiom 1, this becomes

$$P[Y(t + h) = k] = \sum_{j=0}^{k} P[Y(t) = j]P[Y(h) = k - j].$$

To simplify notation, let $P[Y(t) = j] = P_j(t)$. Then,

$$P_k(t + h) = \sum_{j=0}^{k} P_j(t)P_{k-j}(h),$$

and Axiom 4 reduces this to

$$P_k(t + h) = P_{k-1}(t)P_1(h) + P_k(t)P_0(h) + o(h).$$

Axiom 3 gives further reduction to

$$\begin{aligned} P_k(t + h) &= P_{k-1}(t)[\lambda h + o(h)] + P_k(t)[1 - \lambda h - o(h)] \\ &= P_{k-1}(t)(\lambda h) + P_k(t)(1 - \lambda h) + o(h). \end{aligned}$$

We can write this equation as

$$\frac{1}{h}[P_k(t + h) - P_k(t)] = \lambda P_{k-1}(t) - \lambda P_k(t) + \frac{o(h)}{h}$$

and, on taking limits as $h \to 0$, we have

$$\frac{dP_k(t)}{dt} = \lim_{h \to 0} \frac{1}{h}[P_k(t + h) - P_k(t)] = \lambda P_{k-1}(t) - \lambda P_k(t).$$

To solve this differential equation, we take $k = 0$ to obtain

$$\frac{dP_k(t)}{dt} = P_0'(t) = -\lambda P_0(t),$$

since $P_{k-1}(t)$ will be replaced by zero for $k = 0$. Thus, we have

$$\frac{P_0'(t)}{P_0(t)} = -\lambda$$

or, on integrating both sides,

$$\ln P_0(t) = -\lambda t + c.$$

The constant is evaluated by the boundary condition $P_0(0) = P[Y(0) = 0] = 1$ by Axiom 1, which gives $c = 0$. Hence,

$$\ln P_0(t) = -\lambda t$$

or

$$P_0(t) = e^{-\lambda t}.$$

By letting $k = 1$ in the original differential equation, we can obtain

$$P_1(t) = \lambda t e^{-\lambda t},$$

and, recursively,

$$P_k(t) = \frac{1}{k!}(\lambda t)^k e^{-\lambda t}, \qquad k = 0, 1, 2, \ldots.$$

Thus, $Y(t)$ has a Poisson distribution with mean λt.

The notion of a Poisson process is easily extended to points randomly dispersed in a plane where the interval $(0, t)$ is replaced by a planar region (quadrat) of area A. Then, under analogous axioms, $Y(A)$, the number of points in a quadrat of area A, has Poisson distribution with mean λA. Similarly, it could apply to points randomly dispersed in space when the interval $(0, t)$ is replaced by a cube, sphere, or some other three-dimensional figure of volume V.

Example 3.23 Suppose that in a certain area plants are randomly dispersed with a mean density of 20 per square yard. If a biologist randomly locates 100 2-square-yard sampling quadrats on the area, how many of them can he expect to contain no plants?

Solution

Assuming the plant counts per unit area to have a Poisson distribution with a mean of 20 per square yard, the probability of no plants in a 2-square-yard area is

$$P[Y(2) = 0] = e^{-2\lambda}$$
$$= e^{-2(20)} = e^{-40}.$$

If the 100 quadrats are randomly located, the expected number of quadrats containing zero plants is

$$100P[Y(2) = 0] = 100e^{-40}.$$

3.12

MARKOV CHAINS

Consider a system that can be in any of a finite number of states and assume that it moves from state to state according to some prescribed probability law. The system, for example, could involve weather conditions from day to day with the states being clear, partly cloudy, and cloudy. Observation of conditions over a long period of time would allow one to find the probability of it being clear tomorrow given that it is partly cloudy today.

Let X_i denote the state of the system at time point i, and let the possible states be denoted by $S_1, \ldots, S_m$, for a finite integer m. We are not interested in the elapsed time between transitions from one state to another, but only in the states and the probabilities of going from one state to another; that is, the *transition probabilities.* We assume that

$$P(X_i = S_k \mid X_{i-1} = S_j) = p_{jk},$$

where p_{jk} is the transition probability from S_j to S_k; and this probability is independent of i. That is, the transition probabilities do not depend on the time points, but only on the states. The event $(X_i = S_k \mid X_{i-1} = S_j)$ is assumed independent of the past history of the process. Such a process is called a *Markov chain* with stationary transition probabilities. The transition probabilities can conveniently be displayed in a matrix:

$$\boldsymbol{P} = \begin{bmatrix} p_{11} & p_{12} & \cdots & p_{1m} \\ p_{21} & p_{22} & \cdots & p_{2m} \\ \vdots & & & \vdots \\ p_{m1} & p_{m2} & \cdots & p_{mm} \end{bmatrix}.$$

Let X_0 denote the starting state of the system, with probabilities given by

$$p_k^{(0)} = P(X_0 = S_k).$$

Also, let the probability of being in state S_k after n steps be given by $p_k^{(n)}$. These probabilities are conveniently displayed by vectors:

$$\boldsymbol{p}^{(0)} = [p_1^{(0)}, p_2^{(0)}, \ldots, p_m^{(0)}]$$

and

$$\boldsymbol{p}^{(n)} = [p_1^{(n)}, p_2^{(n)}, \ldots, p_m^{(n)}].$$

It can easily be shown that

$$\boldsymbol{p}^{(1)} = \boldsymbol{p}^{(0)}\boldsymbol{P}$$

(note that this is matrix multiplication) and, in general,

$$\boldsymbol{p}^{(n)} = \boldsymbol{p}^{(n-1)}\boldsymbol{P}.$$

$\boldsymbol{P}$ is said to be *regular* if some power of $\boldsymbol{P}$($\boldsymbol{P}^n$ for some n) has all positive entries. Thus, one can get from state S_j to state S_k, eventually, for any pair (j, k). (Note that the condition of regularity rules out certain chains that periodically return to certain states.) If $\boldsymbol{P}$ is regular, the chain has a stationary (or equilibrium) distribution that gives the probabilities of being in the respective states after many transitions have evolved. That is, $p_j^{(n)}$ must have a limit, π_j, as $n \to \infty$. Suppose that these limits exist; then $\boldsymbol{\pi} = (\pi_1, \ldots, \pi_m)$ must satisfy

$$\boldsymbol{\pi} = \boldsymbol{\pi}\boldsymbol{P},$$

because $\boldsymbol{p}^{(n)} = \boldsymbol{p}^{(n-1)}\boldsymbol{P}$.

Example 3.24 A supermarket stocks three brands of coffee, A, B, and C, and it has been observed that customers switch from brand to brand according to the transition matrix

$$\boldsymbol{P} = \begin{bmatrix} 3/4 & 1/4 & 0 \\ 0 & 2/3 & 1/3 \\ 1/4 & 1/4 & 1/2 \end{bmatrix},$$

where S_1 corresponds to a purchase of A, S_2 to B, and S_3 to C. That is, 3/4 of the customers buying A also buy A the next time they purchase coffee, whereas 1/4 of these customers switch to brand B.

(a) Find the probability that a customer who buys brand A today will again purchase A two weeks from today, assuming that he or she purchases coffee once a week.

(b) In the long run, what fraction of customers purchase the respective brands?

Solution

(a) Assuming that the customer is chosen at random, his or her transition probabilities are given by $\boldsymbol{P}$. The given information indicates that $\boldsymbol{p}^{(0)} = (1, 0, 0)$; that is, the customer starts with a purchase of brand A. Then

$$\boldsymbol{p}^{(1)} = \boldsymbol{p}^{(0)}\boldsymbol{P} = \left(\frac{3}{4}, \frac{1}{4}, 0\right)$$

gives the probabilities for next week's purchase. The probabilities for two weeks from now are given by

$$\boldsymbol{p}^{(2)} = \boldsymbol{p}^{(1)}\boldsymbol{P} = \left(\frac{9}{16}, \frac{17}{48}, \frac{1}{12}\right).$$

That is, the chance of purchasing A two weeks from now is only 9/16.

(b) The answer to the long-run frequency ratio is given by $\boldsymbol{\pi}$, the stationary distribution. The equation

$$\boldsymbol{\pi} = \boldsymbol{\pi P}$$

yields the system

$$\pi_1 = \left(\frac{3}{4}\right)\pi_1 + \left(\frac{1}{4}\right)\pi_3$$

$$\pi_2 = \left(\frac{1}{4}\right)\pi_1 + \left(\frac{2}{3}\right)\pi_2 + \left(\frac{1}{4}\right)\pi_3$$

$$\pi_3 = \left(\frac{1}{3}\right)\pi_2 + \left(\frac{1}{2}\right)\pi_3.$$

Combining these equations with the fact that $\pi_1 + \pi_2 + \pi_3 = 1$ yields

$$\boldsymbol{\pi} = \left(\frac{2}{7}, \frac{3}{7}, \frac{2}{7}\right).$$

Thus, the store should stock more brand B coffee than either A or C.

An interesting example of a transition matrix that is not regular is formed by a Markov chain with *absorbing states.* A state, S_i, is said to be *absorbing* if $p_{ii} = 1$ and $p_{ij} = 0$ for $j \neq 1$. That is, once the system is in state S_i, it cannot leave it. The transition matrix for such a chain can always be arranged in a standard form, with the absorbing states listed first. For example, suppose that a chain has five states, of which two are absorbing. Then $\boldsymbol{P}$ can be written

$$\boldsymbol{P} = \begin{bmatrix} 1 & 0 & 0 & 0 & 0 \\ 0 & 1 & 0 & 0 & 0 \\ p_{31} & p_{32} & p_{33} & p_{34} & p_{35} \\ p_{41} & p_{42} & p_{43} & p_{44} & p_{45} \\ p_{51} & p_{52} & p_{53} & p_{54} & p_{55} \end{bmatrix}$$

$$= \begin{bmatrix} \boldsymbol{I} & \boldsymbol{0} \\ \boldsymbol{R} & \boldsymbol{Q} \end{bmatrix},$$

where $\boldsymbol{I}$ is a 2×2 identity matrix and $\boldsymbol{0}$ a matrix of zeros. Many interesting properties of these chains can be expressed in terms of $\boldsymbol{R}$ and $\boldsymbol{Q}$ (see Kemeny et al., 1962).

The following discussion will be restricted to the case in which $\boldsymbol{R}$ and $\boldsymbol{Q}$ are such that it is possible eventually to get to an absorbing state from every other state. In that case, the Markov chain eventually will end up in an absorbing state. Questions of interest then involve the expected number of steps to absorption and the probability of absorption in the various absorbing states.

Let m_{ij} denote the expected (or mean) number of times the system is in state S_j, given that it started in S_i, for nonabsorbing states S_i and S_j. From S_i the system could go to an absorbing state in one step, or it could go to a nonabsorbing state, say S_k, and eventually be absorbed from there. Thus, m_{ij} must satisfy

$$m_{ij} = \delta_{ij} + \sum_k p_{ik} m_{kj},$$

where the summation is over all nonabsorbing states and

$$\delta_{ij} = \begin{cases} 1 & \text{if } i = j \\ 0 & \text{otherwise.} \end{cases}$$

The term δ_{ij} accounts for the fact that, if the chain goes to an absorbing state in one step, it was in state S_i one time.

If we denote the matrix of m_{ij} terms by $\boldsymbol{M}$, the preceding equation can then be generalized to

$$\boldsymbol{M} = \boldsymbol{I} + \boldsymbol{QM}$$

or

$$\boldsymbol{M} = (\boldsymbol{I} - \boldsymbol{Q})^{-1}.$$

(Matrix operations, such as inversion, will not be discussed here. The equations can be solved directly if matrix operations are unfamiliar to the reader.)

The expected number of steps to absorption, from nonabsorbing starting state S_i, will be denoted by m_i and given simply by

$$m_i = \sum_k m_{ik},$$

again summing over nonabsorbing states.

Turning now to the probability of absorption into the various absorbing states, we let a_{ij} denote the probability of the system being absorbed in state S_j, given that it started in state S_i, for nonabsorbing S_i and absorbing S_j. Repeating the preceding argument, the system could move to S_j in one step, or it could move in a nonabsorbing state, S_k, and be absorbed from there. Thus, a_{ij} satisfies

$$a_{ij} = p_{ij} + \sum_k p_{ik} a_{kj}$$

where the summation is over the nonabsorbing states. If we denote the matrix of a_{ij} terms by $\boldsymbol{A}$, the preceding equation then generalizes to

$$\boldsymbol{A} = \boldsymbol{R} + \boldsymbol{QA}$$

or

$$\boldsymbol{A} = (\boldsymbol{I} - \boldsymbol{Q})^{-1}\boldsymbol{R} = \boldsymbol{MR}.$$

The following example illustrates the computations.

Example 3.25 A manager of one section of a plant has employees working at level I and level II. New employees may enter his section at either level. At the end of each year the performances of employees are evaluated; they can either be reassigned to level I or II jobs, terminated, or promoted to level III, in which case they never come back to I or II. The manager can then keep track of employee movement as a Markov chain. The absorbing states are termination (S_1) and employment at level III (S_2), whereas the nonabsorbing states are employment at level I (S_3) and level II (S_4). Records over a long period of time indicate that the following is a reasonable assignment of probabilities;

$$\boldsymbol{P} = \begin{bmatrix} 1 & 0 & 0 & 0 \\ 0 & 1 & 0 & 0 \\ .2 & .1 & .2 & .5 \\ .1 & .3 & .1 & .5 \end{bmatrix}.$$

Thus, if an employee enters at a level I job, the probability is .5 that she will jump to level II work at the end of the year, but she has a probability equal to .2 of being terminated. Find (a) the expected number of evaluations an employee must go through in this section, and (b) the probabilities of being terminated or promoted to level III eventually.

Solution

For the $\boldsymbol{P}$ matrix,

$$\boldsymbol{R} = \begin{bmatrix} .2 & .1 \\ .1 & .3 \end{bmatrix}$$

and

$$\boldsymbol{Q} = \begin{bmatrix} .2 & .5 \\ .1 & .5 \end{bmatrix}.$$

Thus,

$$\boldsymbol{I} - \boldsymbol{Q} = \begin{bmatrix} .8 & .5 \\ -.1 & .5 \end{bmatrix}$$

and

$$M = (I - Q)^{-1} = \begin{bmatrix} 10/7 & 10/7 \\ 2/7 & 16/7 \end{bmatrix}.$$

It follows that

$$m_1 = \frac{20}{7}, m_2 = \frac{18}{7},$$

or, a new employee in this section can expect to remain there through 20/7 evaluations periods if she enters at level I. Also,

$$A = MR = \begin{bmatrix} 3/7 & 4/7 \\ 2/7 & 5/7 \end{bmatrix},$$

which implies that an employee entering at level I has a probability of 4/7 of reaching level III, whereas entering at level II raises this probability to 5/7.

EXERCISES

3.88 During the workday, telephone calls come into an office at the rate of one call every 3 minutes.

(a) Find the probability that no more than one call will come into this office during the next 5 minutes. (Are the assumptions for the Poisson process reasonable here?)

(b) An observer in the office hears two calls come in during the first 2 minutes of a 5-minute visit. Find the probability that no more calls arrive during the visit.

3.89 For a Markov chain, as defined in Section 3.12, show that $P^{(n)} = P^{(n-1)}P$.

3.90 Suppose that two friends, *A* and *B*, toss a balanced coin. If the coin comes up heads, *A* wins \$1 from *B*. If it comes up tails, *B* wins \$1 from *A*. The game ends only when one player has all the other's money. If *A* starts with \$1 and *B* with \$3, find the expected duration of the game and the probability that *A* will win. (Note that this system can be modeled as a Markov chain with absorbing states.)

3.13

ACTIVITIES FOR THE COMPUTER

Computers lend themselves nicely for use in the area of probability. Not only can computers be used to calculate probabilities but also to simulate random variables from specified probability distributions. A simulation, performed on the computer, is used to analyze problems that are both theoretical and applied. A simulated model attempts to copy the behavior of a situation under consideration; practical applications could include models of inventory control problems, queuing systems, production lines, medical systems, and flight patterns of major jets. Simulation also can be used to determine the behavior of a complicated random variable whose precise probability distribution function is difficult to evaluate mathematically.

Generating observations from a probability distribution is based upon random numbers on $[0, 1]$. A random number, R_i, on the interval $[0, 1]$, is chosen on the condition that each number between 0 and 1 has the same probability of being selected. A sequence of numbers that appears to follow a certain pattern or trend would not be considered random. Most computer languages have built-in random generators that will give a number on $[0, 1]$. If a built-in generator is not available, algorithms are available for setting up one. A basic technique used is a congruential method, such as a linear congruential generator (LCG). The generator most commonly used is

$$x_i = (ax_{i-1} + c)(\text{mod}\, m), \qquad i = 1, 2, \ldots;$$

$$x_i,\ a,\ c,\ m \text{ are integers and } 0 \le x_i < m.$$

If $c = 0$, this method is called a multiplicative-congruential sequence. The values of y_i generated by the LCG will be contained in the interval $[0, m)$. To obtain a value in $[0, 1)$, evaluate x_i/m.

It is desirable for a random number generator to have certain characteristics, such as the following:

1. The numbers produced (x_i/m) should appear to be distributed uniformly on $[0, 1]$; that is, the probability function for the numbers is constant over the $[0, 1]$. The numbers also should be independent of each other. Many statistical tests can be applied to check for uniformity and independence, such as comparing the number of occurrences of a digit in a sequence with the expected number of occurrences for that digit.
2. The random number generator should have a long, full period; that is, it is preferable that the sequence of numbers does not begin repeating or "cycling" too quickly. A sequence has a full period (p) if $p = m$. It has been shown that a LCG has a full period if and only if the following are true (Kennedy and Gentle, 1980, p. 137):

   A. the only positive integer that (exactly) divides both m and c is 1;
   B. if q is a prime number that divides m, then q divides $a - 1$;
   C. if 4 divides m, then 4 divides $a - 1$.

   This implies that if c is odd and $a - 1$ is divisible by 4, with such a full-period generator, x_0 can be any integer between 0 and $m - 1$

without affecting the generator's period. A good choice for m is 2^b, where b is the number of bits. Based on the just mentioned relationships of a, c, and m, the following values have been determined to be satisfactory for use on a microcomputer: $a = 25{,}173$; $c = 13{,}849$; $m = 2^{16} = 65{,}536$ (Yang and Robinson, 1986, p. 6).

3. It is beneficial to have the ability to reuse the same random numbers in a different simulation run. The preceding LCG has this ability because all values of x_i are determined by the initial value (seed) x_0. Because of this, the computed random numbers are referred to as *pseudo-random numbers*. Even though these numbers are not truly random, careful selection of a, c, m, and x_0 will result in values for x_i that will behave as random numbers and pass the appropriate statistical tests as described in characteristic 1.
4. The generator should be efficient; that is, fast and in need of little storage.

Given that a random number, R_i, on $[0, 1]$ can be generated, a brief description will now be given for generating discrete random variables for the distributions discussed in this chapter.

BERNOULLI DISTRIBUTION

Let p = probability of success. If $R_i \leq p$, then $X_i = 1$; otherwise, $X_i = 0$.

BINOMIAL DISTRIBUTION

A binomial random variable, X_i, can be expressed as the sum of n independent Bernoulli random variables, Y_j; that is, $X_i = \sum Y_j$, $j = 1, \ldots, n$. Thus, to simulate X_i with parameters n and p, simulate n Bernoulli random variables as stated previously. X_i = the sum of the n Bernoulli variables.

GEOMETRIC DISTRIBUTION

Let X_i = the number of trials necessary for the first success with p = probability of success. $X_i = m$, where m is the number of R_i's generated until $R_i \leq p$.

NEGATIVE BINOMIAL DISTRIBUTION

Let X_i = the number of trials necessary until the rth success with p = probability of success. A negative binomial random variable, X_i, can be expressed as the sum of r independent geometric random variables, Y_j; that is, $X_i = \sum Y_j$, $j = 1, \ldots, r$. Thus, to simulate X_i with parameter p, simulate r

geometric random variables as stated previously. X_i = the sum of the r geometric variables.

POISSON DISTRIBUTION

Generating Poisson random variables will be discussed at the end of Chapter 4.

Let us consider some simple examples of possible uses for simulating discrete random variables. Suppose that n_1 items are to be inspected from one production line and n_2 items from another. Let p_1 = probability of a defective from line 1 and p_2 = probability of a defective from line 2. Let X be a binomial random variable with parameters n_1 and p_1. Let Y be a binomial random variable with parameters n_2 and p_2. A variable of interest is W, which is the total number of defective items observed in both production lines. Let $W = X + Y$. Unless $p_1 = p_2$, the distribution of W will not be binomial. To see how the distribution of W will behave, a simulation could be performed. Useful information could be obtained from the simulation by looking at a histogram of the values of W_i generated and considering the values of the sample mean and sample variance. Let's consider the following random variables X and Y: X is binomial with $n_1 = 7$, $p_1 = .2$, and Y is binomial with $n_1 = 8$, $p_2 = .6$. Defining $W = X + Y$, a simulation produced the following histogram:

SIMULATION

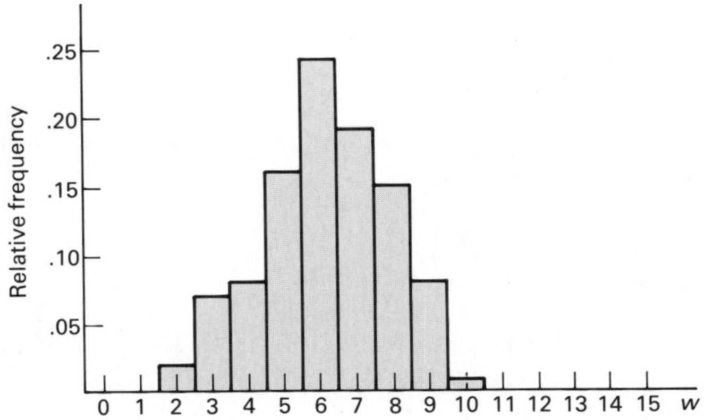

The sample mean was 6.2 with a sample standard deviation of 1.76. In Chapter 5, we will be able to show that these values are very close to the expected value of $\mu_w = 6.2$ and $\sigma_w = 1.74$. (This calculation will be possible after the linear functions of random variables are discussed.) From the histogram, the probability that the total number of defective items is at least 9 is given by .09.

Another example of interest might be the coupon-collector problem, which incorporates the geometric distribution. Suppose that there are n distinct colors of coupons. Each time an individual obtains a coupon, we assume that it is equally likely to be any one of the n colors and the selection of the coupon is independent of a previously obtained coupon. Suppose that one can redeem a set of coupons for a prize if each possible color coupon is represented in the set. We define the random variable X = the number of necessary coupons to be selected to complete a set of each color coupon. Questions of interest might be

1. What is the expected number of coupons needed in order to obtain this complete set; that is, $E(X)$?
2. What is the standard deviation of X?
3. What is the probability that one must select at most x coupons to obtain this complete set?

Instead of answering these questions by deriving the distribution function of X, one might try simulation. Two simulations (2 and 3) follow. The first histogram represents a simulation where n, the number of different color coupons, is equal to 5.

SIMULATION 2

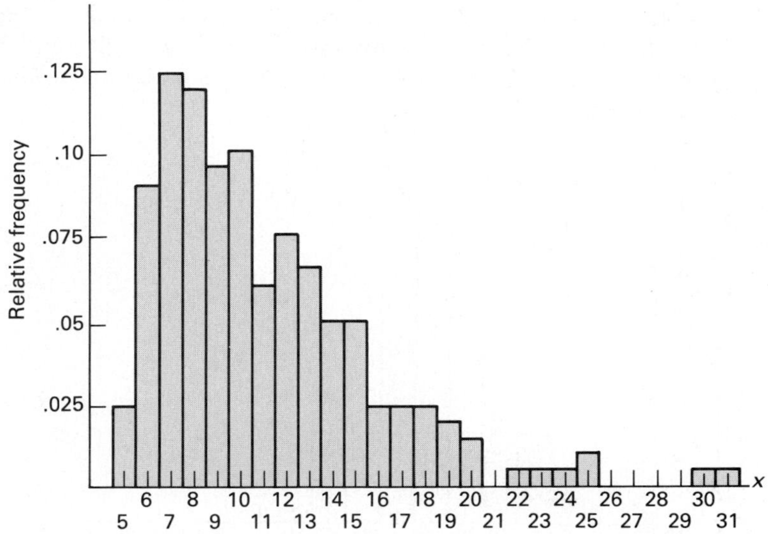

The sample mean was computed to be 11.06, with a sample standard deviation of 4.65. Suppose one is interested in finding $P(X \leq 10)$. Using the results of the simulations, the relative frequency probability is given to be .555. It might also be noted that, from this simulation, the largest number of coupons needed to obtain the complete set was 31.

The second histogram represents a simulation where $n = 10$. The sample mean was computed to be 29.07 with a sample standard deviation of 10.16. The $P(X \leq 20) = .22$ from the simulated values. In this simulation, the largest number of coupons necessary to obtain the complete set was 71.

SIMULATION 3

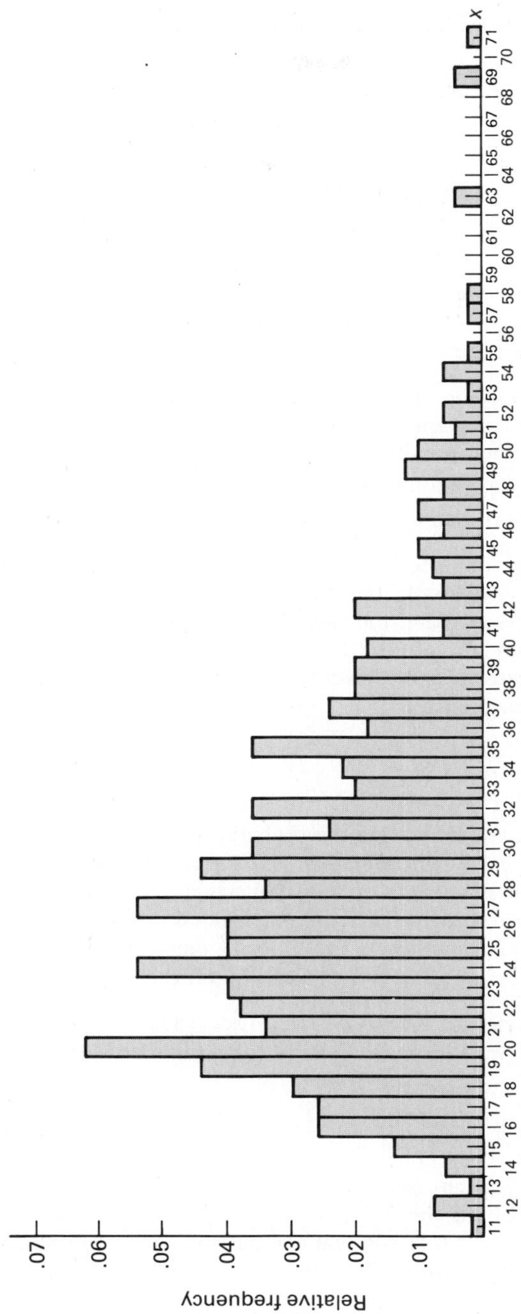

SUPPLEMENTARY EXERCISES

3.91 Construct probability histograms for the binomial probability distribution for $n = 5$, $p = .1$, .5, and .9. (Table 2 of the Appendix will reduce the amount of calculation.) Note the symmetry for $p = .5$ and the direction of skewness for $p = .1$ and $p = .9$.

3.92 Use Table 2 of the Appendix to construct a probability histogram for the binomial probability distribution for $n = 20$ and $p = .5$. Note that almost all the probability falls in the interval $5 \leq y \leq 15$.

3.93 The probability that a single radar set will detect an enemy plane is .9. If we have five radar sets, what is the probability that exactly four sets will detect the plane? (Assume that the sets operate independently of each other.) At least one set?

3.94 Suppose that the four engines of a commercial aircraft were arranged to operate independently and that the probability of in-flight failure of a single engine is .01. What is the probability that, on a given flight,

(a) no failures are observed?

(b) no more than one failure is observed?

3.95 Sampling for defectives from large lots of manufactured product yields a number of defectives, Y, that follows a binomial probability distribution. A sampling plan consists of specifying the number of items to be included in a sample, n, and an acceptance number, a. The lot is accepted if $Y \leq a$ and rejected if $Y > a$. Let p denote the proportion of defectives in the lot. For $n = 5$ and $a = 0$, calculate the probability of lot acceptance if

(a) $p = 0$. (b) $p = .1$.
(c) $p = .3$. (d) $p = .5$.
(e) $p = 1.0$.

A graph showing the probability of lot acceptance as a function of lot fraction defective is called the *operating characteristic curve* for the sample plan. Construct this curve for the plan $n = 5$, $a = 0$. Note that a sampling plan is an example of statistical inference. Accepting or rejecting a lot based on information contained in the sample is equivalent to concluding that the lot is either good or bad, respectively. Good implies that a low fraction is defective, and, therefore, the lot is suitable for shipment.

3.96 Refer to Exercise 3.95. Use Table 2 of the Appendix to construct the operating characteristic curve for a sampling plan with

(a) $n = 10$, $a = 0$.

(b) $n = 10$, $a = 1$.

(c) $n = 10$, $a = 2$.

For each, calculate P(lot acceptance) for $p = 0$, .05, .1, .3, .5, and 1.0. Our intuition suggests that sampling plan (a) would be much less likely to accept bad lots than plans (b) and (c). A visual comparison of the operating characteristic curves will confirm this supposition.

3.97 A quality control engineer wishes to study the alternative sampling plans $n = 5$, $a = 1$

and $n = 25$, $a = 5$. On a sheet of graph paper, construct the operating characteristic curves for both plans; make use of acceptance probabilities at $p = .05, .10, .20, .30$, and .40 in each case.

(a) If you were a seller producing lots with fraction defective ranging from $p = 0$ to $p = .10$, which of the two sampling plans would you prefer?

(b) If you were a buyer wishing to be protected against accepting lots with a fraction defective exceeding $p = .30$, which of the two sampling plans would you prefer?

3.98 For a certain section of a pine forest, the number of diseased trees per acre Y has a Poisson distribution with mean $\lambda = 10$. The diseased trees are sprayed with an insecticide at a cost of \$3.00 per tree, plus a fixed overhead cost for equipment rental of \$50.00. Letting C denote the total spraying cost for a randomly selected acre, find the expected value and standard deviation for C. Within what interval would you expect C to lie with probability at least .75?

3.99 In checking river water samples for bacteria, water is placed in a culture medium to grow colonies of certain bacteria, if present. The number of colonies per dish averages twelve for water samples from a certain river.

(a) Find the probability that the next dish observed will have at least ten colonies.

(b) Find the mean and standard deviation of the number of colonies per dish.

(c) Without calculating exact Poisson probabilities, find an interval in which at least 75 % of the colony count measurements should lie.

3.100 The number of vehicles passing a specified point on a highway averages ten per minute.

(a) Find the probability that at least fifteen vehicles pass this point in the next minute.

(b) Find the probability that at least fifteen vehicles pass this point in the next two minutes.

(c) What assumptions are you making for your answers in (a) and (b) to be valid?

3.101 A production line often produces a variable number N of items each day. Suppose each item produced has the same probability p of not conforming to manufacturing standards. If N has a Poisson distribution with mean λ, then the number of nonconforming items in one day's production Y has a Poisson distribution with mean λp.

The average number of resistors produced by a facility in one day has a Poisson distribution with a mean of 100. Typically, 5 % of the resistors produced do not meet specifications.

(a) Find the expected number of resistors not meeting specifications for a given day.

(b) Find the probability that all resistors will meet the specifications on a given day.

(c) Find the probability that more than five resistors fail to meet specifications on a given day.

3.102 A certain type of bacteria cell divides at a constant rate λ over time. (That is, the probability that a cell will divide in a small interval of time t is approximately λt.) Given that a population starts out at time zero with k cells of this type, and cell divisions are independent of one another, the size of the population at time t, $Y(t)$, has the probability distribution

$$P[Y(t) = n] = \binom{N-1}{k-1} e^{-\lambda k t}(1 - e^{-\lambda t})^{n-k}.$$

(a) Find the expected value of $Y(t)$ in terms of λ and t.

(b) If, for a certain type of bacteria cell, $\lambda = .1$ per second, and the population starts out with two cells at time zero, find the expected population size after 5 seconds.

3.103 The probability that any one vehicle will turn left at a particular intersection is .2. The left-turn lane at this intersection has room for three vehicles. If five vehicles arrive at this intersection while the light is red, find the probability that the left-turn lane will hold all of the vehicles that want to turn left.

3.104 Referring to Exercise 3.103, find the probability that six cars must arrive at the intersection while the light is red to fill up the left-turn lane.

3.105 For any probability function $p(y)$, $\sum_y p(y) = 1$ if the sum is taken over all possible values y that the random variable in question can assume. Show that this is true for

(a) the binomial distribution.

(b) the geometric distribution.

(c) the Poisson distribution.

3.106 The supply office for a large construction firm has three welding units of Brand A in stock. If a welding unit is requested, the probability is .7 that the request will be for this particular brand. On a typical day five requests for welding units come to the office. Find the probability that all three Brand A units will be in use on that day.

3.107 Refer to Exercise 3.106. If the supply office also stocks three welding units that are not Brand A, find the probability that exactly one of these units will be left immediately after the third Brand A unit is requested.

3.108 The probability of a customer arrival at a grocery service counter in any 1 second equals .1. Assume that customers arrive in a random stream and, hence, that the arrival at any 1 second is independent of any other.

(a) Find the probability that the first arrival will occur during the third 1-second interval.

(b) Find the probability that the first arrival will not occur until at least the third 1-second interval.

3.109 Of a population of consumers, 60% is reputed to prefer a particular brand, A, of toothpaste. If a group of consumers are interviewed, what is the probability that exactly five people have to be interviewed before encountering a consumer who prefers brand A? At least five people?

3.110 The mean number of automobiles entering a mountain tunnel per 2-minute period is one. An excessive number of cars entering the tunnel during a brief period of time produces a hazardous situation.

(a) Find the probability that the number of autos entering the tunnel during a 2-minute period exceeds three.

(b) Assume that the tunnel is observed during ten 2-minute intervals, thus giving ten independent observations, $Y_1 Y_2, \ldots, Y_{10}$, on a Poisson random variable. Find the probability that $Y > 3$ during at least one of the ten 2-minute intervals.

3.111 Suppose that 10 % of a brand of microcomputers will fail before their guarantee has expired. If 1000 computers are sold this month, find the expected value and variance of Y, the number that have not failed during the guarantee period. Within what limit would Y be expected to fall? (Hint: Use Tchebysheff's Theorem.)

3.112 (a) Consider a binomial experiment for $n = 20$, $p = .05$. Use Table 2 of the Appendix to calculate the binomial probabilities for $Y = 0, 1, 2, 3, 4$.
(b) Calculate the same probabilities using the Poisson approximation with $\lambda = np$. Compare.

3.113 The manufacturer of a low-calorie dairy drink wishes to compare the taste appeal of a new formula (B) with that of the standard formula (A). Each of four judges is given three glasses in random order, two containing formula A and the other containing formula B. Each judge is asked to choose which glass he most enjoyed. Suppose that the two formulas are equally attractive. Let Y be the number of judges stating a preference for the new formula.

(a) Find the probability function for Y.

(b) What is the probability that at least three of the four judges state a preference for the new formula?

(c) Find the expected value of Y.

(d) Find the variance of Y.

3.114 Show that the hypergeometric probability function approaches the binomial in the limit as $N \to \infty$ and $p = r/N$ remains constant. That is, show that

$$\lim_{\substack{N\to\infty \\ r\to\infty}} \frac{\binom{r}{y}\binom{N-r}{n-y}}{\binom{N}{n}} = \binom{n}{y} p^y q^{n-y}$$

for $p = r/N$ constant.

3.115 A lot of $N = 100$ industrial products contains 40 defectives. Let Y be the number of defectives in a random sample of size 20. Find $p(10)$ using

(a) the hypergeometric probability distribution.

(b) the binomial probability distribution.

Is N large enough so that the binomial probability function is a good approximation to the hypergeometric probability function?

3.116 For simplicity, let us assume that there are two kinds of drivers. The safe drivers, which are 70 % of the population, have a probability of .1 of causing an accident in a year. The rest of the population are accident makers, who have a probability of .5 of causing an accident in a year. The insurance premium is \$400 times one's probability of causing an accident in the following year. A new subscriber has an accident during the first year. What should be her insurance premium for the next year?

3.117 A merchant stocks a certain perishable item. He knows that on any given day he will have a demand for either two, three, or four of these items with probabilities .1, .4, and .5, respectively. He buys the items for \$1.00 each and sells them for \$1.20 each. Any items left at the end of the day represent a total loss. How many items should the merchant stock to maximize his expected daily profit?

3.118 It is known that 5 % of a population have disease A, which can be discovered by a blood test. Suppose that N (a large number) people are to be tested. This can be done in two ways.

1. Each person is tested separately.
2. The blood samples of k people are pooled together and analyzed. (Assume that $N = nk$, with n an integer.) If the test is negative, all of them are healthy (that is, just this one test is needed). If the test is positive, each of the k persons must be tested separately (that is, a total of $k + 1$ tests are needed).

(a) For fixed k, what is the expected number of tests needed in method (2)?
(b) Find the k that will minimize the expected number of tests in method (2).
(c) How many tests does part (b) save in comparison with part (a)?

3.119 Four possible winning numbers for a lottery—AB-4536, NH-7812, SQ-7855, and ZY-3221—are given to you. You will win a prize if one of you numbers matches with one of the winning numbers. You are told that there is one first prize of \$100,000; two second prizes of \$50,000 each; and ten third prizes of \$1000 each. The only thing you need to do is to mail the coupon back. No purchase is required. From the structure of the numbers you have received, it is obvious that the entire list consists of all the permutations of two alphabets followed with four digits. Is the coupon worth mailing back for 25 cents postage?

CHAPTER 4

CONTINUOUS PROBABILITY DISTRIBUTIONS

4.1

CONTINUOUS RANDOM VARIABLES AND THEIR PROBABILITY DISTRIBUTIONS

All of the random variables discussed in Chapter 3 were discrete, assuming only a finite number or countable infinity of values. However, many of the random variables seen in practice have more than a countable collection of possible values. Weights of patients coming into a clinic may be anywhere from, say, 80 to 300 pounds. Diameters of machined rods from a certain industrial process may be anywhere from 1.2 to 1.5 centimeters. Proportions of impurities in ore samples may run from .10 to .80. These random variables can take on any value in an interval of real numbers. That is not to say that every value in the interval can be found in the sample data if one looks long enough; one may never observe a patient weighing exactly 172.38 pounds. Yet, no value can be ruled out as a possible observation; one might have a patient weighing 172.38 pounds, and so this number must be considered in the set of possible

TABLE 4.1 **Life lengths of batteries (in hundreds of hours)**

.406	.685	4.778	1.725	8.223
2.343	1.401	1.507	.294	2.230
.538	.234	4.025	3.323	2.920
5.088	1.458	1.064	.774	.761
5.587	.517	3.246	2.330	1.064
2.563	.511	2.782	6.426	.836
.023	.225	1.514	3.214	3.810
3.334	2.325	.333	7.514	.968
3.491	2.921	1.624	.334	4.490
1.267	1.702	2.634	1.849	.186

outcomes. Since random variables of this type have a continuum of possible values, they are called *continuous random variables*. Probability distributions for continuous random variables are developed in this chapter, and the basic ideas are presented in the context of an experiment on life lengths.

An experimenter is measuring the life length X of a transistor. In this case X can assume an infinite number of possible values. We cannot assign a positive probability to each possible outcome of the experiment because, no matter how small we might make the individual probabilities, they would sum to a value greater than one when accumulated over the entire sample space. However, we can assign positive probabilities to *intervals* of real numbers in a manner consistent with the axioms of probability. To introduce the basic ideas involved here, let us consider a specific example in some detail.

Suppose that we have measured the life lengths of fifty batteries of a certain type, selected from a larger population of such batteries. The observed life lengths are as given in Table 4.1. The relative frequency histogram for these data (Figure 4.1) shows clearly that most of the life lengths are near zero, and the frequency drops off rather smoothly as we look at larger life lengths. Here 32% of the fifty observations fall into the first subinterval and another 22% fall into the second. There is a decline in frequency as we proceed across the subintervals, until the last subinterval (8 or 9) contains a single observation.

This sample relative frequency histogram not only allows us to picture how the sample behaves, but it also gives us some insight into a possible probabilistic model for the random variable X. This histogram of Figure 4.1 looks like it could be approximated quite closely by a negative exponential curve. The particular function

$$f(x) = \frac{1}{2}e^{-x/2} \qquad x > 0$$

is sketched through the histogram in Figure 4.1 and seems to fit reasonably well. Thus, we could take this function as a mathematical model for the

FIGURE 4.1 Relative frequency histogram of data from Table 4.1.

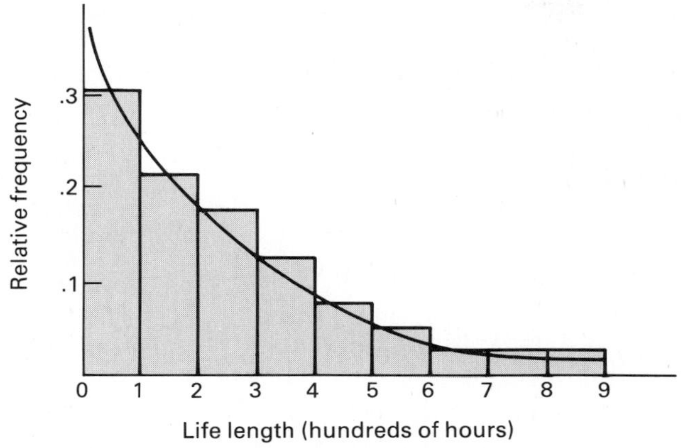

behavior of the random variable X. If we want to use a battery of this type in the future, we might want to know the probability that it will last longer than 400 hours. This probability can be approximated by the area under the curve to the right of the value 4; that is, by

$$\int_4^\infty \frac{1}{2} e^{-x/2}\,dx = .135.$$

Note that this figure is quite close to the observed sample fraction of lifetimes that exceed 4; namely, $(8/50) = .16$. One might suggest that, because the sample fraction .16 is available, we do not really need the model. However, the model would give more satisfactory answers for other questions than could otherwise be obtained. For example, suppose we are interested in the probability that X is greater than 9. Then, the model suggests the answer

$$\int_9^\infty \frac{1}{2} e^{-x/2}\,dx = .011$$

whereas the sample shows no observations in excess of 9. These are quite simple examples, in fact, and we will see many examples of more involved questions for which a model is quite essential.

Why did we choose the exponential function as a model here? Would some others not do just as well? The choice of a model is a fundamental problem, and we will spend considerable time in later sections delving into theoretical and practical reasons for these choices. For this early discussion we will merely suggest some models that look like they might do the job.

The function $f(x)$, which models the relative frequency behavior of X, is called the probability density function.

Definition 4.1

A random variable X is said to be **continuous** if it can take on the infinite number of possible values associated with intervals of real numbers, and there is a function $f(x)$, called the **probability density function,** such that

1. $f(x) \geq 0$, for all x
2. $\int_{-\infty}^{\infty} f(x)\,dx = 1$
3. $P(a \leq X \leq b) = \int_a^b f(x)\,dx.$

Note that for a continuous random variable X,

$$P(X = a) = \int_a^a f(x)\,dx = 0$$

for any specific value a. The need to assign zero probability to any specific value should not disturb us, because X can assume an infinite number of possible values. For example, out of all the possible lengths of the life of a transistor, what is the probability that the transistor we are using will last exactly 497.392 hours? Assigning probability zero to this event does not rule out 497.392 as a possible length, but it does say that the chance of observing this particular length is extremely small.

Example 4.1 The random variable X of the life length example is associated with a probability density function of the form

$$f(x) = \begin{cases} \frac{1}{2} e^{-x/2} & x > 0 \\ 0 & \text{elsewhere.} \end{cases}$$

Find the probability that the life of a particular battery of this type is less than 200 or greater than 400 hours.

Solution

Let A denote the event that X is less than 2 and B the event that X is greater than 4. Then, because A and B are mutually exclusive,

$$\begin{aligned} P(A \cup B) &= P(A) + P(B) \\ &= \int_0^2 \frac{1}{2} e^{-x/2}\,dx + \int_4^{\infty} \frac{1}{2} e^{-x/2}\,dx \\ &= (1 - e^{-1}) + (e^{-2}) \\ &= 1 - .368 + .135 \\ &= .767. \end{aligned}$$

Example 4.2 Refer to Example 4.1. Find the probability that a battery of this type lasts more than 300 hours given that it already has been in use for more than 200 hours.

Solution

We are interested in $P(X > 3 \mid X > 2)$ and, by the definition of conditional probability,

$$P(X > 3 \mid X > 2) = \frac{P(X > 3)}{P(X > 2)}$$

because the intersection of the events $(X > 3)$ and $(X > 2)$ is the event $(X > 3)$. Now

$$\frac{P(X > 3)}{P(X > 2)} = \frac{\int_3^{\infty} \frac{1}{2} e^{-x/2}\, dx}{\int_2^{\infty} \frac{1}{2} e^{-x/2}\, dx} = \frac{e^{-3/2}}{e^{-1}} = e^{-1/2} = .606.$$

Sometimes it is convenient to look at cumulative probabilities of the form $P(X \le b)$. To do this we can make use of the distribution function.

Definition 4.2

The **distribution function** for a random variable X is defined as

$$F(b) = P(X \le b).$$

If X is continuous with probability density function $f(x)$, then

$$F(b) = \int_{-\infty}^{b} f(x)\, dx.$$

Note that $F'(x) = f(x)$.

In the battery example, X has a probability density function given by

$$f(x) = \begin{cases} \frac{1}{2} e^{-x/2} & x > 0 \\ 0 & \text{elsewhere.} \end{cases}$$

FIGURE 4.2 A distribution function for a continuous random variable.

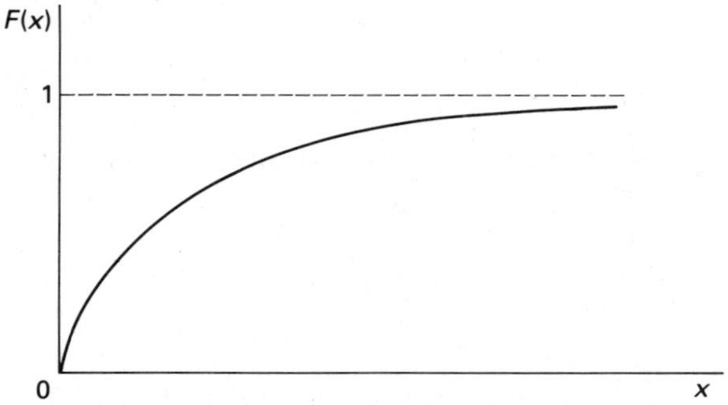

Thus,

$$F(b) = P(X \le b) = \int_0^b \frac{1}{2} e^{-x/2}\,dx$$

$$= -e^{-x/2}\Big|_0^b$$

$$= 1 - e^{-b/2} \qquad b > 0$$

$$= 0 \qquad b \le 0.$$

The function is shown graphically in Figure 4.2

Example 4.3 A supplier of kerosene has a 200-gallon tank filled at the beginning of each week. His weekly demands show a relative frequency behavior that increases steadily up to 100 gallons, and then levels off between 100 and 200 gallons. Letting X denote weekly demand in hundreds of gallons, suppose the relative frequencies for demand are modeled adequately by

$$f(x) = \begin{cases} = 0 & x < 0 \\ = x & 0 \le x \le 1 \\ = 1/2 & 1 < x \le 2 \\ = 0 & x > 2 \end{cases}.$$

This function has the graphical form shown in Figure 4.3. Find $F(b)$ for this random variable. Use $F(b)$ to find the probability that demand will exceed 150 gallons on a given week.

Solution

From the definition

$$F(b) = \int_{-\infty}^{b} f(x)\,dx$$

FIGURE 4.3 $f(x)$ for Example 4.3.

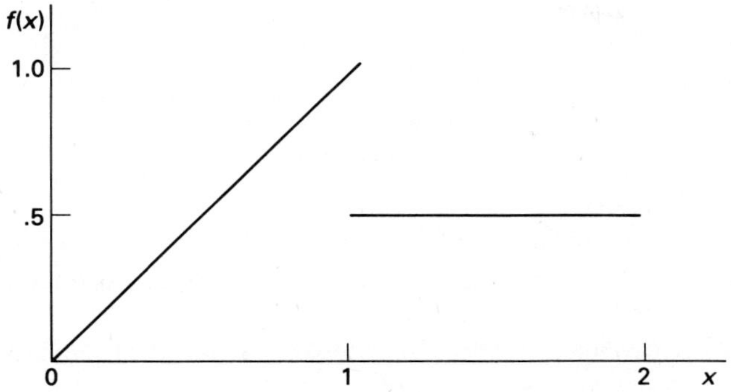

$$
\begin{aligned}
&= 0 && b < 0 \\
&= \int_{-\infty}^{b} x \, dx = \frac{b^2}{2} && 0 \le b \le 1 \\
&= \frac{1}{2} + \int_{1}^{b} \frac{1}{2} \, dx && \\
&= \frac{1}{2} + \frac{b-1}{2} = \frac{b}{2} && 1 < b \le 2 \\
&= 1 && b > 2.
\end{aligned}
$$

This function is graphed in Figure 4.4. Note that $F(b)$ is continuous over the whole real line, even though $f(b)$ has two discontinuities.

FIGURE 4.4 $F(b)$ for Example 4.3.

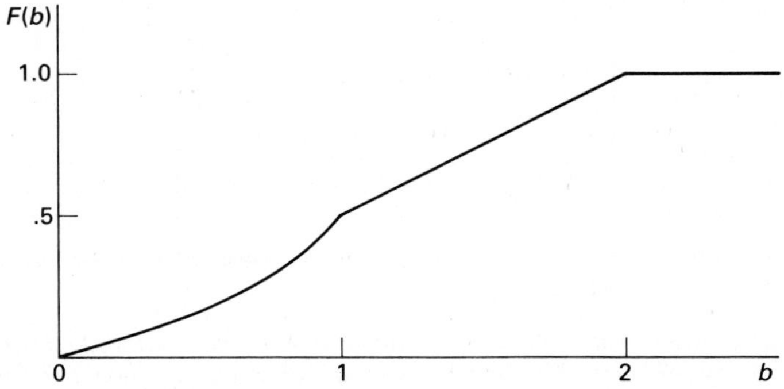

The probability that demand will exceed 150 gallons is given by

$$P(X > 1.5) = 1 - P(X \leq 1.5) = 1 - F(1.5)$$
$$= \frac{1.5}{2} = .75.$$

EXERCISES

4.1 For each of the following situations, define an appropriate random variable and state whether it is continuous or discrete.

(a) An environmental engineer is looking at ten field plots to determine whether they contain a certain type of insect.

(b) A quality control technician samples a continuously produced fabric in square yard sections, and counts the number of defects observed in each sampled section.

(c) A metallurgist counts the number of grains seen in a cross-sectional sample of aluminum.

(d) The metallurgist of (c) measures the area proportion covered by grains of a certain size, rather than simply counting them.

4.2 Suppose a random variable X has a probability density function given by

$$f(x) = \begin{cases} kx(1 - x) & 0 \leq x \leq 1 \\ 0 & \text{elsewhere.} \end{cases}$$

(a) Find the value of k that makes this a probability density function.

(b) Find $P(.4 \leq X \leq 1)$.

(c) Find $P(X \leq .4 \mid X \leq .8)$.

(d) Find $F(b) = P(X \leq b)$, and sketch the graph of this function.

4.3 The effectiveness of solar-energy heating units depends upon the amount of radiation available from the sun. For a typical October, daily total solar radiation in Tampa, Florida, approximately follows the following probability density function (units are hundreds of calories):

$$f(x) = \begin{cases} \frac{3}{32}(x - 2)(6 - x) & 2 \leq x \leq 6 \\ 0 & \text{elsewhere.} \end{cases}$$

(a) Find the probability that solar radiation will exceed 300 calories on a typical October day.

(b) What amount of solar radiation is exceeded on exactly 50% of the October days, according to this model?

4.4 An accounting firm that does not have its own computing facilities rents time from a consulting company. The firm must plan its computing budget carefully and, hence, has studied the weekly use of CPU time quite thoroughly. The weekly use of CPU time approximately follows the probability density function given by (measurements in hours)

$$f(x) = \begin{cases} \frac{3}{64}x^2(4 - x) & 0 \le x \le 4 \\ 0 & \text{elsewhere.} \end{cases}$$

(a) Find the distribution function $F(x)$ for weekly CPU time X.

(b) Find the probability that CPU time used by the firm will exceed 2 hours for a selected week.

(c) The current budget of the firm covers only 3 hours of CPU time per week. How often will the budgeted figure be exceeded?

(d) How much CPU time should be budgeted per week if this figure is to be exceeded with probability only .10?

4.5 The pH level, a measure of acidity, is important in studies of acid rain. For a certain Florida lake, baseline measurements on acidity are made so any changes caused by acid rain can be noted. The pH of water samples from the lake is a random variable X with probability density function

$$f(x) = \begin{cases} \frac{3}{8}(7 - x)^2 & 5 \le x \le 7 \\ 0 & \text{elsewhere.} \end{cases}$$

(a) Sketch the curve of $f(x)$.

(b) Find the distribution function $F(x)$ for X.

(c) Find the probability that the pH will be less than 6 for a water sample from this lake.

(d) Find the probability that the pH of a water sample from this lake will be less than 5.5 given that it is known to be less than 6.

4.6 The "on" temperature of a thermostatically controlled switch for an air conditioning system is set at 60°, but the actual temperature X at which the switch turns on is a random variable having probability density function

$$f(x) = \begin{cases} \frac{1}{2} & 59 \le x \le 61 \\ 0 & \text{elsewhere.} \end{cases}$$

(a) Find the probability that it takes a temperature in excess of 60° to turn the switch on.

(b) If two such switches are used independently, find the probability that they both require a temperature in excess of 60° to turn on.

4.7 The proportion of time, during a 40-hour workweek, that an industrial robot was in operation was measured for a large number of weeks, and the measurements can be modeled by the probability density function

$$f(x) = \begin{cases} 2x & 0 \le x \le 1 \\ 0 & \text{elsewhere.} \end{cases}$$

If X denotes the proportion of time this robot will be in operation during a coming week, find the following:

(a) $P(X > 1/2)$
(b) $P(X > 1/2 \mid X > 1/4)$
(c) $P(X > 1/4 \mid X > 1/2)$
(d) $F(x)$. Graph this function. Is $F(x)$ continuous?

4.8 The proportion of impurities by weight, X, in certain copper ore samples is a random variable having probability density function

$$f(x) = \begin{cases} 12x^2(1-x) & 0 \le x \le 1 \\ 0 & \text{elsewhere.} \end{cases}$$

If four such samples are independently selected, find the probability that

(a) exactly one has a proportion of impurities exceeding .5.
(b) at least one has a proportion of impurities exceeding .5.

4.2

EXPECTED VALUES OF CONTINUOUS RANDOM VARIABLES

As in the discrete case, we often want to summarize the information contained in a probability distribution by calculating expected values for the random variable and certain functions of the random variable.

Definition 4.3

The **expected value** of a continuous random variable X having probability density function $f(x)$ is given by[1]

$$E(X) = \int_{-\infty}^{\infty} xf(x)\,dx.$$

For function of random variables we have the following theorem.

Theorem 4.1 If X is a continuous random variable with probability distribution $f(x)$ and if $g(x)$ is any real valued function of X, then

$$E[g(X)] = \int_{-\infty}^{+\infty} g(x)f(x)\,dx.$$

The proof of Theorem 4.1 will not be given here.

[1] We assume absolute convergence of the integrals.

The definition of variance and standard deviation and the properties given in Theorems 3.2 and 3.3 hold for the continuous case, as well. We illustrate the expectations of continuous random variables in the following examples.

Example 4.4 For a lathe in a machine shop let X denote the percentage of time out of a 40-hour work week that the lathe is actually in use. Suppose X has a probability density function given by

$$f(x) = \begin{cases} 3x^2 & 0 \le x \le 1 \\ 0 & \text{elsewhere.} \end{cases}$$

Find the mean and variance of X.

Solution

From Definition 4.3,

$$\begin{aligned} E(X) &= \int_{-\infty}^{\infty} xf(x)\,dx \\ &= \int_0^1 x(3x^2)\,dx \\ &= \int_0^1 3x^3\,dx \\ &= 3\left[\frac{x^4}{4}\right]_0^1 = \frac{3}{4} = .75. \end{aligned}$$

Thus, on the average, the lathe is in use 75% of the time.

To compute $V(X)$, we first find $E(X^2)$ by

$$\begin{aligned} E(X^2) &= \int_{-\infty}^{\infty} x^2 f(x)\,dx \\ &= \int_0^1 x^2(3x^2)\,dx \\ &= \int_0^1 3x^4\,dx \\ &= 3\left[\frac{x^5}{5}\right]_0^1 = \frac{3}{5} = .60. \end{aligned}$$

Then,

$$\begin{aligned} V(X) &= E(X^2) - \mu^2 \\ &= .60 - (.75)^2 \\ &= .60 - .5625 = .0375. \end{aligned}$$

Example 4.5 The weekly demand, X, for kerosene at a certain supply station has a density function given by

$$f(x) = \begin{cases} x & 0 \le x \le 1 \\ \frac{1}{2} & 1 < x \le 2 \\ 0 & \text{elsewhere.} \end{cases}$$

Find the expected weekly demand.

Solution

Using Definition 4.3 to find $E(X)$, we must now carefully observe that $f(x)$ has different nonzero forms over two disjoint regions. Thus,

$$\begin{aligned} E(X) &= \int_{-\infty}^{\infty} xf(x)\,dx \\ &= \int_0^1 x(x)\,dx + \int_1^2 x\left(\frac{1}{2}\right)dx \\ &= \int_0^1 x^2\,dx + \frac{1}{2}\int_1^2 x\,dx \\ &= \left[\frac{x^3}{3}\right]_0^1 + \frac{1}{2}\left[\frac{x^2}{2}\right]_1^2 \\ &= \frac{1}{3} + \frac{1}{2}\left[2 - \frac{1}{2}\right] \\ &= \frac{1}{3} + \frac{3}{4} = \frac{13}{12} = 1.08. \end{aligned}$$

That is, the expected weekly demand is for 108 gallons.

Tchebysheff's Theorem (Theorem 3.4) holds for continuous random variables, just as it does for discrete ones. Thus, if X is continuous with mean μ and standard deviation σ, then

$$P(|X - \mu| < k\sigma) \ge 1 - \frac{1}{k^2}$$

for any positive number k. We illustrate the use of this result in the next example.

Example 4.6 The weekly amount Y spent for chemicals in a certain firm has a mean of \$445 and a variance of \$236. Within what interval would these weekly costs for chemicals be expected to lie at least 75 % of the time?

Solution

To find an interval guaranteed to contain at least 75 % of the probability mass for Y, we get

$$1 - \frac{1}{k^2} = .75,$$

which gives

$$\frac{1}{k^2} = .25$$

$$k^2 = \frac{1}{.25} = 4$$

or

$$k = 2.$$

Thus, the interval $\mu - 2\sigma$ to $\mu + 2\sigma$ will contain at least 75 % of the probability. This interval is given by

$$445 - 2\sqrt{236} \quad \text{to} \quad 445 + 2\sqrt{236}$$

$$445 - 30.72 \quad \text{to} \quad 445 + 30.72,$$

or

$$414.28 \quad \text{to} \quad 475.72.$$

EXERCISES

4.9 The temperature X at which a thermostatically controlled switch turns on has probability density function

$$f(x) = \begin{cases} \frac{1}{2} & 59 \le x \le 61 \\ 0 & \text{elsewhere.} \end{cases}$$

Find $E(X)$ and $V(X)$.

4.10 The proportion of time X that an industrial robot is in operation during a 40-hour week is a random variable with porbability density function

$$f(x) = \begin{cases} 2x & 0 \le x \le 1 \\ 0 & \text{elsewhere.} \end{cases}$$

(a) Find $E(X)$ and $V(X)$.

(b) For the robot under study, the profit Y for a week is given by $Y = 200X - 60$. Find $E(Y)$ and $V(Y)$.

(c) Find an interval in which the profit should lie for at least 75 % of the weeks that the robot is in use.

4.11 Daily total solar radiation for a certain location in Florida in October has probability density function

$$f(x) = \begin{cases} \frac{3}{32}(x-2)(6-x) & 2 \le x \le 6 \\ 0 & \text{elsewhere} \end{cases}$$

with measurements in hundreds of calories. Find the expected daily solar radiation for October.

4.12 Weekly CPU time used by an accounting firm has probability density function (measured in hours)

$$f(x) = \begin{cases} \frac{3}{64}x^2(4-x) & 0 \le x \le 4 \\ 0 & \text{elsewhere.} \end{cases}$$

(a) Find the expected value and variance of weekly CPU time.
(b) The CPU time costs the firm \$200 per hour. Find the expected value and variance of the weekly cost for CPU time.
(c) Would you expect the weekly cost to exceed \$600 very often? Why?

4.13 The pH of water samples from a specific lake is a random variable X with probability density function

$$f(x) = \begin{cases} \frac{3}{8}(7-x)^2 & 5 \le x \le 7 \\ 0 & \text{elsewhere.} \end{cases}$$

(a) Find $E(X)$ and $V(X)$.
(b) Find an interval shorter than $(5, 7)$ in which at least $3/4$ of the pH measurements must lie.
(c) Would you expect to see a pH measurement below 5.5 very often? Why?

4.14 A retail grocer has a daily demand X for a certain food sold by the pound, such that X (measured in hundred of pounds) has probability density function

$$f(x) = \begin{cases} 3x^2 & 0 \le x \le 1 \\ 0 & \text{elsewhere.} \end{cases}$$

The grocer, who cannot stock over 100 pounds, wants to order $100k$ pounds of food on a certain day. He buys the food at 6 cents per pound and sells it at 10 cents per pound. What value of k will maximize his expected daily profit? (There is no salvage value for food not sold.)

4.3

THE UNIFORM DISTRIBUTION

We now move from a general discussion of continuous random variables to discussions of specific models found useful in practice. Consider an experiment that consists of observing events in a certain time frame, such as buses arriving at a bus stop or telephone calls coming into a switchboard during a specified period. Suppose we know that one such event has occurred in the time interval (a, b) (a bus arrived between 8:00 and 8:10). It may then be of interest to place a probability distribution on the actual time of occurrence of the event under observation, which we will denote by X. A very simple model assumes that X is equally likely to lie in any small subinterval, say, of length d, no matter where that subinterval lies within (a, b). This assumption leads to the *uniform* probability distribution, which has probability density function given by

$$\begin{aligned} f(x) &= \frac{1}{b - a} \qquad a \le x \le b \\ &= 0 \qquad \text{elsewhere.} \end{aligned}$$

This density function is graphed in Figure 4.5.

The distribution function for a uniformly distributed X is given by

$$\begin{aligned} F(x) &= 0, && x < a, \\ F(x) &= \int_a^x \frac{1}{b - a}\,dx = \frac{x - a}{b - a}, && a \le x \le b, \\ F(x) &= 1, && x > b. \end{aligned}$$

FIGURE 4.5 The uniform probability density function.

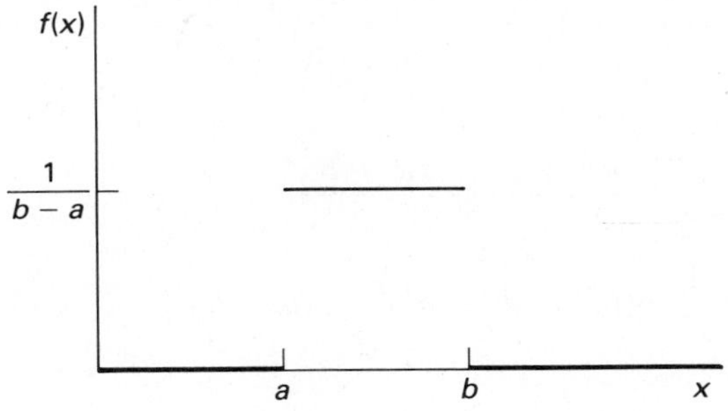

If we consider a subinterval $(c, c + d)$ contained entirely within (a, b), we have

$$P(c \le X \le c + d) = F(c + d) - F(c)$$
$$= \frac{(c + d) - a}{b - a} - \frac{c - a}{b - a}$$
$$= \frac{d}{b - a}.$$

Note that this probability does not depend on the location, c, but only on the length, d, of the subinterval.

A relationship exists between the uniform distribution and the Poisson distribution, which was introduced in Section 3.7. Suppose the number of events occurring in an interval, say $(0, t)$, has a Poisson distribution. If exactly one of these events is known to have occurred in the interval (a, b), with $a \ge 0$ and $b \le t$, then the conditional probability distribution of the actual time of occurrence for this event is uniform over (a, b).

Paralleling the material presented in Chapter 3, we now look at the mean and variance of the uniform distribution. From Definition 4.3,

$$E(X) = \int_{-\infty}^{\infty} xf(x)\,dx = \int_a^b x\left(\frac{1}{b - a}\right) dx$$
$$= \left(\frac{1}{b - a}\right)\left(\frac{b^2 - a^2}{2}\right) = \frac{a + b}{2}.$$

It is intuitively reasonable that the mean value of a uniformly distributed random variable should lie at the midpoint of the interval.

Recalling from Theorem 3.3 that $V(X) = E(X - \mu)^2 = E(X^2) - \mu^2$, we have, for the uniform case,

$$E(X^2) = \int_{-\infty}^{\infty} x^2 f(x)\,dx$$
$$= \int_a^b x^2\left(\frac{1}{b - a}\right) dx$$
$$= \left(\frac{1}{b - a}\right)\left(\frac{b^3 - a^3}{3}\right) = \frac{b^2 + ab + a^2}{3}.$$

Thus,

$$V(X) = \frac{b^2 + ab + a^2}{3} - \left(\frac{a + b}{2}\right)^2$$
$$= \frac{1}{12}[4(b^2 + ab + a^2) - 3(a + b)^2]$$
$$= \frac{1}{12}(b - a)^2.$$

This result may not be intuitive, but we see that the variance depends only upon the length of the interval (a, b).

Example 4.7 The failure of a circuit board interrupts work by a computing system until a new board is delivered. Delivery time X is uniformly distributed over the interval 1 to 5 days. The cost C of this failure and interruption consists of a fixed cost c_0 for the new part and a cost that increases proportional to X^2, so that

$$C = c_0 + c_1X^2.$$

(a) Find the probability that the delivery time is 2 or more days.
(b) Find the expected cost of a single failure, in terms of c_0 and c_1.

Solution

(a) The delivery time X is distributed uniformly from 1 to 5 days, which gives

$$f(x) = \begin{cases} \frac{1}{4} & 1 \le x \le 5 \\ 0 & \text{elsewhere.} \end{cases}$$

Thus,

$$\begin{aligned} P(X \ge 2) &= \int_2^5 \left(\frac{1}{4}\right) dx \\ &= \frac{1}{4}(5 - 2) = \frac{3}{4}. \end{aligned}$$

(b) We know that

$$E(C) = c_0 + c_1E(X^2),$$

so it remains to find $E(X^2)$. This could be found directly from the definition or by using the variance and the fact that

$$E(X^2) = V(X) + \mu^2.$$

Using the latter approach,

$$\begin{aligned} E(X^2) &= \frac{(b - a)^2}{12} + \left(\frac{a + b}{2}\right)^2 \\ &= \frac{(5 - 1)^2}{12} + \left(\frac{1 + 5}{2}\right)^2 = \frac{31}{3}. \end{aligned}$$

Thus,

$$E(C) = c_0 + c_1\left(\frac{31}{3}\right).$$

We now summarize the properties of uniform distribution.

The Uniform Distribution

$$f(x) = \frac{1}{b - a} \qquad a \le x \le b$$

$$= 0 \qquad \text{elsewhere}$$

$$E(X) = \frac{a + b}{2} \qquad V(X) = \frac{(b - a)^2}{12}$$

EXERCISES

4.15 Suppose X has a uniform distribution over the interval (a, b).

(a) Find $F(x)$.

(b) Find $P(X > c)$, for some point c, between a and b.

(c) If $a \le c \le d \le b$, find $P(X > d \mid X > c)$.

4.16 Upon studying low bids for shipping contracts, a microcomputer manufacturing firm finds that intrastate contracts have low bids uniformly distributed between 20 and 25, in units of thousands of dollars. Find the probability that the low bid on the next intrastate shipping contract is

(a) below \$22,000.

(b) in excess of \$24,000.

Now find the average cost of low bids on contracts of this type.

4.17 If a point is *randomly* located in an interval (a, b) and X denotes its distance from a, then X will be assumed to have a uniform distribution over (a, b). A plant efficiency expert randomly picks a spot along a 500-foot assembly line from which to observe work habits. Find the probability that she is

(a) within 25 feet of the end of the line.

(b) within 25 feet of the beginning of the line.

(c) closer to the beginning of than the end of the line.

4.18 A bomb is to be dropped along a mile-long line that stretches across a practice target. The target center is at the midpoint of the line. The target will be destroyed if the bomb falls within a tenth of a mile to either side of the center. Find the probability that the target is destroyed if the bomb falls at a random location along the line.

4.19 A telephone call arrived at a switchboard at a random time within a 1-minute interval. The switchboard was fully busy for 15 seconds into this 1-minute period. Find the probability that the call arrived when the switchboard was not fully occupied.

4.20 Beginning at 12:00 midnight, a computer center is up for 1 hour and down for 2 hours on a regular cycle. A person who does not know the schedule dials the center at a random time between 12:00 midnight and 5:00 A.M. What is the probability that the center will be operating when this call comes in?

4.21 The number of defective circuit boards among those coming out of a soldering machine follows a Poisson distribution. On a particular 8-hour day, one defective board is found.

(a) Find the probability that it was produced the first hour of operation of that day.

(b) Find the probability that it was produced during the last hour of operation for that day.

(c) Given that no defective boards were seen during the first 4 hours of operation, find the probability that the defective board was produced during the fifth hour.

4.22 In determinging the range of an acoustic source by triangulation, the time at which the spherical wave front arrives at a receiving sensor must be measured accurately. According to Perruzzi and Hilliard (1984), measurement errors in these times can be modeled as having uniform distributions. Suppose measurement errors are uniformly distributed from $-.05$ to $+.05$ microsecond.

(a) Find the probability that a particular arrival time measurement will be in error by less than .01 microsecond.

(b) Find the mean and variance of these measurement errors.

4.23 In the setting of Exercise 4.22, suppose the measurement errors are distributed uniformly from $-.02$ to $+.05$ microsecond.

(a) Find the probability that a particular arrival time measurement will be in error by less than .01 microsecond.

(b) Find the mean and variance of these measurement errors.

4.24 According to Y. Zimmels (1983), the sizes of particles used in sedimentation experiments often have uniform distributions. It is important to study both the mean and variance of particle sizes because, in sedimentation with mixtures of various-sized particles, the larger particles hinder the movements of the smaller ones.

Suppose spherical particles have diameters uniformly distributed between .01 and .05 centimeters. Find the mean and variance of the *volumes* of these particles. (Recall that the volume of a sphere is $(4/3)\pi r^3$.)

4.25 Arrivals of customers at a certain checkout counter follows a Poisson distribution. During a given 30-minute period, one customer arrived at the counter. Find the probability that he arrived during the last 5 minutes of the period.

4.26 Using the conditions given in Exercise 4.25, find the conditional probability that the customer arrived during the last 5 minutes of the 30-minute period if there were no arrivals during the first 10 minutes of the period.

4.27 In tests of stopping distances for automobiles, those automobiles traveling at 30 miles per hour before the brakes were applied tended to travel distances that appear uniformly distributed between two points, a and b. Find the probability that one of these automobiles

(a) stops closer to a than b.

(b) stops so that the distance to a is more than three times the distance to b.

4.28 Suppose three automobiles are used in a test of the type discussed in Exercise 4.27. Find the probability that exactly one of the three travels past the midpoint between a and b.

4.29 The cycle time for trucks hauling concrete to a highway construction site is uniformly distributed over the interval 50 to 70 minutes.

(a) Find the expected value and variance for these cycle times.

(b) How many trucks would you expect to schedule for this job so that a truckload of concrete can be dumped at the site every 15 minutes?

4.4

THE EXPONENTIAL DISTRIBUTION

The life-length data of Section 4.1 displayed a probabilistic behavior that was not uniform, but rather one in which the probability over intervals of constant length decreased as the intervals moved further and further to the right. We saw that an exponential curve seemed to fit this data rather well and now discuss the exponential probability distribution in more detail. In general the exponential density function is given by

$$\begin{aligned} f(x) &= \frac{1}{\theta} e^{-x/\theta} \qquad && x \geq 0 \\ &= 0 && \text{elsewhere} \end{aligned}$$

where the parameter θ is a constant ($\theta > 0$) that determines the rate at which the curve decreases.

An exponential density function with $\theta = 2$ was sketched in Figure 4.1, and in general, the exponential functions have the form shown in Figure 4.6. Many

FIGURE 4.6 An exponential probability density function.

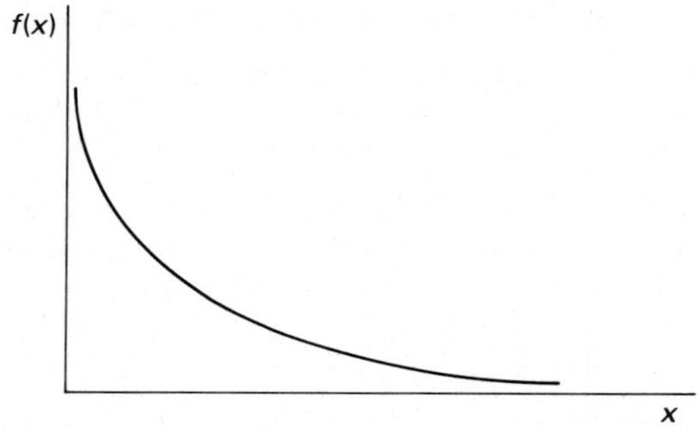

FIGURE 4.7 Interarrival times of vehicles on a one-directional roadway.

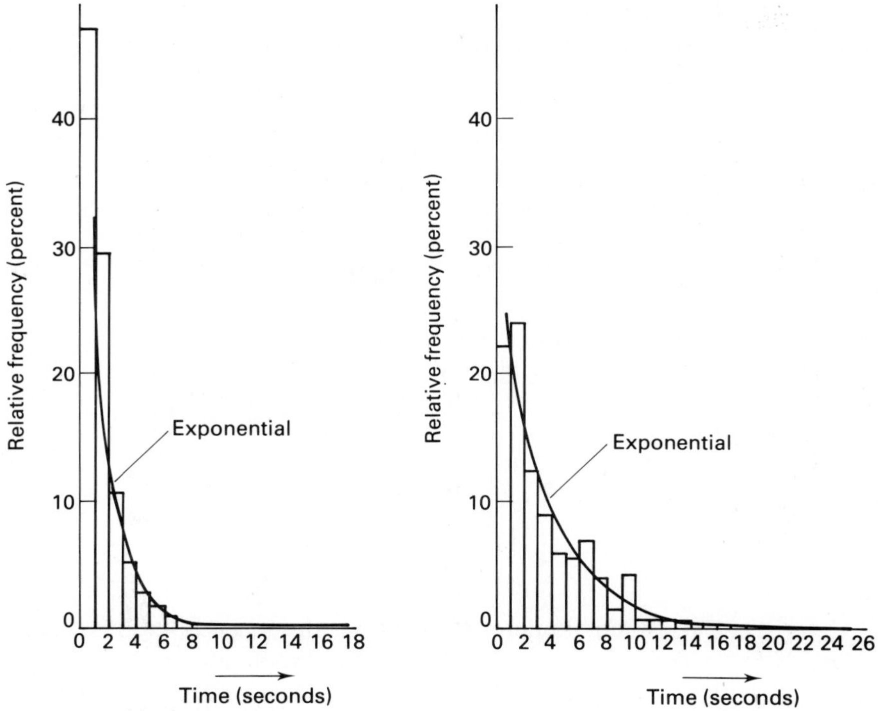

Source: D. Mahalel, and A. S. Hakkert, "Traffic Arrival Patterns on a Cross Section of a Multilane Highway," *Transportation Research* 17A, no. 4 (1983): 267. Used by permission.

random variables in engineering and the sciences can be modeled appropriately as having exponential distributions. Figure 4.7 shows two examples of relative frequency distributions for times between arrivals (interarrival times) of vehicles at a fixed point on a one-directional roadway. Both of these relative frequency histograms can be modeled quite nicely by exponential functions. Note that the higher traffic density causes shorter interarrival times to be more frequent.

Finding expected values for the exponential case requires the evaluation of a certain type of integral called a gamma (Γ) function. The function $\Gamma(n)$ is defined by

$$\Gamma(n) = \int_0^\infty x^{n-1}e^{-x}\, dx.$$

Upon integrating by parts, it can be shown that $\Gamma(n) = (n - 1)\Gamma(n - 1)$ and hence that, when n is a positive integer,

$$\Gamma(n) = (n - 1)!.$$

It is useful to note that $\Gamma(1/2) = \sqrt{\pi}$.

Suppose that a constant term θ appears in the exponent and we wish to evaluate

$$\int_0^\infty x^{n-1}e^{-x/\theta}\,dx.$$

Integrating by substitution ($y = x/\theta$), we can show that

$$\int_0^\infty x^{n-1}e^{-x/\theta}\,dx = \Gamma(n)\theta^n.$$

Using this result, we see that, for the exponential distribution,

$$\begin{aligned} E(X) &= \int_{-\infty}^\infty xf(x)\,dx = \int_0^\infty x\left(\frac{1}{\theta}\right)e^{-x/\theta}\,dx \\ &= \frac{1}{\theta}\int_0^\infty xe^{-x/\theta}\,dx \\ &= \frac{1}{\theta}\Gamma(2)\theta^2 = \theta. \end{aligned}$$

Thus, the parameter θ actually is the mean of the distribution.

To evaluate the variance of the exponential distribution, first we can find

$$\begin{aligned} E(X^2) &= \int_0^\infty x^2\left(\frac{1}{\theta}\right)e^{-x/\theta}\,dx \\ &= \frac{1}{\theta}\Gamma(3)\theta^3 = 2\theta^2. \end{aligned}$$

It follows that

$$\begin{aligned} V(X) &= E(X^2) - \mu^2 \\ &= 2\theta^2 - \theta^2 = \theta^2 \end{aligned}$$

and θ becomes the standard deviation as well as the mean.

The distribution function for the exponential case has a simple form, seen to be

$$F(t) = 0 \qquad \text{for } t < 0$$

$$\begin{aligned} F(t) &= P(X \le t) = \int_0^t \frac{1}{\theta}e^{-x/\theta}\,dx \\ &= -e^{-x/\theta}\Big|_0^t = 1 - e^{-t/\theta} \qquad \text{for } t \ge 0. \end{aligned}$$

Example 4.8 A sugar refinery has three processing plants, all receiving raw sugar in bulk. The amount of sugar that one plant can process in one day can be modeled as having an exponential distribution with a mean of 4 (measurements in tons) for each of the three plants. If the plants operate independently, find the probability that exactly two of the three plants process more than 4 tons on a given day.

Solution

The probability that any given plant processes more than 4 tons, with X denoting the amount used, is

$$P(X > 4) = \int_4^\infty f(x)\, dx = \int_4^\infty \frac{1}{4} e^{-x/4}\, dx$$
$$= -e^{-x/4}\Big|_4^\infty = e^{-1} = .37.$$

(Note: $P(X > 4) \neq .5$, even though the mean of X is 4.)

Knowledge of the distribution function could allow us to evaluate this immediately as

$$P(X > 4) = 1 - P(X \leq 4) = 1 - (1 - e^{-4/4})$$
$$= e^{-1}.$$

Assuming the three plants operate independently, the problem is to find the probability of two successes out of three tries, where the probability of success is .37. This is a binomial problem, and the solution is

$$P(\text{exactly two plants use more than 4 tons}) = \binom{3}{2}(.37)^2(.63)$$
$$= 3(.37)^2(.63)$$
$$= .26.$$

Example 4.9 Consider a particular plant in Example 4.8. How much raw sugar should be stocked for that plant each day so that the chance of running out of product is only .05?

Solution

Let a denote the amount to be stocked. Since the amount to be used, X, has an exponential distribution,

$$P(X > a) = \int_a^\infty \frac{1}{4} e^{-x/4}\, dx = e^{-a/4}.$$

We want to choose a so that

$$P(X > a) = e^{-a/4} = .05$$

and solving this equation yields

$$a = 11.98.$$

As in the uniform case, there is a relationship between the exponential distribution and the Poisson distribution. Suppose events are occurring in time according to a Poisson distribution with a rate of λ events per hour. Thus, in t hours, the number of events, say, Y, will have a Poisson distribution with mean value λt. Suppose we start at time zero and ask the question, How long do I have to wait to see the first event occur? Let X denote the length of time until this first event. Then,

$$\begin{aligned} P(X > t) &= P[Y = 0 \text{ on the interval } (0, t)] \\ &= (\lambda t)^0 e^{-\lambda t}/0 = e^{-\lambda t} \end{aligned}$$

and

$$P(X \leq t) = 1 - P(X > t) = 1 - e^{-\lambda t}.$$

We see that $P(X \leq t) = F(t)$, the distribution function for X, has the form of an exponential distribution function with $\lambda = (1/\theta)$. Upon differentiating, we see that the probability density function of X is given by

$$\begin{aligned} f(t) &= \frac{dF(t)}{dt} = \frac{d(1 - e^{-\lambda t})}{dt} \\ &= \lambda e^{-\lambda t} \\ &= \frac{1}{\theta} e^{-t/\theta} \qquad t > 0 \end{aligned}$$

and X has an exponential distribution. Actually, we need not start at at time zero, for it can be shown that the waiting time from the occurrence of any one event until the occurrence of the next will have an exponential distribution for events occurring according to a Poisson distribution.

Besides probability density and distribution functions, another function is of use in studying properties of continuous distributions, especially life-length data. Suppose X denotes the life length of a component with density function $f(x)$ and distribution function $F(x)$. The *failure rate function*, $r(t)$, is defined as

$$r(t) = \frac{f(t)}{1 - F(t)} \qquad t > 0, F(t) < 1.$$

For an intuitive look at what $r(t)$ is measuring, suppose dt denotes a very small interval around the point t. Then $f(t)\, dt$ is approximately the probability that X

takes on a value in $(t, t + dt)$. Also $1 - F(t) = P(X > t)$. Thus,

$$r(t)\,dt = \frac{f(t)\,dt}{1 - F(t)}$$
$$\approx P[X \in (t, t + dt) \mid X > t].$$

In other words, $r(t)\,dt$ represents the probability of failure during the time interval $(t, t + dt)$ given that the component has survived to time t.

For the exponential case,

$$r(t) = \frac{f(t)}{1 - F(t)} = \frac{\frac{1}{\theta}e^{-t/\theta}}{e^{-t/\theta}} = \frac{1}{\theta}$$

or X has a *constant* failure rate. It is unlikely that many individual conponents have a constant failure rate over time (most fail more frequently as they age), but it may be true of some systems that undergo regular preventive maintenance.

The Exponential Distribution

$$f(x) = \frac{1}{\theta}e^{-x/\theta} \qquad x > 0$$
$$= 0 \qquad \text{elsewhere}$$
$$E(X) = \theta \qquad V(X) = \theta^2$$

EXERCISES

4.30 Suppose Y has an exponential density function with mean θ. Show that $P(Y > a + b \mid Y > a) = P(Y > b)$. This is referred to as the *memoryless property* of the exponential distribution.

4.31 The magnitudes of earthquakes recorded in a region of North America can be modeled by an exponential distribution with mean of 2.4, as measured on the Richter scale. Find the probability that the next earthquake to strike this region will

(a) exceed 3.0 on the richter scale.

(b) fall between 2.0 and 3.0 on the Richter scale.

4.32 Referring to Exercise 4.31, out of the next ten earthquakes to strike this region, find the probability that at least one will exceed 5.0 on the Richter scale.

4.33 A pumping station operator observes that the demand for water at a certain hour of the day can be modeled as an exponential random variable with a mean of 100 cfs (cubic feet per second).

(a) Find the probability that the demand will exceed 200 cfs on a randomly selected day.

(b) What is the maximum water-producing capacity that the station should keep on line for this hour so that the demand will exceed this production capacity with a probability of only .01?

4.34 Suppose customers arrive at a certain checkout counter at the rate of two every minute.

(a) Find the mean and variance of the waiting time between successive customer arrivals.

(b) If a clerk takes 3 minutes to serve the first customer arriving at the counter, what is the probability that at least one more customer is waiting when the service of the first customer is completed?

4.35 The length of time X to complete a certain key task in house construction is an exponentially distributed random variable with a mean of 10 hours. The cost C of completing this task is related to the square of the time to completion by the formula

$$C = 100 + 40X + 3X^2$$

(a) Find the expected value and variance of C.

(b) Would you expect C to exceed 2000 very often?

4.36 The interaccident times (times between accidents) for all fatal accidents on scheduled American domestic passenger airflights, 1948–1961 were found to follow an exponential distribution with a mean of approximately 44 days (Pyke, 1968, p. 426).

(a) If one of those accidents occurred on July 1, find the probability that another one occurred in that same month.

(b) Find the variance of the interaccident times.

(c) What does this information suggest about the clumping of airline accidents?

4.37 The life lengths of automobile tires of a certain brand, under average driving conditions, are found to follow an exponential distribution with a mean of 30 (in thousands of miles). Find the probability that one of these tires, bought today, will last

(a) over 30,000 miles.

(b) over 30,000 miles given that it already has gone 15,000 miles.

4.38 The dial-up connections from remote terminals come into a computing center at the rate of 4 per minute. The calls follow a Poisson distribution. If a call arrives at the beginning of a 1-minute period, find the probability that a second call will not arrive in the next 20 seconds.

4.39 The breakdowns of an industrial robot follow a Poisson distribution with an average of .5 breakdowns per 8-hour workday. If this robot is placed in service at the beginning of the day, find the probability that

(a) it will not break down during the day.

(b) it will work for at least 4 hours without breaking down.

Does what happened the day before have any effect on your answers? Why?

4.40 Air samples from a large city are found to have 1-hour carbon monoxide concentrations in an exponential distribution with a mean of 3.6 ppm (Zamurs, 1984, p. 637).

(a) Find the probability that a concentration will exceed 9 parts per million (ppm).

(b) A traffic control strategy reduced the mean to 2.5 ppm. Now find the probability that a concentration will exceed 9 ppm.

4.41 The weekly rainfall totals for a section of the midwestern United States follow an exponential distribution with a mean of 1.6 inches.

(a) Find the probability that a weekly rainfall total in this section will exceed 2 inches.

(b) Find the probability that the weekly rainfall totals will not exceed 2 inches in either of the next two weeks.

4.42 The service times at teller windows in a bank were found to follow an exponential distribution with a mean of 3.2 minutes. A customer arrives at a window at 4:00 P.M.

(a) Find the probability that he will still be there at 4:02 P.M.

(b) Find the probability that he will still be there at 4:04 given that he was there at 4:02.

4.43 In deciding how many customer service representatives to hire and in planning their schedules, it is important for a firm marketing electronic typewriters to study repair times for the machines. Such a study revealed that repair times have, approximately, an exponential distribution with a mean of 22 minutes.

(a) Find the probability that a repair time will last less than 10 minutes.

(b) The charge for typewriter repairs is $50 for each half hour, or part thereof, for labor. What is the probability that a repair job will result in a charge for labor of $100?

(c) In planning schedules, how much time should be allowed for each repair so that the chance of any one repair time exceeding this allowed time is only .01?

4.44 Explosive devices used in a mining operation cause nearly circular craters to form in a rocky surface. The radii of these craters are exponentially distributed with a mean of 10 feet. Find the mean and variance of the area covered by such a crater.

4.5 THE GAMMA DISTRIBUTION

Many sets of data, of course, will not have relative frequency curves with the smooth decreasing trend found in the exponential model. It perhaps is more common to see distributions that have low probabilities for intervals close to zero, with the probability increasing for a while as the interval moves to the right (in the positive direction) and then decreasing as the interval moves out even further. That is, the relative frequency curves appear as in Figure 4.8. In the case of electronic components, for example, few will have very short life

FIGURE 4.8 A common relative frequency curve.

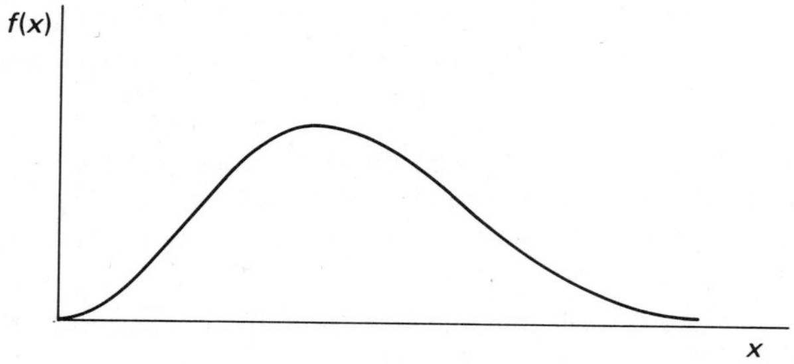

lengths, many will have something close to an average life length, and very few will have extraordinarily long life lengths.

A class of functions that serve as good models for this type of behavior is the *gamma* class. The gamma probability density function is given by

$$f(x) = \frac{1}{\Gamma(\alpha)\beta^{\alpha}} x^{\alpha-1} e^{-x/\beta} \qquad x \geq 0$$
$$= 0 \qquad \text{elsewhere}$$

where α and β are parameters that determine the specific shape of the curve. Note immediately that the gamma density reduces to the exponential when $\alpha = 1$. The parameters α and β must be positive, but need not be integers. The symbol $\Gamma(\alpha)$ is defined by

$$\Gamma(\alpha) = \int_0^{\infty} x^{\alpha-1} e^{-x}\, dx.$$

The integral

$$\int_0^{\infty} x^{\alpha-1} e^{-x/\beta}\, dx$$

can be evaluated by making the transformation $y = x/\beta$, or $x = \beta y$, $dx = \beta\, dy$. We have, then,

$$\int_0^{\infty} (\beta y)^{\alpha-1} e^{-y} \beta\, dy = \beta^{\alpha} \int_0^{\infty} y^{\alpha-1} e^{-y}\, dy = \beta^{\alpha}\Gamma(\alpha),$$

which makes $\int_0^{\infty} f(x)\, dx = 1$. Some typical gamma densities are shown in Figure 4.9.

An example of a real data set that closely follows a gamma distribution is shown in Figure 4.10. The data are 6-week summer rainfall totals for Ames, Iowa. Notice that many totals are from 2 to 8 inches, but occasionally a rainfall total goes well beyond 8 inches. Of course, no rainfall measurements can be negative.

FIGURE 4.9 The gamma density function, $\beta = 1$.

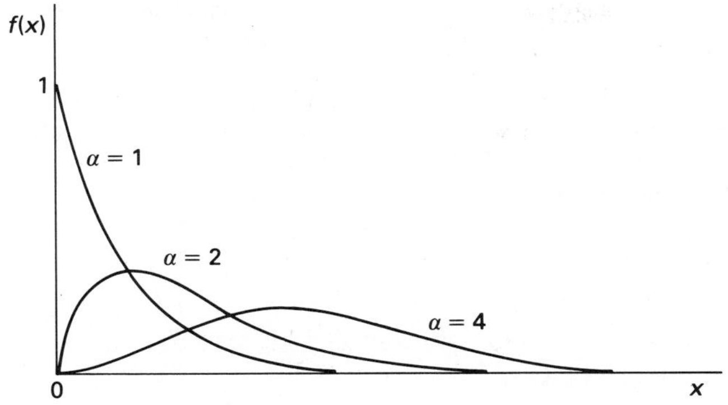

The derivation of expectations here is very similar to the exponential case of Section 4.4. We have

$$E(X) = \int_{-\infty}^{\infty} xf(x)\,dx = \int_0^{\infty} x\frac{1}{\Gamma(\alpha)\beta^\alpha}x^{\alpha-1}e^{-x/\beta}\,dx$$

$$= \frac{1}{\Gamma(\alpha)\beta^\alpha}\int_0^{\infty} x^\alpha e^{-x/\beta}\,dx$$

$$= \frac{1}{\Gamma(\alpha)\beta^\alpha}\Gamma(\alpha+1)\beta^{\alpha+1} = \alpha\beta.$$

FIGURE 4.10 Summer rainfall (6-week totals) for Ames, Iowa.

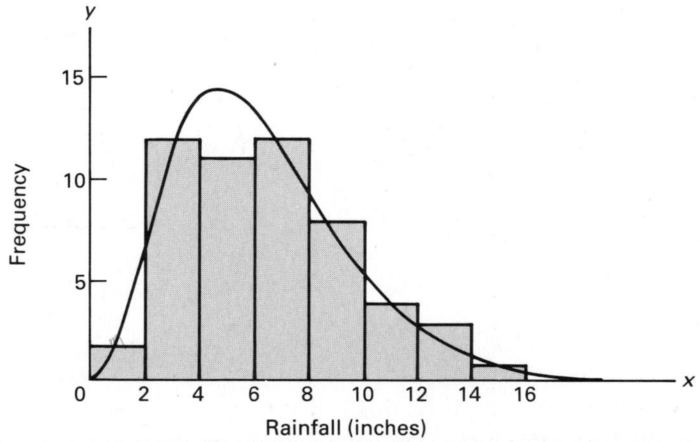

Source: G. L. Barger and H. C. S. Thom, "Evolution of Drought Hazard," *Agronomy Journal* 41, no. 11 (November 1949): 521. Reproduced by permission of American Society of Agronomy. Inc.

Similar manipulations yield $E(X^2) = \alpha(\alpha + 1)\beta^2$ and hence,

$$V(X) = E(X^2) - \mu^2$$
$$= \alpha(\alpha + 1)\beta^2 - \alpha^2\beta^2 = \alpha\beta^2.$$

A simple and often used property of sums of identically distributed, independent gamma random variables will be stated, but not proved, at this point. Suppose $X_1, X_2, \ldots, X_n$ represent independent gamma random variables with parameters α and β, as just used. If

$$Y = \sum_{i=1}^{n} X_i$$

then Y also has a gamma distribution with parameters $n\alpha$ and β. Thus, one can immediately see that

$$E(Y) = n\alpha\beta$$

and

$$V(Y) = n\alpha\beta^2.$$

Example 4.10 A certain electronic system having life length X_1 with an exponential distribution and a mean of 400 hours is supported by an identical backup system with a life length of X_2. The backup system takes over immediately when the primary system fails. If the systems operate independently, find the probability distribution and expected value for the total life length of the primary and backup system.

Solution

Letting Y denote the total life length, we have $Y = X_1 + X_2$, where X_1 and X_2 are independent exponential random variables, each with a mean $\beta = 400$. By the results stated earlier, Y then will have a gamma distribution with $\alpha = 2$ and $\beta = 400$; that is,

$$f_Y(y) = \frac{1}{\Gamma(2)(400)^2} y e^{-y/400} \qquad y > 0.$$

The mean value is given by

$$E(Y) = \alpha\beta = 2(400) = 800,$$

which is intuitively reasonable.

Example 4.11 Suppose that the length of time, Y, to conduct a periodic maintenance check on a dictating machine (known from previous experience) follows a gamma distribution with $\alpha = 3$ and $\beta = 2$ (minutes). Suppose that a new repair person requires 20 minutes to check a machine. Does this time to perform a maintenance check seem to disagree with prior experience?

Solution

The mean and variance for the length of maintenance times (prior experience) are

$$\mu = \alpha\beta \text{ and } \sigma^2 = \alpha\beta^2.$$

Then, for our example,

$$\mu = \alpha\beta = (3)(2) = 6, \quad \sigma^2 = \alpha\beta^2 = (3)(2)^2 = 12, \quad \sigma = \sqrt{12} = 3.46$$

and the observed deviation $(Y - \mu)$ is $20 - 6 = 14$ minutes.

For our example, $y = 20$ minutes exceeds the mean, $\mu = 6$, by $k = 14/3.46$ standard deviations. Then, from Tchebysheff's Theorem,

$$P(|Y - \mu| \geq k\sigma) \leq \frac{1}{k^2}$$

or

$$P(|Y - 6| \geq 14) \leq \frac{1}{k^2} = \frac{(3.46)^2}{(14)^2} = .06.$$

Note that this probability is based on the assumption that the distribution of maintenance times has not changed from prior experience. Then observing that $P(Y \geq 20 \text{ minutes})$ is small, we must conclude that either our new maintenance person has encountered a machine needing lengthy maintenance time, which occurs with low probability, or is somewhat slower than previous repairers. Noting the low probability for $P(Y \geq 20)$, we would be inclined to favor the latter view.

The failure rate function $r(t)$ for the gamma case is not easily displayed because $F(t)$ does not have a simple closed form. However, for $\alpha > 1$, this function will increase but is always bounded above by $1/\beta$. A typical form is shown in Figure 4.11. The properties of the gamma distribution are summarized as follows.

The Gamma Distribution

$$f(x) = \frac{1}{\Gamma(\alpha)\beta^\alpha} x^{\alpha-1} e^{-x/\beta} \qquad x > 0$$

$$= 0 \qquad \text{elsewhere}$$

$$E(X) = \; = \alpha\beta \qquad V(X) = \alpha\beta^2$$

FIGURE 4.11 The failure rate function for the gamma distribution ($\alpha > 1$).

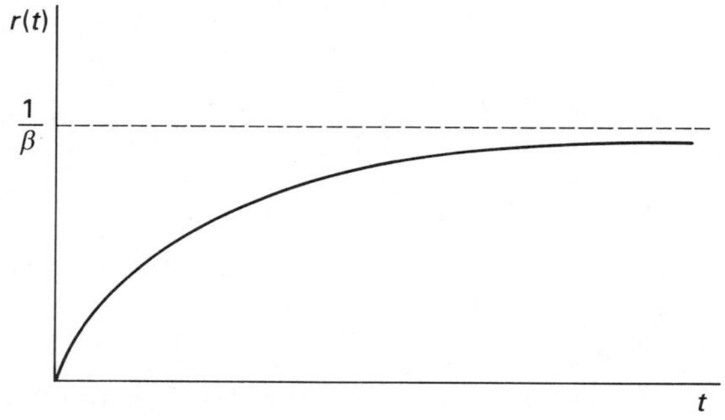

EXERCISES

4.45 Four-week summer rainfall totals in a certain section of the midwestern United States have a relative frequency histogram that appears to fit closely to a gamma distribution with $\alpha = 1.6$ and $\beta = 2.0$.

(a) Find the mean and variance of this distribution of 4-week rainfall totals.

(b) Find an interval that will include the rainfall total for a selected 4-week period with probability at least .75.

4.46 Annual incomes for engineers in a certain industry have approximately a gamma distribution with $\alpha = 600$ and $\beta = 50$.

(a) Find the mean and variance of these incomes.

(b) Would you expect to find many engineers in this industry with an annual income exceeding $35,000?

4.47 The weekly downtime Y (in hours) for a certain industrial machine has approximately a gamma distribution with $\alpha = 3$ and $\beta = 2$. The loss, in dollars, to the industrial operation as a result of this downtime is given by

$$L = 30Y + 2Y^2.$$

(a) Find the expected value and variance of L.

(b) Find an interval that will contain L on approximately 89 % of the weeks that the machine is in use.

4.48 Customers arrive at a checkout counter according to a Poisson process with a rate of two per minute. Find the mean, variance, and probability density function of the waiting time between the opening of the counter and

(a) the arrival of the second customer.

(b) the arrival of the third customer.

4.49 Suppose two houses are to be built, and each involves the completion of a certain key task. The task has an exponentially distributed time to completion with a mean of 10 hours. Assuming the completion times are independent for the two houses, find the expected value and variance of

(a) the total time to complete both tasks.

(b) the average time to complete the two tasks.

4.50 The total sustained load on the concrete footing of a planned building is the sum of the dead load plus the occupancy load. Suppose the dead load, X_1, has a gamma distribution with $\alpha_1 = 50$ and $\beta_1 = 2$ whereas the occupancy load, X_2, has a gamma distribution with $\alpha_2 = 20$ and $\beta_2 = 2$. (Units are in kips, or thousands of pounds.)

(a) Find the mean, variance, and probability density function of the total sustained load on the footing.

(b) Find a value for the sustained load that should be exceeded only with probability less than 1/16.

4.51 A 40-year history of annual maximum river flows for a certain small river in the United States shows a relative frequency histogram that can be modeled by a gamma density function with $\alpha = 1.6$ and $\beta = 150$ (measurements in cubic feet per second).

(a) Find the mean and standard deviation of the annual maximum river flows.

(b) Within what interval will the maximum annual flow be contained with probability at least 8/9?

4.52 The time intervals between dial-up connections to a computer center from remote terminals are exponentially distributed with a mean of 15 seconds. Find the mean, variance, and probability distribution of the waiting time from the opening of the computer center until the fourth dial-up connection from a remote terminal.

4.53 If service times at a teller window of a bank are exponentially distributed with a mean of 3.2 minutes, find the probability distribution, mean, and variance of the time taken to serve three waiting customers.

4.54 The response times at an online terminal have, approximately, a gamma distribution with a mean of 4 seconds and a variance of 8. Write the probability density function for these response times.

4.6

THE NORMAL DISTRIBUTION

The most widely used continuous probability distribution is referred to as the *normal distribution.* The normal probability density function has the familiar symmetric "bell" shape as in Figure 4.12. The curve is centered at the mean value μ and its spread, of course, is measured by the variance σ^2. These two parameters, μ and σ^2, completely determine the shape and location of the normal denstiy function, whose functional form is given by

$$f(x) = \frac{1}{\sqrt{2\pi}\sigma} e^{-(x-\mu)^2/2\sigma^2} \qquad -\infty < x < \infty.$$

FIGURE 4.12 A normal density function.

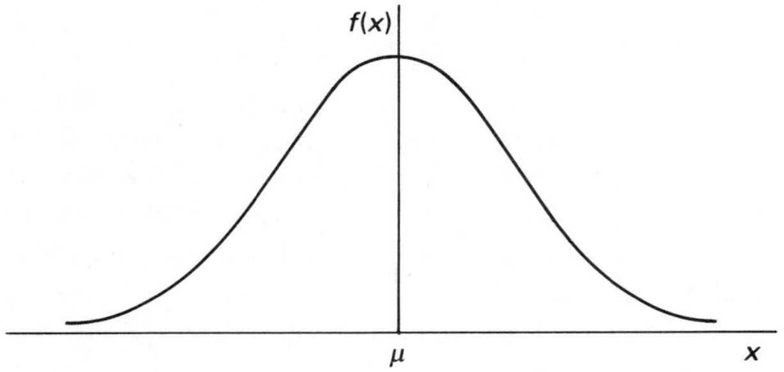

The basic reason that the normal distribution works well as a model for many different types of measurements generated in real experiments will be discussed in some detail in Chapter 7. For now we say simply that, any time responses tend to be averages of independent quantities, the normal distribution quite likely will provide a reasonably good model for their relative frequency behavior. Many naturally occurring measurements tend to have relative frequency distributions closely resembling the normal curve, probably because nature tends to "average out" the effects of the many variables that relate to a particular response. For example, heights of American men tend to have a distribution that shows many measurements clumped closely about a mean height, with relatively few very short or very tall men in the population. In other words, the relative frequency distribution is close to normal.

In contrast, life lengths of biological organisms or electronic components tend to have relative frequency distributions that are not normal or close to normal. This often is because life-length measurements are a product of "extreme" behavior, not "average" behavior. A component may fail because of one extremely hard shock rather than the average effect of many shocks. Thus, the normal distribution often is not used to model life length; consequently, we will not discuss the failure rate function for this distribution.

A naturally occurring example of the normal distribution is seen in Michelson's measures of the speed of light. A histogram of these measurements is given in Figure 4.13. The distribution is not perfectly symmetrical but still exhibits an approximately normal shape. A very important property of the normal distribution, which will be proved in Section 4.9, is that any linear function of a normally distributed random variable also is normally distributed. That is, if X has a normal distribution with mean μ and variance σ^2, and $Y = aX + b$, for constants a and b, then Y also is normally distributed. It is easily seen that

$$E(Y) = a\mu + b, \qquad V(Y) = a^2\sigma^2.$$

Suppose Z has a normal distribution with $\mu = 0$ and $\sigma = 1$. This random

FIGURE 4.13 Michelson's 100 measures of the speed of light in air (+299,000).

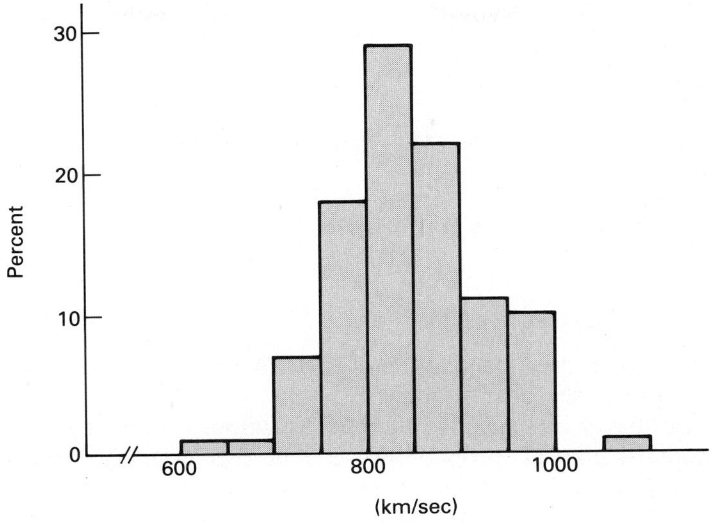

Source: Michelson, *Astronomical Papers* (1881): 231.

variable, Z, is said to have a *standard normal distribution.* Direct integration will show that $E(Z) = 0$ and $V(Z) = 1$. We have

$$E(Z) = \int_{-\infty}^{\infty} z \frac{1}{\sqrt{2\pi}} e^{-z^2/2}\, dz$$
$$= \frac{1}{\sqrt{2\pi}} \int_{-\infty}^{\infty} e^{-z^2/2} z\, dz$$
$$= \frac{1}{\sqrt{2\pi}} [-e^{-z^2/2}]_{-\infty}^{\infty} = 0.$$

Similarly,

$$E(Z^2) = \int_{-\infty}^{\infty} \frac{1}{\sqrt{2\pi}} z^2 e^{-z^2/2}\, dz$$
$$= \frac{1}{\sqrt{2\pi}} (2) \int_0^{\infty} z^2 e^{-z^2/2}\, dz.$$

On making the transformation $u = z^2$, the integral becomes

$$\frac{1}{\sqrt{2\pi}} \int_0^{\infty} u^{1/2} e^{-u/2}\, du = \frac{1}{\sqrt{2\pi}} \Gamma\left(\frac{3}{2}\right)(2)^{3/2}$$
$$= \frac{1}{\sqrt{\pi}} (2)\left(\frac{1}{2}\right)\Gamma\left(\frac{1}{2}\right) = 1,$$

because $\Gamma(1/2) = \sqrt{\pi}$. Therefore, $E(Z) = 0$ and $V(Z) = E(Z^2) - \mu^2 = E(Z^2) = 1$.

For any normally distributed random variable X, with parameters μ and σ^2,

$$Z = \frac{X - \mu}{\sigma}$$

will have a standard normal distribution. Then,

$$X = Z\sigma + \mu,$$
$$E(X) = \sigma E(Z) + \mu = \mu,$$

and

$$V(X) = \sigma^2 V(Z) = \sigma^2.$$

This shows that the parameters μ and σ^2 do indeed measure the mean and variance of the distribution.

Because any normally distributed random variable can be transformed to the standard normal, probabilities can be evaluated for any normal distribution simply by using a table of standard normal integrals, as in Table 4 of the Appendix. Table 4 gives numerical values for

$$P(0 \le Z \le z) = \int_0^z \frac{1}{\sqrt{2\pi}} e^{-x^2/2}\, dx.$$

Values of the integral are given for z between .00 and 3.09. We will now show how to use Table 4 to find $P(-.5 \le Z \le 1.5)$ for a standard normal variable Z. Figure 4.14 will aid in visualizing the necessary areas.

We first must write the probability in terms of intervals to the left and right of zero, the mean of the distribution. This produces

$$P(-.5 \le Z \le 1.5) = P(0 \le Z \le 1.5) + P(-.5 \le Z \le 0).$$

Now $P(0 \le Z \le 1.5) = A_1$ in Figure 4.14, found by looking up $z = 1.5$ in Table 4. The result is $A_1 = .4332$. Similarly, $P(-.5 \le Z \le 0) = A_2$ in Figure 4.14, found by looking up $z = .5$ in Table 4. Areas under the standard

FIGURE 4.14 A standard normal density function.

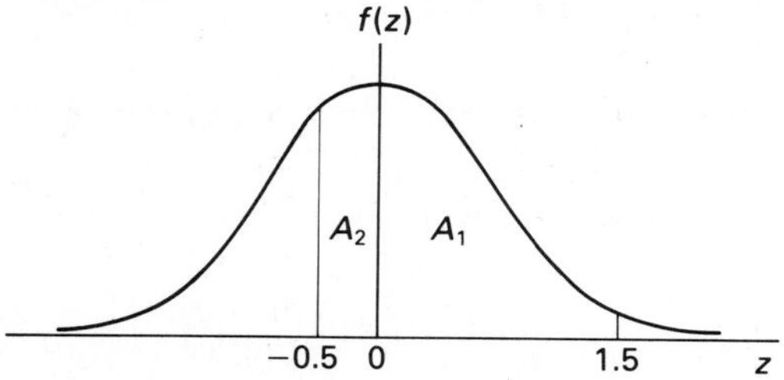

normal curve for negative z values are equal to those for corresponding positive z values because the curve is symmetric around zero. We find $A_2 = .1915$. It follows that

$$P(-.5 \leq Z \leq 1.5) = A_1 + A_2 = .4332 + .1915 = .6247.$$

Example 4.12 If Z denotes a standard normal variable, find

(a) $P(Z \leq 1)$. (b) $P(Z > 1)$.
(c) $P(Z < -1.5)$ (d) $P(-1.5 \leq Z \leq .5)$.

Also find a value of z, say z_0, such that $P(0 \leq Z \leq z_0) = .49$.

Solution

This example provides practice in reading Table 4. We see that

$$\text{(a) } P(Z \leq 1) = P(Z \leq 0) + P(0 \leq Z \leq 1) = .5 + .3413 = .8413.$$

$$\text{(b) } P(Z > 1) = .5 - P(0 \leq Z \leq 1) = .5 - .3413 = .1587.$$

$$\text{(c) } P(Z < -1.5) = P(Z > 1.5) = .5 - P(0 \leq Z \leq 1.5) = .5 - .4332 = .0668.$$

$$\text{(d) } P(1.5 \leq Z \leq .5) = P(-1.5 \leq Z \leq 0) + P(0 \leq Z \leq .5) = P(0 \leq Z \leq 1.5) + P(0 \leq Z \leq .5) = .4332 + .1915 = .6247.$$

To find the value of z_0, we must look for the given probability of .49 on the area side of Table 4. The closest we can come is at .4901, which corresponds to a z value of 2.33. Hence, $z_0 = 2.33$.

The next example illustrates how the standardization allows Table 4 to be used for any normally distributed random variable.

Example 4.13 A firm that manufactures and bottles apple juice has a machine that automatically fills 16-ounce bottles. There is some variation, however, in the amount of liquid dispensed into each bottle by the machine. Over a long period of time the average amount dispensed into the bottles was 16 ounces, but with a standard deviation of 1 ounce in the measurements. If the amounts per bottle can be assumed to be normally distributed, find the probability that the machine will dispense more than 17 ounces of liquid in any one bottle.

Solution

Let X denote the ounces of liquid dispensed into one bottle by the filling machine. Then X is assumed to be normally distributed with a mean of 16 and standard deviation 1. Hence,

$$P(X > 17) = P\left(\frac{X - \mu}{\sigma} > \frac{17 - \mu}{\sigma}\right) = P\left(Z > \frac{17 - 16}{1}\right)$$
$$= P(Z > 1) = .1587.$$

The answer is found from Table 4 because $Z = (X - \mu)/\sigma$ has a *standard* normal distribtion.

Example 4.14 Suppose that another machine, similar to the one of Example 4.13, operates so that the amounts have a mean equal to the dial setting for "amount of liquid," but have a standard deviation of 1.2 ounces. Find the proper setting for the dial so that 17-ounce bottles will overflow only 5 % of the time. Assume that the amounts dispensed have a normal distribution.

Solution

Letting X denote the amount of liquid dispensed, we now look for a μ-value such that

$$P(X > 17) = .05.$$

Now,

$$P(X > 17) = P\left(\frac{X - \mu}{\sigma} > \frac{17 - \mu}{\sigma}\right)$$
$$= P\left(Z > \frac{17 - \mu}{1.2}\right).$$

From Table 4 we know that if

$$P(Z > z_0) = .05,$$

then $z_0 = 1.645$. Thus, it must be that

$$\frac{17 - \mu}{1.2} = 1.645$$

and

$$\mu = 17 - 1.2(1.645) = 15.026.$$

As mentioned earlier, we will make much more use of the normal distribution in later chapters. The properties of the normal distribution are summarized as follows.

The Normal Distribution

$$f(x) = \frac{1}{\sqrt{2\pi}\sigma} e^{-(x-\mu)^2/2\sigma^2} \qquad -\infty < x < \infty$$

$$E(X) = \mu \qquad V(X) = \sigma^2$$

EXERCISES

4.55 Use Table 4 of the Appendix to find the following probabilities for a standard normal random variable Z.

(a) $P(0 \leq Z \leq 1.2)$
(b) $P(-.9 \leq Z \leq 0)$
(c) $P(.3 \leq Z \leq 1.56)$
(d) $P(-.2 \leq Z \leq .2)$
(e) $P(-2.00 \leq Z \leq 1.56)$

4.56 For a standard normal random variable Z, use Table 4 of the Appendix to find a number z_0 such that

(a) $P(Z \leq z_0) = .5$.
(b) $P(Z \leq z_0) = .8749$.
(c) $P(Z \geq z_0) = .117$.
(d) $P(Z \geq z_0) = .617$.
(e) $P(-z_0 \leq Z \leq z_0) = .90$.
(f) $P(-z_0 \leq Z \leq z_0) = .95$.

4.57 The weekly amount spent for maintenance and repairs in a certain company has an approximately normal distribution with a mean of \$400 and a standard deviation of \$20. If \$450 is budgeted to cover repairs for next week, what is the probability that the actual costs will exceed the budgeted amount?

4.58 In the setting of Exercise 4.57, how much should be budgeted weekly for maintenance and repairs so that the budgeted amount will be exceeded with probability only .1?

4.59 A machining operation produces steel shafts having diameters that are normally distributed with a mean of 1.005 inches and a standard deviation of .01 inch. Specifications call for diameters to fall within the interval 1.00 ± .02 inches. What percentage of the output of this operation will fail to meet specifications?

4.60 Referring to Exercise 4.59, what should be the mean diameter of the shafts produced to minimize the fraction not meeting specifications?

4.61 Wires manufactured for a certain computer system are specified to have resistances between .12 and .14 ohms. The actual measured resistances of the wires produced by

Company A have a normal probability distribution with a mean of .13 ohms and a standard deviation of .005 ohms.

(a) What is the probability that a randomly selected wire from Company A's production will meet the specifications?

(b) If four such wires are used in the system and all are selected from Company A, what is the probability that all four will meet the specifications?

4.62 At a temperature of 25°C the resistances of a type of thermistor are normally distributed with a mean of 10,000 ohms and a standard deviation of 4000 ohms. The thermistors are to be sorted, with those having resistances between 8000 and 15,000 ohms shipped to a vendor. What fraction of these thermistors will be shipped?

4.63 A vehicle driver gauges the relative speed of a preceding vehicle by the speed with which the image of the width of that vehicle varies. This speed is proportional to the speed, X, of variation of the angle at which the eye subtends this width. According to P. Ferrani and others (1984, pp. 50, 51), a study of many drivers revealed X to be normally distributed with a mean of zero and a standard deviation of 10 (10^{-4} radian per second). What fraction of these measurements is more than 5 units away from zero? What fraction is more than 10 units away from zero?

4.64 A type of capacitor has resistances that vary according to a normal distribution with a mean of 800 megohms and a standard deviation of 200 megohms (Nelson, 1967, pp. 261–268, for a more thorough discussion). A certain application specifies capacitors with resistances between 900 and 1000 megohms.

(a) What proportion of these capacitors will meet this specification?

(b) If two capacitors are randomly chosen from a lot of capacitors of this type, what is the probability that both will satisfy the specification?

4.65 Sick leave time used by employees of a firm in one month has, approximately, a normal distribution with a mean of 200 hours and a variance of 400.

(a) Find the probability that total sick leave for next month will be less than 150 hours.

(b) In planning schedules for next month, how much time should be budgeted for sick leave if that amount is to be exceeded with a probability of only .10?

4.66 The time until first failure of a brand of ink jet printers are approximately normally distributed with a mean of 1500 hours and a standard deviation of 200 hours.

(a) What fraction of these printers will fail before 1000 hours?

(b) What should be the guarantee time for these printers if the manufacturer wants only 5 % to fail within the guarantee period?

4.67 A machine for filling cereal boxes has a standard deviation of 1 ounce on ounces of fill per box. What setting of the mean ounces of fill per box will allow 16-ounce boxes to overflow only 1 % of the time? Assume that the ounces of fill per box are normally distributed.

4.68 Referring to Exercise 4.67, suppose the standard deviation σ is not known but can be fixed at certain levels by carefully adjusting the machine. What is the largest value of σ

that will allow the actual value dispensed to be within 1 ounce of the mean with probability at least .95?

4.7

THE BETA DISTRIBUTION

Except for the uniform distribution of Section 4.3, the continuous distributions discussed thus far have density functions that are positive over an infinite interval. It is of value to have another class of distributions that can be used to model phenomena constrained to a finite interval of possible values. One such class, the beta distributions, is very useful for modeling the probabilistic behavior of certain random variables, such as proportions, constrained to fall in the interval $(0, 1)$. [Actually, any finite interval can be transformed to $(0, 1)$.] The beta distribution has the functional form

$$f(x) = \frac{\Gamma(\alpha + \beta)}{\Gamma(\alpha)\Gamma(\beta)} x^{\alpha-1}(1 - x)^{\beta-1} \qquad 0 < x < 1$$
$$= 0 \qquad \text{elsewhere}$$

where α and β are positive constants. The constant term in $f(x)$ is necessary so that

$$\int_0^1 f(x)\,dx = 1.$$

In other words,

$$\int_0^1 x^{\alpha-1}(1 - x)^{\beta-1}\,dx = \frac{\Gamma(\alpha)\Gamma(\beta)}{\Gamma(\alpha + \beta)}$$

for positive α and β. This is a handy result to keep in mind. The graphs of some common beta density functions are shown in Figure 4.15.

One measurement of interest in the process of sintering copper is the proportion of the volume that is solid, rather than made up of voids. (The proportion due to voids is sometimes called the *porosity of the solid.*) Figure 4.16 shows a relative frequency histogram of proportions of solid copper in samples from a sintering process. This distribution could be modeled with a beta distribution having a large α and small β. The expected value of a beta random variable is easily found because

$$E(X) = \int_0^1 x \frac{\Gamma(\alpha + \beta)}{\Gamma(\alpha)\Gamma(\beta)} x^{\alpha-1}(1 - x)^{\beta-1}\,dx$$
$$= \frac{\Gamma(\alpha + \beta)}{\Gamma(\alpha)\Gamma(\beta)} \int_0^1 x^{\alpha}(1 - x)^{\beta-1}\,dx$$

FIGURE 4.15 The beta density function.

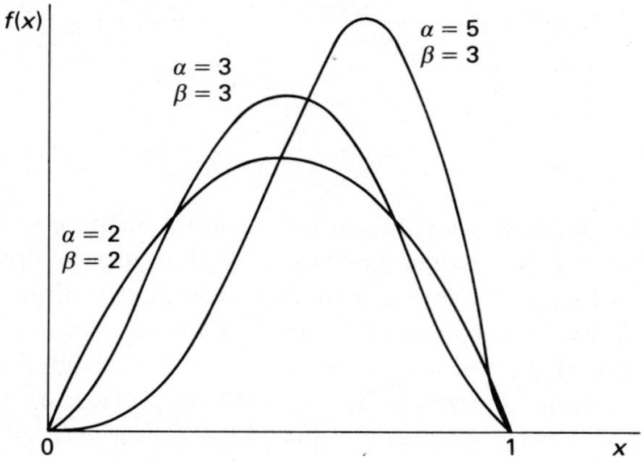

$$= \frac{\Gamma(\alpha + \beta)}{\Gamma(\alpha)\Gamma(\beta)} \times \frac{\Gamma(\alpha + 1)\Gamma(\beta)}{\Gamma(\alpha + \beta + 1)}$$

$$= \frac{\alpha}{\alpha + \beta}.$$

[Recall that $\Gamma(n + 1) = n\Gamma(n)$.] Similar manipulations reveal that

$$V(X) = \frac{\alpha\beta}{(\alpha + \beta)^2(\alpha + \beta + 1)}.$$

We illustrate the use of this density function in an example.

FIGURE 4.16 Solid mass in sintered linde copper.

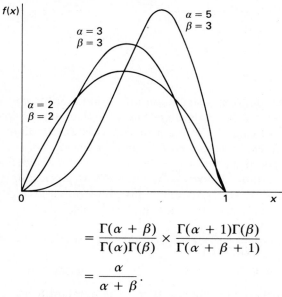

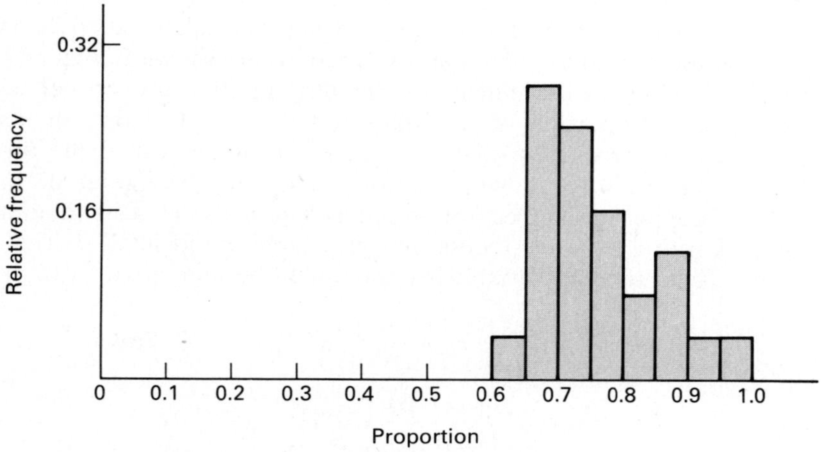

Source: Department of Materials Science, University of Florida.

Example 4.15 A gasoline wholesale distributor has bulk storage tanks holding a fixed supply. The tanks are filled every Monday. Of interest to the wholesaler is the proportion of this supply sold during the week. Over many weeks this proportion has been observed to be modeled fairly well by a beta distribution with $\alpha = 4$ and $\beta = 2$. Find the expected value of this proportion. Is it highly likely that the wholesaler will sell at least 90 % of the stock in a given week?

Solution

By the results given earlier, with X denoting the proportion of the total supply sold in a given week,

$$E(X) = \frac{\alpha}{\alpha + \beta} = \frac{4}{6} = \frac{2}{3}.$$

For the second part we are interested in

$$\begin{aligned} P(X > .9) &= \int_{.9}^{1} \frac{\Gamma(4 + 2)}{\Gamma(4)\Gamma(2)} x^3(1 - x)\, dx \\ &= 20\int_{.9}^{1} (x^3 - x^4)\, dx \\ &= 20(.004) = .08. \end{aligned}$$

It is not very likely that 90% of the stock will be sold in a given week.

The basic properties of the beta distribution are summarized as follows.

The Beta Distribution

$$f(x) = \frac{\Gamma(\alpha + \beta)}{\Gamma(\alpha)\Gamma(\beta)} x^{\alpha-1}(1 - x)^{\beta-1} \qquad 0 < x < 1$$

$$= 0 \qquad \text{elsewhere}$$

$$E(X) = \frac{\alpha}{\alpha + \beta} \qquad V(X) = \frac{\alpha\beta}{(\alpha + \beta)^2(\alpha + \beta + 1)}$$

EXERCISES

4.69 Suppose X has a probability density function given by

$$f(x) = \begin{cases} kx^3(1 - x)^2 & 0 \le x \le 1 \\ 0 & \text{elsewhere.} \end{cases}$$

(a) Find the value of k that makes this a probability density function.

(b) Find $E(X)$ and $V(X)$.

4.70 If X has a beta distribution with parameters α and β, show that

$$V(X) = \frac{\alpha\beta}{(\alpha + \beta)^2(\alpha + \beta + 1)}.$$

4.71 During an 8-hour shift, the proportion of time X that a sheet-metal stamping machine is down for manintenance or repairs has a beta distribution with $\alpha = 1$ and $\beta = 2$. That is,

$$f(x) = \begin{cases} 2(1 - x) & 0 \le x \le 1 \\ 0 & \text{elsewhere.} \end{cases}$$

The cost (in hundreds of dollars) of this downtime, due to lost production and cost of maintenance and repair, is given by

$$C = 10 + 20X + 4X^2.$$

(a) Find the mean and variance of C.

(b) Find an interval in which C will lie with probability at least .75.

4.72 The percentage of impurities per batch in a certain type of industrial chemical is a random variable X having the probability density function

$$f(x) = \begin{cases} 12x^2(1 - x) & 0 \le x \le 1 \\ 0 & \text{elsewhere.} \end{cases}$$

(a) Suppose a batch with more than 40% impurities cannot be sold. What is the probability that a randomly selected batch will not be allowed to be sold?

(b) Suppose the dollar value of each batch is given by $V = 5 - .5X$. Find the expected value and variance of V.

4.73 To study the disposal of pollutants emerging from a power plant, the prevailing wind direction was measured for a large number of days. The direction is measured on a scale of 0° to 360°, but by dividing each daily direction by 360, the measurements can be rescaled to the interval $(0, 1)$. These rescaled measurements X are found to follow a beta distribution with $\alpha = 4$ and $\beta = 2$. Find $E(X)$. To what angle does this mean correspond?

4.74 Errors in measuring the arrival time of a wave front from an acoustic source can sometimes be modeled by a beta distribution (Perruzzi and Hilliard, 1984, p. 197). Suppose these errors have a beta distribution with $\alpha = 1$ and $\beta = 2$, with measurements in microseconds.

(a) Find the probability that such a measurement error will be less than .5 microseconds.

(b) Find the mean and standard deviation of these error measurements.

4.75 The proper blending of fine and coarse powders prior to copper sintering is necessary for uniformity in the finished product. One way to check the blending is to select many small samples of the blended powders and measure the weight fractions of the fine particles. These measurements should be relatively constant if good blending has been achieved.

(a) Suppose the weight fractions have a beta distribution with $\alpha = \beta = 3$. Find their mean and variance.
(b) Repeat (a) for $\alpha = \beta = 2$.
(c) Repeat (a) for $\alpha = \beta = 1$.
(d) Which case, (a), (b), or (c), would exemplify the best blending?

4.76 The proportion of pure iron in certain ore samples has a beta distribution with $\alpha = 3$ and $\beta = 1$.
(a) Find the probability that one of these samples will have more than 50% pure iron.
(b) Find the probability that two out of three samples will have less than 30% pure iron.

4.8

THE WEIBULL DISTRIBUTION

We have suggested that the gamma distribution often can serve as a probabilistic model for life lengths of components or systems. However, the failure rate function for the gamma distribution has an upper bound that limits its applicability to real systems. For this and other reasons, other distributions often provide better models for life-length data. One such distribution is the *Weibull,* which is explored in this section.

A Weibull density function has the form

$$f(x) = \frac{\gamma}{\theta} x^{\gamma-1} e^{-x^{\gamma}/\theta} \qquad x > 0$$
$$= 0 \qquad \text{elsewhere}$$

for positive parameters θ and γ. For $\gamma = 1$, this becomes an exponential density. For $\gamma > 1$, the functions look something like the gamma functions of Section 4.5 but have somewhat different mathematical properties. We can integrate directly to see that

$$F(x) = \int_0^x \frac{\gamma}{\theta} t^{\gamma-1} e^{-t^{\gamma}/\theta}\, dt$$
$$= -e^{-t^{\gamma}/\theta}\Big|_0^x = 1 - e^{-x^{\gamma}/\theta} \qquad x > 0.$$

A convenient way to look at properties of the Weibull density is to use the transformation $Y = X^{\gamma}$. Then,

$$F_Y(y) = P(Y \le y) = P(X^{\gamma} \le y) = P(X \le y^{1/\gamma})$$
$$= F_X(y^{1/\gamma}) = 1 - e^{-(y^{1/\gamma})^{\gamma}/\theta}$$
$$= 1 - e^{-y/\theta} \qquad y > 0.$$

Hence,

$$f_Y(y) = \frac{dF_Y(y)}{dy} = \frac{1}{\theta}e^{-y/\theta} \qquad y > 0$$

and Y has the familiar exponential density.

If we want to find $E(X)$ for an X having the Weibull distribution, then

$$\begin{aligned} E(X) = E(Y^{1/\gamma}) &= \int_0^\infty y^{1/\gamma}\frac{1}{\theta}e^{-y/\theta}\,dy \\ &= \frac{1}{\theta}\int_0^\infty y^{1/\gamma}e^{-y/\theta}\,dy \\ &= \frac{1}{\theta}\Gamma\left(1 + \frac{1}{\gamma}\right)\theta^{(1+1/\gamma)} = \theta^{1/\gamma}\Gamma\left(1 + \frac{1}{\gamma}\right). \end{aligned}$$

This result follows from recognizing the integral to be of the gamma type.

If we let $\gamma = 2$ in the Weibull density, we see that $Y = X^2$ has an exponential distribution. To reverse the idea just outlined, if we start with an exponentially distributed random variable Y, then the square root of Y will have a Weibull distribution with $\gamma = 2$. We can illustrate this empirically by taking the square roots of the data from an exponential distribution, given in Table 4.1. These square roots are given in Table 4.2. A relative frequency histogram for these data is given in Figure 4.17. Note that the exponential form has now disappeared and that the curve given by the Weibull density with $\gamma = 2$, $\theta = 2$ (seen in Figure 4.18) is a much more plausible model for these observations.

The Weibull distribution is used commonly as a model for life lengths because of the properties of its failure rate function. In this case,

$$\gamma(t) = \frac{f(t)}{1 - F(t)} = \frac{\frac{\gamma}{\theta}t^{\gamma-1}e^{-t^\gamma/\theta}}{e^{-t^\gamma/\theta}} = \frac{\gamma}{\theta}t^{\gamma-1}$$

TABLE 4.2 **Square roots of the life lengths of Table 4.1**

.637	.828	2.186	1.313	2.868
1.531	1.184	1.223	.542	1.459
.733	.484	2.006	1.823	1.709
2.256	1.207	1.032	.880	.872
2.364	1.305	1.623	1.360	.431
1.601	.719	1.802	1.526	1.032
.152	.715	1.668	2.535	.914
1.826	.474	1.230	1.793	.617
1.868	1.525	.577	2.746	.984
1.126	1.709	1.274	.578	2.119

FIGURE 4.17 Relative frequency histogram for the data of Table 4.2.

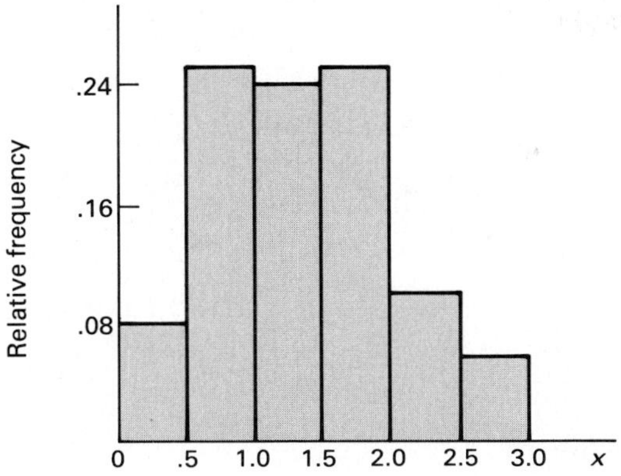

FIGURE 4.18 The Weibull density function $\gamma = 2$, $\theta = 2$.

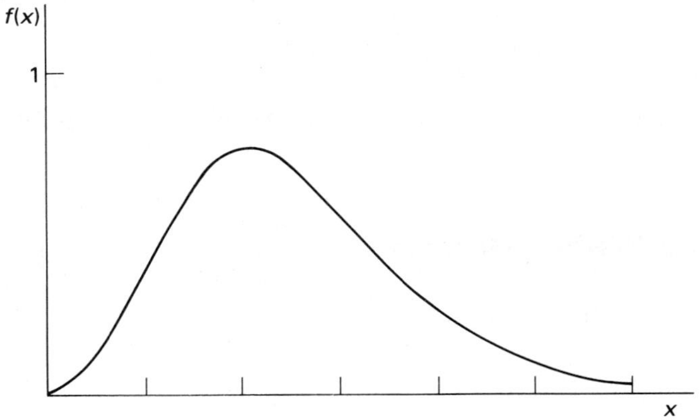

which, for $\gamma > 1$, is a monotonically increasing function with no upper bound. By appropriate choice of γ this failure rate function can be made to model many sets of life-length data seen in practice. We illustrate the use of the Weibull distribution in the following example.

Example 4.16 The length of service time during which a certain type of thermistor produces resistances within its specifications has been observed to follow a Weibull distribution with $\theta = 50$ and $\gamma = 2$ (measurements in thousands of hours)

(a) Find the probability that one of these thermistors, to be installed in a system today, will function properly for over 10,000 hours.

(b) Find the expected life length for thermistors of this type.

Solution

(a) The Weibull distribution has a closed-form expression for $F(x)$. Thus, if X represents the life length of the thermistor in question,

$$\begin{aligned}P(X > 10) &= 1 - F(10) \\ &= 1 - [1 - e^{-(10)^2/50)}] \\ &= e^{-(10)^2/50} = e^{-2} = .14\end{aligned}$$

because $\theta = 50$ and $\gamma = 2$.

(b) We know that

$$\begin{aligned}E(X) &= \theta^{1/\gamma}\Gamma\left(1 + \frac{1}{\gamma}\right) \\ &= (50)^{1/2}\Gamma\left(\frac{3}{2}\right) \\ &= (50)^{1/2}\frac{1}{2}\Gamma\left(\frac{1}{2}\right) \\ &= (50)^{1/2}\left(\frac{1}{2}\right)\sqrt{\pi} = 6.27.\end{aligned}$$

Thus, the average service time for these thermistors is 6270 hours.

The Weibull Distribution

$$\begin{aligned}f(x) &= \frac{\gamma}{\theta}x^{\gamma-1}e^{-x^{\gamma}/\theta} \qquad && x > 0 \\ &= 0 && \text{elsewhere}\end{aligned}$$

$$E(X) = \theta^{1/\gamma}\Gamma\left(1 + \frac{1}{\gamma}\right)$$

$$V(X) = \theta^{2/\gamma}\left\{\Gamma\left(1 + \frac{2}{\gamma}\right) - \left[\Gamma\left(1 + \frac{1}{\gamma}\right)\right]^2\right\}$$

EXERCISES

4.77 Fatigue life, in hundreds of hours, for a certain type of bearings has approximately a Weibull distribution with $\gamma = 2$ and $\theta = 4$.

(a) Find the probability that a bearing of this type fails in less than 200 hours.

(b) Find the expected value of the fatigue life for these bearings.

4.78 The maximum flood levels, in millions of cubic feet per second, for a certain United States river have a Weibull distribution with $\gamma = 1.5$ and $\theta = .6$ (Cohen, Whitten, and Ding, 1984, p. 165, for more details). Find the probability that the maximum flood level for next year

(a) will exceed .5.

(b) will be less than .8.

4.79 The time necessary to achieve proper blending of copper powders before sintering was found to have a Weibull distribution with $\gamma = 1.1$ and $\theta = 2$ (measurements in minutes). Find the probability that a proper blending will take less than 2 minutes.

4.80 The ultimate tensile strength of steel wire used to wrap concrete pipe was found to have a Weibull distribution with $\gamma = 1.2$ and $\theta = 270$ (measurements in thousands of pounds). Pressure in the pipe at a certain point may require an ultimate tensile strength of at least 300,000 pounds. What is the probability that a wire will possess this strength?

4.81 The yield strengths of certain steel beams have a Weibull distribution with $\gamma = 2$ and $\theta = 3600$ (measurement in pounds per square inch). Two such beams are used in a construction project, which calls for yield strengths in excess of 70,000 psi. Find the probability that both beams meet the specifications for the project.

4.82 The pressure, in thousands of psi, exerted on the tank of a steam boiler has a Weibull distribution with $\gamma = 1.8$ and $\theta = 1.5$. The tank is built to withstand pressures of 2000 psi. Find the probability that this limit will be exceeded.

4.83 Resistors being used in the construction of an aircraft guidance system have life lengths that follow a Weibull distribution with $\gamma = 2$ and $\theta = 10$ (measurements in thousands of hours).

(a) Find the probability that a randomly selected resistor of this type has a life length that exceeds 5000 hours.

(b) If three resistors of this type are operating independently, find the probability that exactly one of the three burns out prior to 5000 hours of use.

4.84 Referring to Exercise 4.83, find the mean and variance of the life length of a resistor similar to the type discussed.

4.85 Maximum wind-gust velocities in summer thunderstorms were found to follow a Weibull distribution with $\gamma = 2$ and $\theta = 400$ (measurements in feet per second). Engineers designing structures in the areas in which these thunderstorms are found are interested in finding a gust velocity that will be exceeded only with a probability of .01. Find such a value.

4.86 The velocities of gas particles can be modeled by the Maxwell distribution, with the probability density function given by

$$f(v) = 4\pi\left(\frac{m}{2\pi KT}\right)^{3/2} v^2 e^{-v^2(m/2KT)} \qquad v > 0$$

where m is the mass of the particle, K is Boltzmann's constant, and T is the absolute temperature.

(a) Find the mean velocity of these particles.

(b) The kinetic energy of a particle is given by $(1/2)mV^2$. Find the mean kinetic energy for a particle.

4.9

MOMENT-GENERATING FUNCTIONS FOR CONTINUOUS RANDOM VARIABLES

As in the case of discrete distributions, the moment-generating functions of continuous random variables help us find expected values and identify certain properties of probability distributions. We now present a short discussion of moment-generating functions for continuous random variables.

The moment-generating function of a continuous random variable X with a probability density function of $f(x)$ is given by

$$M(t) = E(e^{tX}) = \int_{-\infty}^{\infty} e^{tx} f(x)\,dx$$

when the integral exists.

For the exponential distribution this becomes

$$\begin{aligned} M(t) &= \int_0^{\infty} e^{tx} \frac{1}{\theta} e^{-x/\theta}\,dx \\ &= \frac{1}{\theta} \int_0^{\infty} e^{-x(1/\theta - t)}\,dx \\ &= \frac{1}{\theta} \int_0^{\infty} e^{-x(1-\theta t)/\theta}\,dx \\ &= \frac{1}{\theta} \Gamma(1)\left(\frac{\theta}{1-\theta t}\right) = (1 - \theta t)^{-1}. \end{aligned}$$

We can now use $M(t)$ to find $E(X)$, because

$$M'(0) = E(X)$$

and for the exponential distribution we have

$$\begin{aligned} E(X) = M'(0) &= [-(1 - \theta t)^{-2}(-\theta)]_{t=0} \\ &= \theta. \end{aligned}$$

Similarly, we could find $E(X^2)$ and then $V(X)$ by using the moment-generating function.

An argument analogous to the one used for the exponential distribution will show that, for the gamma distribution,

$$M(t) = (1 - \beta t)^{-\alpha}.$$

From this we can see that, if X has a gamma distribution,

$$\begin{aligned} E(X) = M'(0) &= [-\alpha(1-\beta t)^{-\alpha-1}(-\beta)]_{t=0} \\ &= \alpha\beta, \\ E(X^2) = M^{(2)}(0) &= [\alpha\beta(-\alpha-1)(1-\beta t)^{-\alpha-2}(-\beta)]_{t=0} \\ &= \alpha(\alpha+1)\beta^2, \\ E(X^3) = M^{(3)}(0) &= [\alpha(\alpha+1)\beta^2(-\alpha-2)(1-\beta t)^{-\alpha-3}(-\beta)]_{t=0} \\ &= \alpha(\alpha+1)(\alpha+2)\beta^3, \end{aligned}$$

and so on.

Example 4.17 The force, h, exerted by a mass, m, moving at velocity, v, is

$$h = \frac{mv^2}{2}.$$

Consider a device that fires a serrated nail into concrete at a mean velocity of 500 feet per second, where V, the random velocity, possesses a density function

$$f(v) = \frac{v^3 e^{-v/b}}{b^4 \Gamma(4)} \qquad b = 500,\ v \geq 0.$$

If each nail possesses mass m, find the expected force exerted by a nail.

Solution

$$E(H) = E\left(\frac{mV^2}{2}\right) = \frac{m}{2}E(V^2).$$

The density function for V is a gamma-type function with $\alpha = 4$, $\beta = 500$. Therefore,

$$\begin{aligned} E(V^2) &= \alpha(\alpha+1)\beta^2 \\ &= (4)(5)(500)^2 \\ &= 5{,}000{,}000 \end{aligned}$$

and

$$\begin{aligned} E(H) &= \frac{m}{2}E(V^2) \\ &= \frac{m}{2}(5{,}000{,}000) = 2{,}500{,}000m. \end{aligned}$$

Two important properties of moment-generating functions are as follows:

1. If a random variable X has moment-generating function $M_X(t)$, then

$Y = aX + b$, for constants a and b, has moment-generating function

$$M_Y(t) = e^{tb}M_X(at).$$

2. Moment-generating functions are unique. That is, two random variables that have the same moment-generating function have the same probability distribution.

Property 1 is easily shown (see Exercise 3.85). The proof of property 2 is beyond the scope of this textbook, but we make use of it in identifying distributions, as will be demonstrated.

The normal random variable is used so commonly that we should investigate its moment-generating function. We do so by first finding the moment-generating function of $X - \mu$, where X is normally distributed with a mean of μ and a variance of σ^2. Now,

$$E(e^{t(X-\mu)}) = \int_{-\infty}^{\infty} e^{t(x-\mu)} \frac{1}{\sigma\sqrt{2\pi}} e^{-(x-\mu)^2/2\sigma^2}\,dx.$$

Letting $y = x - \mu$, the integral becomes

$$\frac{1}{\sigma\sqrt{2\pi}}\int_{-\infty}^{\infty} e^{ty-y^2/2\sigma^2} = \frac{1}{\sigma\sqrt{2\pi}}\int_{-\infty}^{\infty} \exp\left[-\frac{1}{2\sigma^2}(y^2 - 2\sigma^2 ty)\right] dy.$$

Upon the completing of the square in the exponent, we have

$$\begin{aligned} -\frac{1}{2\sigma^2}(y^2 - 2\sigma^2 ty) &= -\frac{1}{2\sigma^2}(y^2 - 2\sigma^2 ty + \sigma^4 t^2) + \frac{1}{2}t^2\sigma^2 \\ &= -\frac{1}{2\sigma^2}(y + \sigma^2 t)^2 + \frac{1}{2}t^2\sigma^2 \end{aligned}$$

so that the integral becomes

$$e^{t^2\sigma^2/2}\frac{1}{\sigma\sqrt{2\pi}}\int_{-\infty}^{\infty} \exp\left[-\frac{1}{2\sigma^2}(y - \sigma^2 t)^2\right] dy = e^{t^2\sigma^2/2},$$

because the remaining integrand forms a normal probability density that integrates to unity.

If $Y = X - \mu$ has moment-generating function given by

$$M_Y(t) = e^{t^2\sigma^2/2},$$

then

$$X = Y + \mu$$

has moment-generating function given by

$$\begin{aligned} M_X(t) &= e^{t\mu}M_Y(t) \\ &= e^{t\mu}e^{t^2\sigma^2/2} \\ &= e^{t\mu + t^2\sigma^2/2}, \end{aligned}$$

by property 1 stated earlier.

For this same normal random variable X, let

$$Z = \frac{X - \mu}{\sigma} = \frac{1}{\sigma}X - \frac{\mu}{\sigma}.$$

Then, by property 1,

$$M_Z(t) = e^{-\mu t/\sigma} e^{\mu t/\sigma + t^2\sigma^2/2\sigma^2}$$
$$= e^{t^2/2}.$$

The normal generating function of Z has the form of a moment-generating function for a normal random variable with a mean of zero and a variance of one. Thus, by property 2, Z must have that distribution.

EXERCISES

4.87 Show that a gamma distribution with parameters α and β has moment-generating function

$$M(t) = (1 - \beta t)^{-\alpha}.$$

4.88 Using the moment-generating function for the exponential distribution with mean θ, find $E(X^2)$. Use this result to show that $V(X) = \theta^2$.

4.89 Let Z denote a standard normal random variable. Find the moment-generating function of Z directly from the definition.

4.90 Let Z denote a standard normal random variable. Find the moment-generating function of Z^2. What does the uniqueness property of the moment-generating function tell you about the distribution of Z^2?

4.10

EXPECTATIONS OF DISCONTINUOUS FUNCTIONS AND MIXED PROBABILITY DISTRIBUTIONS

Problems in probability and statistics frequently involve functions that are partly continuous and partly discrete in one of two ways. First, we may be interested in the properties, perhaps the expectation, of a random variable $g(X)$ that is a discontinuous function of a discrete or continuous random variable X. Second, the random variable of interest may itself have a probability distribution made up of isolated points having discrete probabilities and intervals having continuous probability. We illustrate the first of these two situations with the following example.

Example 4.18 A certain retailer for a petroleum product sells a random amount, X, each day. Suppose that X, measured in hundreds of gallons, has the probability density function

$$f(x) = \begin{cases} \left(\frac{3}{8}\right)x^2 & 0 \le x \le 2 \\ 0 & \text{elsewhere.} \end{cases}$$

The retailer's profit turns out to be \$5 for each 100 gallons sold (5 cents per gallon) if $X \le 1$ and \$8 per 100 gallons if $X > 1$. Find the retailer's expected profit for any given day.

Solution

Let $g(X)$ denote the retailer's daily profit. Then,

$$g(X) = \begin{cases} 5X & \text{if } 0 \le X \le 1 \\ 8X & \text{if } 1 < X \le 2. \end{cases}$$

We want to find expected profit, and

$$\begin{aligned} E[g(X)] &= \int_{-\infty}^{\infty} g(x)f(x)\,dx \\ &= \int_0^1 5x\left[\left(\frac{3}{8}\right)x^2\right]dx + \int_1^2 8x\left[\left(\frac{3}{8}\right)x^2\right]dx \\ &= \frac{15}{(8)(4)}[x^4]_0^1 + \frac{24}{(8)(4)}[x^4]_1^2 \\ &= \frac{15}{32}(1) + \frac{24}{32}(15) \\ &= \frac{(15)(25)}{32} = 11.72. \end{aligned}$$

Thus, the retailer can expect to profit by \$11.72 on the daily sale of this particular product.

A random variable X that has some of its probability at discrete points and the remainder spread over intervals is said to have a *mixed distribution.* Let $F(x)$ denote a distribution function representing a mixed distribution. For all practical purposes, any mixed distribution function $F(x)$ can be written uniquely as

$$F(x) = c_1F_1(x) + c_2F_2(x),$$

where $F_1(x)$ is a step distribution function, $F_2(x)$ is a continuous distribution function, c_1 is the accumulated probability of all discrete points, and $c_2 = 1 - c_1$ is the accumulated probability of all continuous portions. The following example gives an illustration of a mixed distribution.

Example 4.19 Let X denote the life length (in hundreds of hours) of a certain type of electronic component. These components frequently fail immediately upon insertion into a system. It has been observed that the probability of immediate failure is 1/4. If a component does not fail immediately, its life-length distribution has the exponential density

$$f(x) = \begin{cases} e^{-x} & x > 0 \\ 0 & \text{elsewhere.} \end{cases}$$

Find the distribution function for X and evaluate $P(X > 10)$.

Solution

There is only one discrete point, $X = 0$, and this point has probability 1/4. Hence, $c_1 = 1/4$ and $c_2 = 3/4$. It follows that X is a mixture of two random variables, X_1 and X_2, where X_1 has a probability of one at the point zero and X_2 has the given exponential density. That is,

$$F_1(x) = \begin{cases} 0 & \text{if } x < 0 \\ 1 & \text{if } x \geq 0 \end{cases}$$

and

$$\begin{aligned} F_2(x) &= \int_0^x e^{-y}\,dy \\ &= 1 - e^{-x} \qquad x > 0. \end{aligned}$$

Now,

$$F(x) = \left(\frac{1}{4}\right)F_1(x) + \left(\frac{3}{4}\right)F_2(x).$$

Hence,

$$\begin{aligned} P(X > 10) &= 1 - P(X \leq 10) \\ &= 1 - F(10) \\ &= 1 - \left[\frac{1}{4} + \left(\frac{3}{4}\right)(1 - e^{-10})\right] \\ &= \left(\frac{3}{4}\right)[1 - (1 - e^{-10})] = \left(\frac{3}{4}\right)e^{-10}. \end{aligned}$$

An easy method for finding expectations of random variables having mixed distributions is given in Definition 4.4.

Definition 4.4

Let X have the mixed distribution function

$$F(x) = c_1F_1(x) + c_2F_2(x)$$

and suppose that X_1 is a discrete random variable having distribution function $F_1(x)$ and X_2 is a continuous random variable having distribution function $F_2(x)$. Let $g(X)$ denote a function of X. Then

$$E[g(X)] = c_1E[g(X_1)] + c_2E[g(X_2)].$$

Example 4.20 Find the mean and variance of the random variable defined in Example 4.19.

Solution

With all definitions as in Example 4.19, it follows that

$$E(X_1) = 0$$

and

$$E(X_2) = \int_0^\infty ye^{-y}\,dy = 1.$$

Therefore,

$$\mu = E(X) = \left(\frac{1}{4}\right)E(X_1) + \left(\frac{3}{4}\right)E(X_2)$$

$$= \frac{3}{4}.$$

Also,

$$E(X_1^2) = 0$$

and

$$E(X_2^2) = \int_0^\infty y^2e^{-y}\,dy = 2.$$

Therefore,

$$E(X^2) = \left(\frac{1}{4}\right)E(X_1^2) + \left(\frac{3}{4}\right)E(X_2^2)$$

$$= \left(\frac{1}{4}\right)(0) + \left(\frac{3}{4}\right)(2) = \frac{3}{2}.$$

Then,

$$V(X) = E(X^2) - \mu^2$$
$$= \frac{3}{2} - \left(\frac{3}{4}\right)^2 = \frac{15}{16}.$$

For any nonnegative random variable X, it is useful to know that the mean can be expressed as

$$E(X) = \int_0^\infty [1 - F(t)]\,dt,$$

where $F(t)$ is the distribution function for X. To see this, write

$$\int_0^\infty [1 - F(t)]\,dt = \int_0^\infty \left(\int_t^\infty f(x)\,dx\right) dt.$$

Upon changing the order of integration, the integral becomes

$$\int_0^\infty \left(\int_0^x f(x)\,dt\right) dx = \int_0^\infty xf(x)\,dx = E(X).$$

Employing this result for the mixed distribution of Examples 4.19 and 4.20,

$$1 - F(x) = \frac{3}{4}e^{-x} \qquad \text{for } x > 0,$$

and

$$E(X) = \int_0^\infty [1 - F(t)]\,dt = \int_0^\infty \frac{3}{4}e^{-t}\,dt$$
$$= \frac{3}{4}[-e^t]_0^\infty = \frac{3}{4}.$$

Thus, the mean can be found without resorting to the two-part calculation of Definition 4.4.

4.11

ACTIVITIES FOR THE COMPUTER

When generating random variables from a continuous distribution with a (cumulative) distribution function (CDF) $F(x)$, the inverse transformation method or inverse CDF method may be considered. Using this method, the procedure is as follows:

1. Generate R_i, which is uniform over the interval $(0, 1)$. This is denoted as $R_i \sim U(0, 1)$.

2. Let $R_i = F(X)$, where $F(x)$ is the CDF for the distribution of values of x.
3. Evaluate step 2 for X, which gives $X = F^{-1}(R)$.

UNIFORM DISTRIBUTION

The probability density function (pdf) for uniform distribution is defined as $f(x) = 1/(b - a)$, $a \le x \le b$. The CDF is given by $F(x) = (x - a)/(b - a)$. Using the inverse CDF method to generate uniform random numbers, x_i, let $R_i = (x - a)/(b - a)$. Solving for x, we obtain $x_i = a + (b - a)R_i$.

EXPONENTIAL DISTRIBUTION

The pdf for exponential distribution is defined as $f(x) = (1/\theta)e^{-x/\theta}$, $x \ge 0$. The CDF is given by $F(x) = 1 - e^{-x/\theta}$. Using the inverse CDF method, we can generate exponential random numbers, x_i, by letting $R_i = 1 - e^{-x/\theta}$. Solving for x, $e^{-x/\theta} = 1 - R_i$. Taking the ln on both sides, we obtain $x = -\theta \ln(1 - R_i)$. Because $(1 - R_i) \sim U(0, 1)$, we let $x = -\theta \ln R_i$.

GAMMA DISTRIBUTION

If x_i is a random variable from the gamma distribution with parameters (α, β) and α is an integer n, then x_i is the convolution (sum) of n exponential random variables, y_j, with parameter β. Thus, to simulate a gamma random variable, x_i, generate n exponential random variables as stated previously. Thus, $X_i = \sum Y_j$, where $j = 1, 2, \ldots, n$.

POISSON DISTRIBUTION

It was stated in Chapter 3 that the method of generating Poisson random variables would be given in this chapter. The procedure that follows requires that we make use of exponential random variables. If the number of events in an interval of a specified unit length follows the Poisson distribution, then the waiting time between the successive events in the interval follows an exponential distribution with $\theta = 1/\lambda$, where λ is the average value of the Poisson random variable. Thus, we can generate exponential random variables as many times as necessary until the sum of these exponential random variables is greater than the length of the interval (L) under consideration. Let n = the number of exponential random variables generated. The Poisson random variable will be given by $Y_j = n - 1$. For example, suppose we desire to generate Poisson random variables from a distribution with a mean rate of $\lambda = 2$ arrivals per day. Let $L = 2.5$ days. We would proceed to generate exponential random variables with mean $\theta = 1/2$ until the sum of the exponentials exceeded $L = 2.5$. The Poisson random variable Y_j would be given by $n - 1$. We can simplify this procedure. Let $x_i = -(1/\lambda) \ln R_i$. The random variable Y_j is defined as the $(n - 1)$ such that $\sum_{i=1}^{n-1} x_i \leq L < \sum_{i=0}^{n} x_i$. Replacing x_i, we have $-\sum_{i=0}^{n-1} \ln R_i < \lambda(L) < -\sum_{i=0}^{n} \ln R_i$. Recall that $\sum \ln R_i = \ln \prod R_i$. Multiplying by (-1), we have

$$\ln \prod_{i=0}^{n-1} R_i \geq -\lambda(L) > \ln \prod_{i=0}^{n} R_i, \qquad \text{or} \qquad \prod_{i=0}^{n-1} R_i \geq e^{-\lambda(L)} > \prod_{i=0}^{n} R_i.$$

Therefore, the method for generating Poisson random variables can be stated as

1. Let $w = e^{-\lambda(L)}$, $z = 1$, $i = 0$.
2. Generate $R_i \sim U(0, 1)$. Replace z by zR_i. If $z < w$, then output $Y = i$ and return to step 1; otherwise, proceed to step 3.
3. Replace i by $i + 1$ and return to step 2.

NORMAL DISTRIBUTION

In attempting to apply the inverse CDF method to generate normal random variables, a problem arises in that there is no explicit closed-form expression for the CDF of x, $F(x)$, or the inverse function, $F^{-1}(R)$. Methods of approximation have been developed that work well in simulating normal random variables. A few of these will be discussed briefly.

Box–Muller Polar Method for Standard Normal Variables

This method will generate a pair of standard normal random variables. To use this method,

1. Generate R_1 and R_2, which are two uniform numbers on (0, 1).
2. Evaluate $Z_1 = (\sqrt{-2 \ln R_i}) \cos(2\pi R_2)$ and $Z_2 = (\sqrt{-2 \ln R_1}) \sin(2\pi R_2)$.
3. The normal random variables X_1 and X_2 with mean μ_x and σ_x will be given by $X_1 = \mu_x + Z_1\sigma_x$, $X_2 = \mu_x + Z_2\sigma_x$.

Central Limit Theorem Method

Discussion of this method will be deferred until Chapter 7.

General Method for Simulating Continuous Random Variables When Inverse CDF Method Is Not Acceptable

If the pdf $f(x)$ of a continuous random variable, X, is bounded (in closed form) such that $a < x < b$, the *Von Neumann (acceptance) rejection method* can be utilized. This method is applied as follows:

1. Let c be a constant, where $f(x) \le c$ on (a, b).
2. Generate random numbers R_1 and R_2. Let $X_i = a + (b - a)R_1$, where X_i is uniform on (a, b). Let $Y_i = cR_2$, where Y_i is uniform on $(0, c)$.
3. Evaluate $f(X_i)$. Then compare Y_i to $f(X_i)$. If $Y_i \le f(X_i)$, then accept X_i as a random variable for the desired distribution. If $Y_i > f(X_i)$, then reject X_i as an acceptable random variable.
4. Repeat steps 1–3 until the desired number of random variables, X_i, are generated.

This method can be visualized as follows:

The value of c is the maximum functional value of $f(x)$ on (a, b). Construct a rectangular region with height c that encloses the area under $f(x)$ on (a, b). Thus, the point (X_i, Y_i) is a random point in this rectangular region. If $Y_i \le f(X_i)$, then (X_i, Y_i) lies under the curve given by $f(x)$. This method will be more efficient if the area under the curve differs from the area of the rectangular region by a small amount. If the ratio of these two areas is close to one, then the number of values of X_i that must be generated to obtain m acceptable values for the random variable will be slightly greater than the sample size, n. In order to make this method more efficient, it may be desirable to construct more than one rectangular region over the curve given by $f(x)$. Once the rectangles have been constructed, one would randomly select a rectangle and proceed by generating a point as stated previously.

To illustrate the Von Neumann rejection method, consider the density function $f(x) = (1/8)(x^3 - 6x^2 + 9x + 1)$ on the interval $[0, 2]$ and bounded by the x axis. To generate random variables from this density function, we must first determine the maximum functional value for $f(x)$ on $(0, 2)$. Taking the first derivative, $f'(x)$, we find $x = 1$ and $x = 3$ as critical values. Since $x = 3$ does not lie in $(0, 2)$, we consider only $x = 1$ as a possibility for a maximum value. We also must consider the end points of the interval, which are $x = 0$ and $x = 2$. Evaluating, we find that $f(0) = 1/8$, $f(1) = 5/8$, and $f(2) = 3/8$. We proceed with the following steps:

1. Let $c = 5/8$.
2. Generate random numbers R_1 and R_2. Let $X_i = 2R_i$, where X_i is uniform on $(0, 2)$. Let $Y_i = (5/8)R_2$, where Y_i is uniform on $(0, 5/8)$.
3. Evaluate $f(x_i) = (1/8)(x_i^3 - 6x_i^2 + 9x_i + 1)$. Compare y_i to $f(x_i)$. If $y_i < f(x_i)$ then accept x_i as a realization of the random variable in question.
4. Repeat steps 1–3 for the desired number of random variables.

Suppose we are interested in evaluating the $P(0 \le X \le 1)$ for the preceding $f(x)$. This is the same as evaluating the area under the curve between 0 and 1. The area for this $f(x)$ can be found easily by integration. However, let us assume that we cannot evaluate the mentioned probability by integration techniques. Utilizing steps 1–4 mentioned previously, we modify step 2 to let $X_i = R_i$ and add the following steps.

5. Keep a running total of the number of accepted random variables divided by the total number of points (X_i, Y_i) generated, which will be denoted as $\hat{p}$.
6. Compute $\hat{A} = B\hat{p}$, where $B = c(b - a)$. $\hat{A}$ is a relative frequency probability estimate of the specified interval.

The method just described is often referred to as the *Monte Carlo hit or miss method.*

The rejection method can be used to generate normal random variables. Because the space for the random variable X is infinite $(-\infty < x < \infty)$, using a rectangular region can present problems. A rectangle works well for enclosing the density function $f(x)$ if the range of $f(x)$ is finite. If a rectangular region is chosen, it would be most practical to truncate each end of the space. The upper and lower bounds should be chosen so that their tail probabilities are close to zero.

However, a more general technique to generate X is to enclose the density function $f(x)$ by another function $w(x)$, where $f(x) \le w(x)$ for all values of x in the spaces S. The function $w(x)$ is chosen because it is well known, and, at the same time, we can easily simulate points from it. We proceed to simulate points under $w(x)$ and then accept the ones that fall under $f(x)$. As it is not always possible to choose $w(x)$ so that it completely encloses $f(x)$, we may have to modify $w(x)$ to obtain the function $g(x) = cw(x)$, where c is a constant. To apply the acceptance-rejection method, we proceed as follows:

1. Simulate x from $w(x)$.
2. Simulate R, a $U(0, 1)$ random number, and then let $y = Rg(x)$.
3. Accept x if and only if $y < f(x)$.

Using this technique to generate standard normal random variables, a suitable choice for the function used to enclose the density function $f(x)$ of the normal distribution might be the exponential function. We know how to easily simulate points from the exponential function. The exponential function can enclose only half of the area given by the density function for the normal distribution, thus consideration will be given to

$$f(x) = (2)[(1/\sqrt{2\pi})e^{-x^2/2}], \qquad x \ge 0,$$

which simplifies to $f(x) = (\sqrt{2/\pi})e^{-x^2/2}$. Let $w(x) = e^{-x}$, $g(x) = ce^{-x}$. The next step is to solve for c. Graphically, we can see that we would like to find the point of intersection for $f(x)$ and $g(x)$, shown in Figure 4.19. We see that in solving for c, we want $cw(x) \ge f(x)$ for all x. The value of c we desire will be given by the $\max_x [f(x)/w(x)]$. By taking the first and second derivatives of

FIGURE 4.19 Approximating $f(x)$ by a simpler function, $g(x)$.

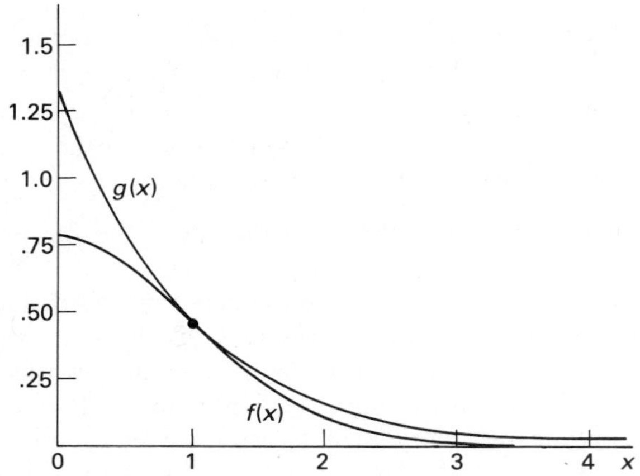

$[f(x)/w(x)]$, we can see if this maximum exists. For our example,

$$[f(x)/w(x)] = \frac{\sqrt{(2/\pi)}e^{x^2/2}}{e^{-x}} = \sqrt{2/\pi}\, e^{x-x^2/2}.$$

Taking the ln of both sides yields $y = \ln[f(x)/w(x)] = \ln\sqrt{2/\pi} + x - (x^2/2)$. The first derivative of y w.r.t. x is $y' = 1 - x$. Setting $1 - x = 0$ and solving for critical points results in $x = 1$ as a possible maximum. The second derivative of y w.r.t. x is $y'' = -1$. Because the second derivative is always less than zero, we know that $x = 1$ is a maximum. Solving $f(x)$ at $x = 1$ gives $f(1) = .485$, which equals $g(x) = ce^{-x}$ at $x = 1$. Thus, we can solve for c, which will be $.485/e^{-1} \approx 1.315$.

We then use the previously stated algorithm to generate the normal random variables. Recall that we are dealing with only the half-normal distribution. Therefore, once a random variable is generated from the half-normal distribution, we must convert this to a standard normal random variable. One way to do this is to generate R, a $U(0, 1)$, and check to see if R is less than .5. If it is, let the simulated half-normal random variable be $-x$. If R is not less than .5, x remains as is. (Note that x represents what we usually denote as z.)

Let us now consider where simulation might be useful in regard to some of the distributions mentioned earlier. First, we look at the discrete Poisson distribution. Suppose a local business firm is interested in comparing the number of local telephone calls to the number of long distance calls coming into their switchboard. Assuming independence, we define X = the number of local calls during a 15-minute period and Y = the number of long distance calls during a 15-minute period. A random variable of interest would be $W = X/Y$, where X and Y represent Poisson random variables with parameters λ_1 and λ_2, respectively. How will the distribution of W behave, given

SIMULATION 1

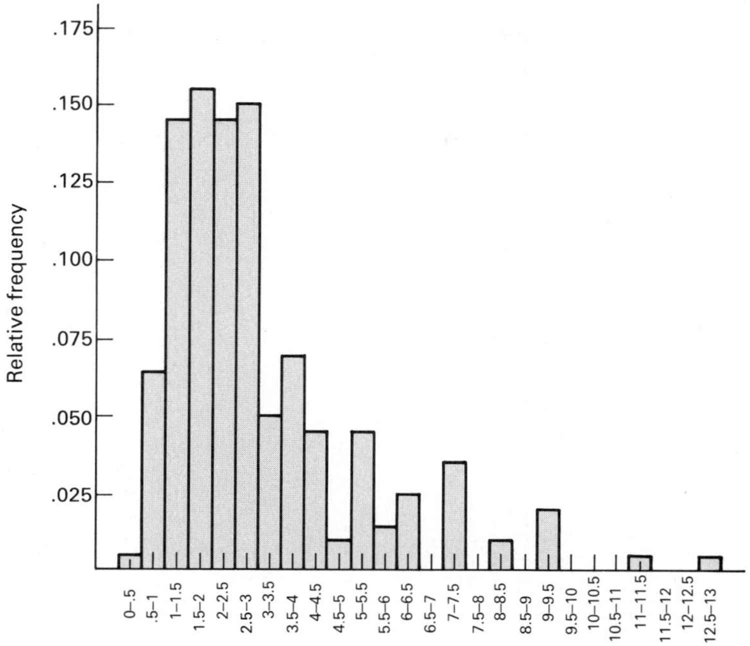

the restriction that Y cannot take on the value of 0? The results of two simulations are presented. In Simulation 1, the random variable X has a mean of 6 and the random variable Y has a mean of 3; the histogram represents 200 random values of W. The sample mean of W was 2.76 with a sample standard deviation of 2.12.

In Simulation 2, the mean of the random variable X is 2 and the mean of the random Y is 5. The histogram represents 200 random values of W. Here, the sample mean for W was .53 with a sample standard deviation of .57.

Let us now consider a situation involving exponential random variables. In engineering, we often are concerned with a parallel system. A system is defined as parallel if it functions when at least one of the components of the system is working. Let us consider a parallel system that has two components. We define the random variable X = time until failure for component 1 and the random variable Y = time until failure for component 2. X and Y are independent exponential random variables with mean for $X = \theta_x$ and the mean for $Y = \theta_y$. A variable of interest is the maximum of X and Y; that is, let $W = \max\{X, Y\}$. In Simulations 3–5, we can see how the distribution of W will behave.

SIMULATION 2

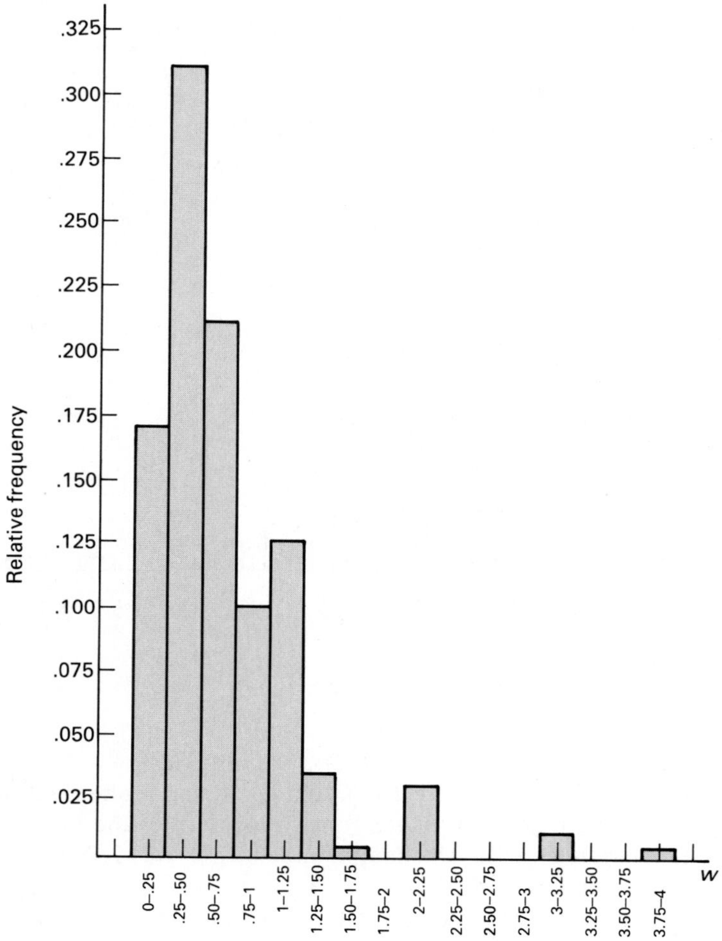

As a final example, we consider a situation that involves normal random variables. A business is interested in looking at their profits versus their total income before expenses are paid. Let X = amount of profit, and let Y = amount of expenses. Suppose that X and Y both are normally distributed. A variable of interest is $W = X/(X + Y)$. Simulations 6–8 show the form of W for different settings of the parameters underlying X and Y.

SIMULATION 3

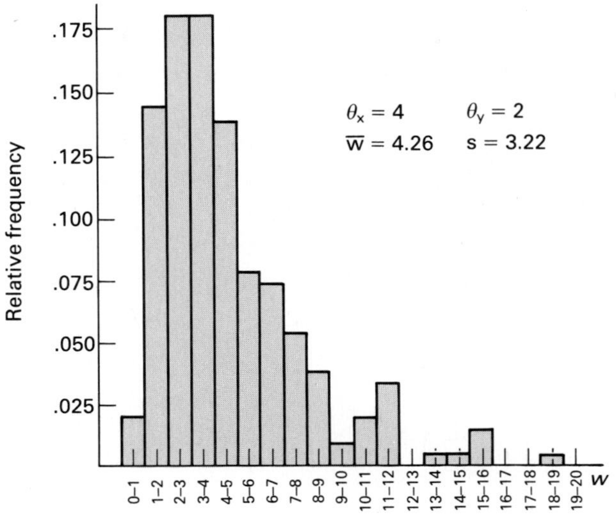

SIMULATION 4

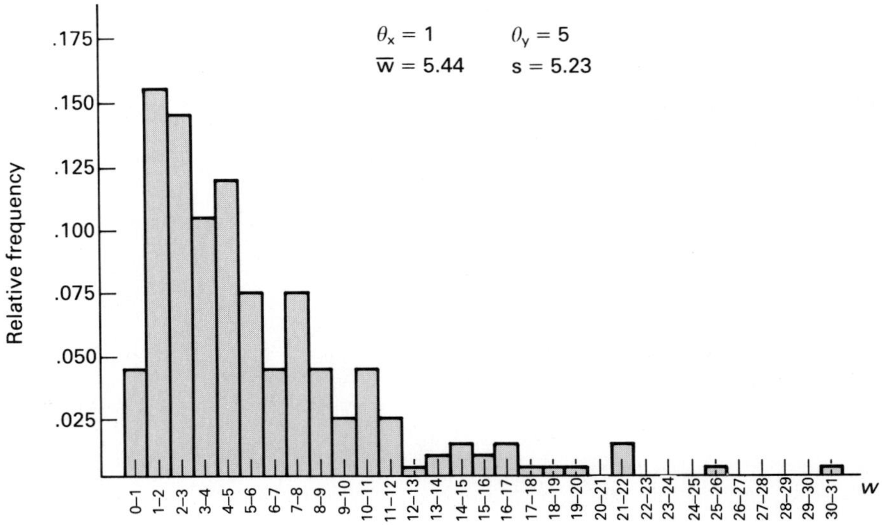

SIMULATION 5

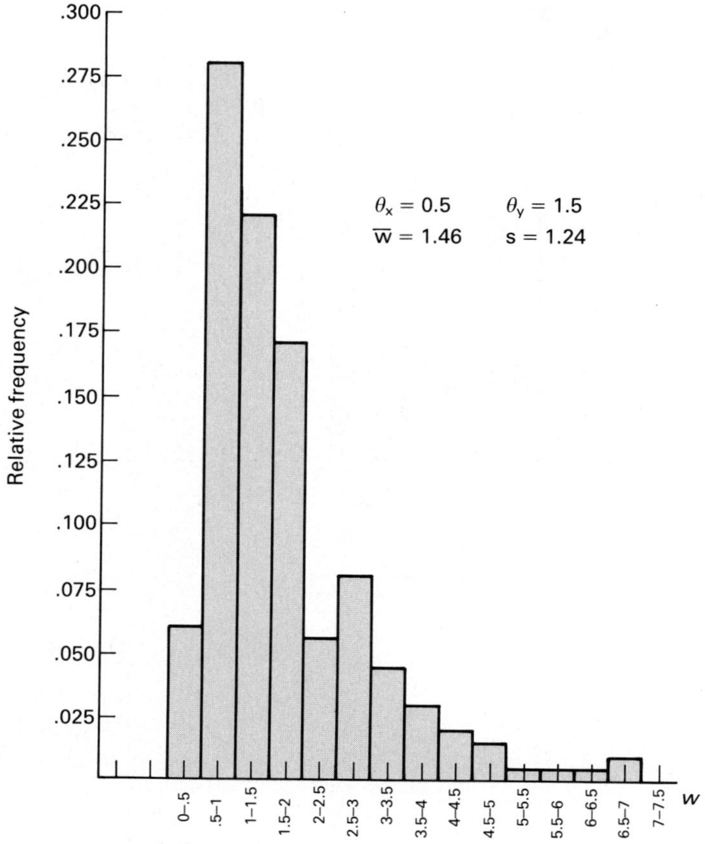

SIMULATION 6

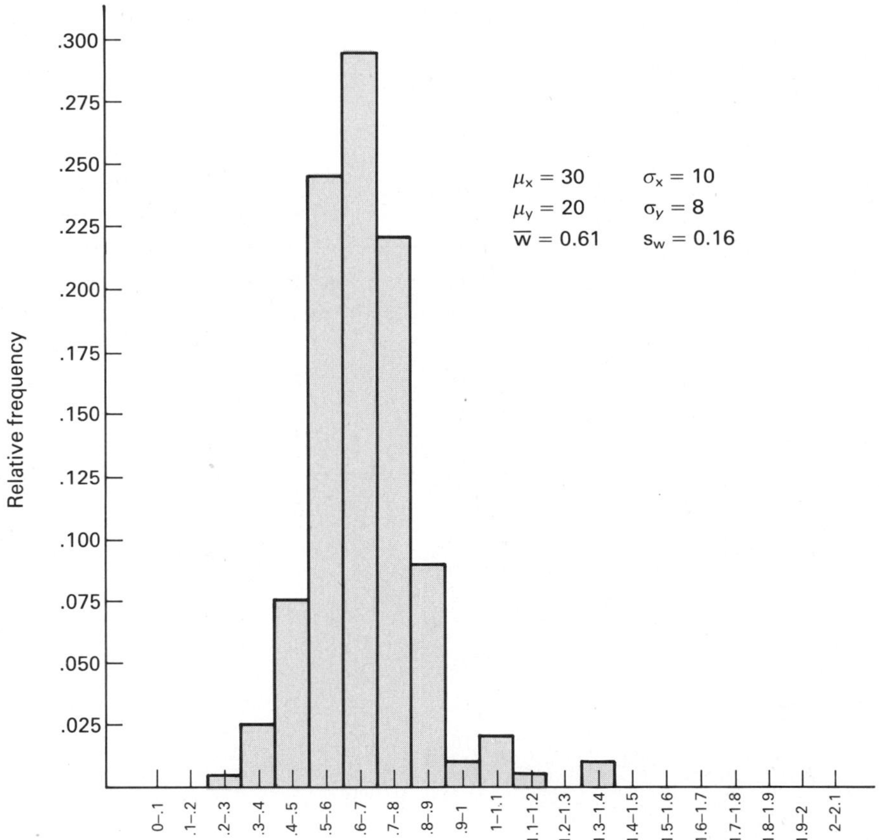

SIMULATION 8

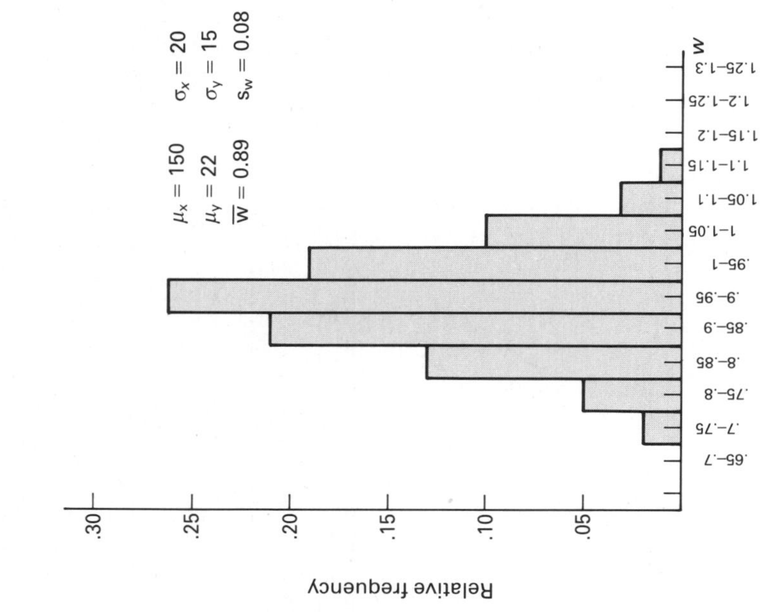

SIMULATION 7

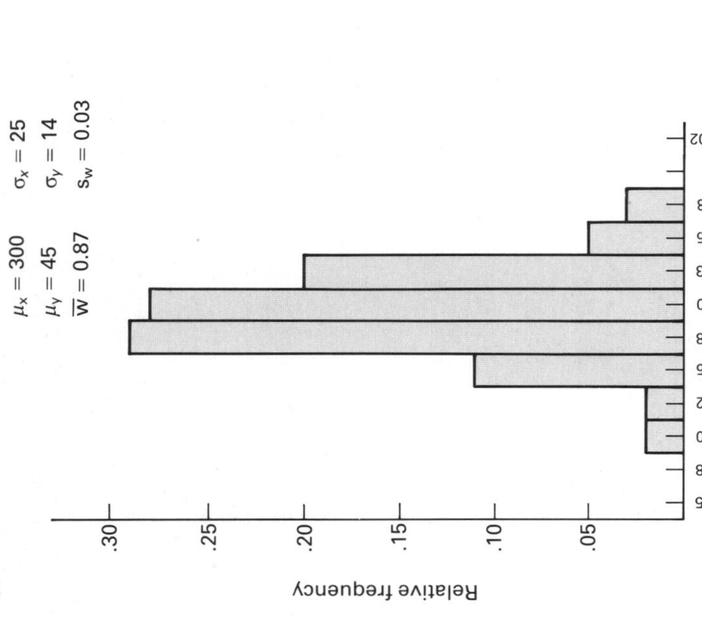

SUPPLEMENTARY EXERCISES

4.91 Let Y possess a density function

$$f(y) = \begin{cases} cy & 0 \le y \le 2 \\ 0 & \text{elsewhere.} \end{cases}$$

(a) Find c.
(b) Find $F(y)$.
(c) Graph $f(y)$ and $F(y)$.
(d) Use $F(y)$ in part (b) to find $P(1 \le Y \le 2)$.
(e) Use the geometric figure for $f(y)$ to calculate $P(1 \le Y \le 2)$.

4.92 Let Y have the density function given by

$$f(y) = \begin{cases} cy^2 + y & 0 \le y \le 2 \\ 0 & \text{elsewhere.} \end{cases}$$

(a) Find c.
(b) Find $F(y)$.
(c) Graph $f(y)$ and $F(y)$.
(d) Use $F(y)$ in part (b) to find $F(-1)$, $F(0)$, and $F(1)$.
(e) Find $P(0 \le Y \le .5)$.
(f) Find the mean and variance of Y.

4.93 Let Y have the density function given by

$$f(y) = \begin{cases} .2 & -1 < y \le 0 \\ .2 + cy & 0 < y \le 1 \\ 0 & \text{elsewhere.} \end{cases}$$

Answer parts (a) through (f), Exercise 4.92.

4.94 The grade-point averages of a large population of college students are approximately normally distributed with a mean equal to 2.4 and a standard deviation equal to .5. What fraction of the students will possess a grade-point average in excess of 3.0?

4.95 Referring to Exercise 4.94, if students possessing a grade-point average equal to or less than 1.9 are dropped from college, what percentage of the students will be dropeed?

4.96 Referring to Exercise 4.94, suppose that three students are selected randomly from the student body. What is the probability that all three will possess a grade-point average in excess of 3.0?

4.97 A machine operation produces bearings with a diameter that is normally distributed with mean and standard deviation equal to 3.005 and .001, respectively. Customer specifications require the bearing diameters to lie in the interval $3.000 \pm .0020$. Those outside the interval are considered scrap and must be remachined or used as stock for smaller bearings. With the existing machine setting, what fraction of total production will be scrap?

4.98 Referring to Exercise 4.97, suppose that five bearings are drawn from production. What is the probability that at least one will be defective?

4.99 Let Y have density function

$$f(y) = \begin{cases} cye^{-2y} & 0 \le y \le \infty \\ 0 & \text{elsewhere.} \end{cases}$$

(a) Give the mean and variance for Y.

(b) Give the moment-generating function for Y.

(c) Find the value of c.

4.100 Find $E(X^k)$ for the beta random variable. Then find the mean and variance for the beta random variable.

4.101 The yield force of a steel reinforcing bar of a certain type is found to be normally distributed with a mean of 8500 pounds and a standard deviation of 80 pounds. If three such bars are to be used on a certain project, find the probability that all three will have yield forces in excess of 8700 pounds.

4.102 An engineer has observed that the next gap times between vehicles passing a certain point on a highway have an exponential distribution with a mean of 10 seconds.

(a) Find the probability that the next gap observed will be no longer than 1 minute.

(b) Find the probability density function for the sum of the next four gap times to be observed.

What assumptions are necessary for this answer to be correct?

4.103 The proportion of time, per day, that all checkout counters in a supermarket are busy is a random variable X having probability density function

$$f(x) = \begin{cases} kx^2(1-x)^4 & 0 \le x \le 1 \\ 0 & \text{elsewhere.} \end{cases}$$

(a) Find the value of k that makes this a probability density function.

(b) Find the mean and variance of X.

4.104 If the life length X for a certain type of battery has a Weibull distribution with $\gamma = 2$ and $\theta = 3$ (with measurements in years), find the probability that the battery lasts less than 4 years given that it is now 2 years old.

4.105 The time (in hours) it takes a manager to interview an applicant has an exponential distribution with $\theta = 1/2$. Three applicants arrive at 8:00 A.M. and the interviews begin. A fourth applicant arrives at 8:45 A.M. What is the probability that the latecomer has to wait before seeing the manager?

4.106 The weekly repair cost, Y, for a certain machine has a probability density function given by

$$f(y) = \begin{cases} 3(1-y)^2 & 0 < y < 1 \\ 0 & \text{elsewhere} \end{cases}$$

with measurements in hundreds of dollars. How much money should be budgeted each week for repair cost so that the actual cost will exceed the budgeted amount only 10 % of the time?

4.107 A house builder has to order some supplies that have a waiting time for delivery Y uniformly distributed over the interval 1 to 4 days. Because the builder can get by without them for 2 days, the cost of the delay is a fixed \$100 for any waiting time up to 2 days. However, after 2 days, the cost of the delay becomes \$100 plus \$20 per day for each additional day. That is, if the waiting time is 3.5 days, the cost of the delay is \$100 + \$20(1.5) = \$130. Find the expected value of the builder's cost due to waiting for supplies.

4.108 There is a relationship between incomplete gamma integrals and sums of Poisson probabilities given by

$$\frac{1}{\Gamma(\alpha)}\int_{\lambda}^{\infty} y^{\alpha-1}e^{-y}\,dy = \sum_{y=0}^{\alpha-1}\frac{\lambda^y e^{-\lambda}}{y!}$$

for integer values of α.

If Y has a gamma distribution with $\alpha = 2$ and $\beta = 1$, find $P(Y > 1)$ by using this equality and Table 3 of the Appendix.

4.109 The weekly downtime, in hours, for a certain production line has a gamma distribution with $\alpha = 3$ and $\beta = 2$. Find the probability that the downtime for a given week will not exceed 10 hours.

4.110 Suppose that plants of a certain species are randomly dispersed over a region, with a mean density of λ plants per unit area. That is, the number of plants in a region of area A has a Poisson distribution with mean λA. For a randomly selected plant in this region, let R denote the distance to the nearest neighboring plant.

(a) Find the probability density function for R. [Hint: Note that $P(R > r)$ is the same as the probability of seeing no plants in a circle of radius r.]

(b) Find $E(R)$.

4.111 A random variable X is said to have a log normal distribution if $Y = \ln(X)$ has a normal distribution. (The symbol *ln* denotes natural logarithm.) In this case, X must not be negative. The shape of the log normal probability density function is similar to that of the gamma distribution, with long tails to the right. The equation of the log normal density function is

$$f(x) = \frac{1}{\sqrt{2\pi}\,\sigma x}e^{-(\ln(x)-\mu)^2/2\sigma^2} \qquad x > 0$$
$$= 0 \qquad \text{elsewhere.}$$

Since $\ln(x)$ is a monotonic function of x,

$$P(X \le x) = P[\ln(X) \le \ln(x)] = P[Y \le \ln(x)]$$

where Y has a normal distribution with a mean of μ and a variance of σ^2. Thus, probabilities in the log normal case can be found by transforming them to the normal case. If X has a log normal distribution with $\mu = 4$ and $\sigma^2 = 1$, find

(a) $P(X \le 4)$.

(b) $P(X > 8)$.

4.112 If X has a log normal distribution with parameters μ and σ^2, then it can be shown that

$$E(X) = e^{\mu+\sigma^2/2}$$

and

$$V(X) = e^{2\mu+\sigma^2}(e^{\sigma^2} - 1).$$

The grains composing polycrystalline metals tend to have weights that follow a log normal distribution. For a certain type of aluminum, grain weights have a log normal distribution with $\mu = 3$ and $\sigma = 4$ (in units of 10^{-2} gram).

(a) Find the mean and variance of the grain weights.

(b) Find an interval in which at least 75% of the grain weights should lie (use Tchebysheff's Theorem).

(c) Find the probability that a randomly chosen grain weighs less than the mean grain weight.

4.113 Let Y denote a random variable with probability density function given by

$$f(y) = \left(\frac{1}{2}\right)e^{-|y|}, \qquad -\infty < y < \infty.$$

Find the moment-generating function of Y and use it to find $E(Y)$.

4.114 The life length Y of a certain component in a complex electronic system is known to have an exponential density with a mean of 100 hours. The component is replaced at failure or at age 200 hours, whichever comes first.

(a) Find the distribution function for X, the length of time that the component is in use.

(b) Find $E(X)$.

4.115 We can show that the normal density function integrates to unity by showing that

$$\frac{1}{\sqrt{2\pi}}\int_{-\infty}^{\infty} e^{-(1/2)uy^2}\,dy = \frac{1}{\sqrt{u}}.$$

This, in turn, can be shown by considering the product of two such integrals,

$$\frac{1}{2\pi}\left(\int_{-\infty}^{\infty} e^{-(1/2)uy^2}\,dy\right)\left(\int_{-\infty}^{\infty} e^{-(1/2)ux^2}\,dx\right) = \frac{1}{2\pi}\int_{-\infty}^{\infty}\int_{-\infty}^{\infty} e^{-(1/2)u(x^2+y^2)}\,dx\,dy.$$

By transforming to polar coordinates, show that the double integral is equal to $1/u$.

4.116 The function $\Gamma(u)$ is defined by

$$\Gamma(u) = \int_0^{\infty} y^{u-1}e^{-y}\,dy.$$

Integrate by parts to show that

$$\Gamma(u) = (u - 1)\Gamma(u - 1).$$

Hence, if n is a positive integer, it follows that $\Gamma(u) = (n - 1)!$.

4.117 Show that $\Gamma(1/2) = \sqrt{\pi}$ by writing

$$\Gamma\left(\frac{1}{2}\right) = \int_0^{\infty} y^{-1/2}e^{-y}\,dy,$$

making the transformation $y = (1/2)x^2$, and employing the result of Exercise 4.116.

4.118 The function $B(\alpha, \beta)$ is defined by

$$B(\alpha, \beta) = \int_0^1 y^{\alpha-1}(1 - y)^{\beta-1}\, dy.$$

(a) Let $y = \sin^2 \theta$, and show that

$$B(\alpha, \beta) = 2\int_0^{\pi/2} \sin^{2\alpha-1} \theta \cos^{2\beta-1} \theta\, d\theta.$$

(b) Write $\Gamma(\alpha)\Gamma(\beta)$ as a double integral, transform to polar coordinates, and conclude that

$$B(\alpha, \beta) = \frac{\Gamma(\alpha)\Gamma(\beta)}{\Gamma(\alpha + \beta)}.$$

4.119 The lifetime, X, of a certain electronic component is a random variable with density function

$$f(x) = \begin{cases} \left(\dfrac{1}{100}\right)e^{-x/100} & x > 0 \\ 0 & \text{elsewhere.} \end{cases}$$

Three of these components operate independently in a piece of equipment. The equipment fails if at least two of the components fail. Find the probability that the equipment operates for at least 200 hours without failure.

C H A P T E R 5

MULTIVARIATE PROBABILITY DISTRIBUTIONS

5.1

BIVARIATE AND MARGINAL PROBABILITY DISTRIBUTIONS

Chapters 3 and 4 dealt with experiments that produced a single numerical response, or random variable, of interest. We discussed, for example, the life length, X, of a battery or the strength, Y, of a steel casing. Often, however, we want to study the joint behavior of two random variables, such as the joint behavior of life length *and* casing strength for these batteries. Perhaps, in such a study, we can identify a region in which a combination of life length and casing strength will be optimal in terms of balancing the cost of manufacturing with customer satisfaction. To proceed with such a study, we must know how to handle joint probability distributions. When only two random variables are involved, these joint distributions are called *bivariate distributions*. We will discuss the bivariate case in some detail, as extensions to more than two variables follow along similar lines.

Other situations in which bivariate probability distributions are important come to mind easily. The physician studies the joint behavior of pulse and amount of exercise, or blood pressure and weight. The educator studies the joint behavior of grades and time devoted to study, or the interrelationship of pretest and posttest scores. The economist studies the joint behavior of business volume and profits. In fact, most real problems we come across will have more than one underlying random variable of interest.

To simplify the basic ideas of joint distributions, we consider a bivariate example in some detail. Suppose two light bulbs are to be drawn sequentially from a box containing four good bulbs and one defective bulb. The drawings are made without replacement; that is, the first one is *not* put back into the box before the second one is drawn. Let X_1 denote the number of defectives observed on the first draw and X_2 the number of defectives observed on the second draw. Note that X_1 and X_2 can take on only the values zero and one. How might we calculate the probability of the event ($X_1 = 0$ and $X_2 = 0$)? Five choices are available for the first bulb drawn, and then only four choices for the second bulb. Thus there are $5 \times 4 = 20$ possible ordered outcomes for the drawing of two bulbs. Because the draws supposedly were made at random, a reasonable probabilistic model for this experiment would be one that assigns equal probability of 1/20 to each of the twenty possible outcomes. Now how many of these twenty outcomes are in the event $X_1 = 0$, $X_2 = 0$? Four good bulbs are in the box, any one of which could occur on the first draw. After that one good bulb is removed, three are left for the second draw. Thus, there are $4 \times 3 = 12$ outcomes that result in no defectives on either draw. The probability of the event in question is now apparent:

$$P(X_1 = 0, X_2 = 0) = \frac{12}{20} = .6$$

After similar calculations for the other possible values of X_1 and X_2, we can arrange the probabilities as in Table 5.1. The entries in Table 5.1 represent the joint probability distribution of X_1 and X_2. Sometimes we write

$$P(X_1 = x_1, X_2 = x_2) = p(x_1, x_2)$$

and call $p(x_1, x_2)$ the joint probability function of (X_1, X_2). From these probabilities we can extract the probabilities that represent the behavior of X_1, or X_2, by itself. Note that

$$\begin{aligned} P(X_1 = 0) &= P(X_1 = 0, X_2 = 0) + P(X_1 = 0, X_2 = 1) \\ &= .6 + .2 = .8. \end{aligned}$$

This entry (.8) is found on Table 5.1 as the total of the entries in the first column (under $X_1 = 0$). In similar fashion, one finds $P(X_1 = 1) = .2$, the total of the second-column entries. These values, .8 and .2, are placed conveniently on the margin of the table and hence are termed the *marginal probabilities* of X_1. The marginal probability distribution for X_2 is found by looking at the row

TABLE 5.1 **Joint distribution of X_1 and X_2**

X_2 \ X_1	0	1	Row totals (marginal probabilities for X_2)
0	.6	.2	.8
1	.2	0	.2
Column totals (marginal probabilities for X_1)	.8	.2	1.0

totals, which show $P(X_2 = 0) = .8$ and $P(X_2 = 1) = .2$. The distributions considered in previous chapters, in fact, were marginal probability distributions.

In addition to joint and marginal probabilities, we may be interested in conditional probabilities, such as the probability that the second bulb turns out to be good given that we have found the first one to be good, or $P(X_2 = 0 \mid X_1 = 0)$. Looking back at our probabilistic model for this experiment, we see that, after it is known that the first bulb is good, there are only three good bulbs among the four remaining. Thus, we should have

$$P(X_2 = 0 \mid X_1 = 0) = \frac{3}{4}.$$

The definition of conditional probability gives

$$P(X_2 = 0 \mid X_1 = 0) = \frac{P(X_1 = 0, X_2 = 0)}{P(X_1 = 0)}$$

$$= \frac{.6}{.8} = \frac{3}{4}$$

so the result agrees with our earlier intuitive assessment. The bivariate discrete case is summarized in Definition 5.1.

Definition 5.1

Let X_1 and X_2 be discrete random variables. The **joint probability distribution** of X_1 and X_2 is given by

$$p(x_1, x_2) = P(X_1 = x_1, X_2 = x_2)$$

defined for all real numbers x_1 and x_2. The function $p(x_1, x_2)$ is called the *joint probability function* of X_1 and X_2.

The marginal probability functions of X_1 and X_2, respectively, are given by

$$p_1(x_1) = \sum_{x_2} p(x_1, x_2)$$

and

$$p_2(x_2) = \sum_{x_1} p(x_1, x_2).$$

Example 5.1 Three checkout counters are in operation at a local supermarket. Two customers arrive at the counters at different times, when the counters are serving no other customers. It is assumed that the customers then choose a checkout station at random and independent of one another. Let X_1 denote the number of times counter A is selected and X_2 the number of times counter B is selected by the two customers. Find the joint probability distribution of X_1 and X_2. Find the probability that one of the customers visits counter B given that one of the customers is known to have visited counter A.

Solution

For convenience let us introduce X_3, defined as the number of customers visiting counter C. Now the event ($X_1 = 0$ and $X_2 = 0$) is equivalent to the event ($X_1 = 0$, $X_2 = 0$, and $X_3 = 2$). It follows that

$$\begin{aligned} P(X_1 = 0, X_2 = 0) &= P(X_1 = 0, X_2 = 0, X_3 = 2) \\ &= P(\text{customer I selects counter C and} \\ &\quad \text{customer II selects counter C}) \\ &= \text{P}(\text{customer I selects counter C}) \\ &\quad \times P(\text{customer II selects counter C}) \\ &= \frac{1}{3} \times \frac{1}{3} = \frac{1}{9} \end{aligned}$$

because customers' choices are independent and each customer makes a random selection from among the three available counters.

It is slightly more complicated to calculate $P(X_1 = 1, X_2 = 0) = P(X_1 = 1, X_2 = 0, X_3 = 1)$. In this event, customer I could select counter A and customer II could select counter C, or I could select C and II could select A. Thus,

$$\begin{aligned} P(X_1 = 1, X_2 = 0) &= P(\text{I selects A})P(\text{II selects C}) \\ &\quad + P(\text{I selects C})P(\text{II selects A}) \\ &= \frac{1}{3} \times \frac{1}{3} + \frac{1}{3} \times \frac{1}{3} = \frac{2}{9}. \end{aligned}$$

Similar arguments will allow one to derive the results in Table 5.2. The second statement asks for

$$\begin{aligned} P(X_2 = 1 \mid X_1 = 1) &= \frac{P(X_1 = 1, X_2 = 1)}{P(X_1 = 1)} \\ &= \frac{2/9}{4/9} = \frac{1}{2}. \end{aligned}$$

Does this answer, 1/2, agree with your intuition?

TABLE 5.2 Joint distribution of X_1 and X_2 for Example 5.1

X_2 \ X_1	0	1	2	Marginal probabilities for X_2
0	1/9	2/9	1/9	4/9
1	2/9	2/9	0	4/9
2	1/9	0	0	1/9
Marginal probabilities for X_1	4/9	4/9	1/9	1

As we move to the continuous case, let us first quickly review the situation in one dimension. If $f(x)$ denotes the probability density function of a random variable X, then $f(x)$ represents a relative frequency curve, and probabilities, such as $P(a \le X \le b)$, are represented as areas under this curve. That is,

$$P(a \le X \le b) = \int_a^b f(x)\,dx.$$

Now, suppose we are interested in the joint behavior of two continuous random variables, say X_1 and X_2, where X_1 and X_2 might represent the amounts of two different hydrocarbons found in an air sample taken for a pollution study, for example. The relative frequency of these two random variables can be modeled by a bivariate function, $f(x_1, x_2)$, which forms a probability, or relative frequency, surface in three dimensions. Figure 5.1 shows such a surface. The probability that X_1 lies in one interval and X_2 lies in another interval is then represented as a volume under this surface. Thus,

$$P(a_1 \le X_1 \le a_2, b_1 \le X_2 \le b_2) = \int_{b_1}^{b_2}\int_{a_1}^{a_2} f(x_1, x_2)\,dx_1\,dx_2.$$

FIGURE 5.1 A bivariate density function.

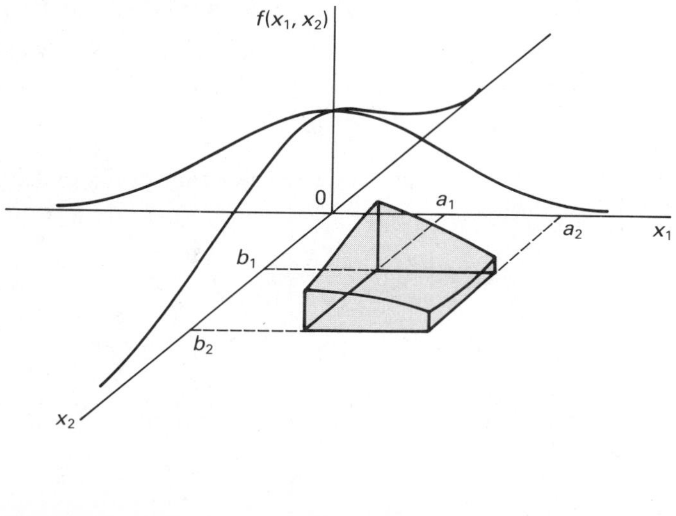

Note that the above integral simply gives the volume under the surface and over the shaded region in Figure 5.1. We will illustrate the actual computations involved in such a bivariate problem with a very simple example.

Example 5.2 A certain process for producing an industrial chemical yields a product containing two predominant types of impurities. For a certain volume of sample from this process, let X_1 denote the proportion of total impurities in the sample, and let X_2 denote the proportion of type I impurity among all impurities found. Suppose, after investigation of many such samples, the joint distribution of X_1 and X_2 can be adequately modeled by the following function:

$$f(x_1, x_2) = \begin{cases} 2(1 - x_1) & 0 \leq x_1 \leq 1, 0 \leq x_2 \leq 1 \\ 0 & \text{elsewhere.} \end{cases}$$

This function graphs as the surface given in Figure 5.2. Calculate the probability that X_1 is less than .5 and that X_2 is between .4 and .7.

FIGURE 5.2 Probability density function for Example 5.2.

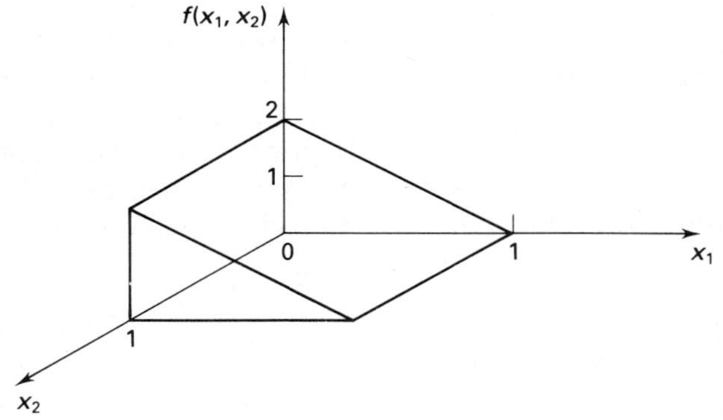

Solution

From the preceding discussion, we see that

$$\begin{aligned} P(0 \leq x_1 \leq .5, .4 \leq x_2 \leq .7) &= \int_{.4}^{.7} \int_{0}^{.5} 2(1 - x_1)\, dx_1\, dx_2 \\ &= \int_{.4}^{.7} [-(1 - x_1)^2]_0^{.5}\, dx_2 \\ &= \int_{.4}^{.7} (.75)\, dx_2 \\ &= .75(.7 - .4) = .75(.3) = .225. \end{aligned}$$

Thus, the fraction of such samples having less than 50 % impurities and a relative proportion of type I impurities between 40 % and 70 % is .225.

Just as the univariate, or marginal, probabilities were computed by summing over rows or columns in the discrete case, the univariate density function for X_1 in the continuous case can be found by integrating ("summing") over values of X_2. Thus, the marginal density function of X_1, $f_1(x_1)$, is given by

$$f_1(x_1) = \int_{-\infty}^{\infty} f(x_1, x_2)\, dx_2.$$

Similarly, the marginal density function of X_2, $f_2(x_2)$, is given by

$$f_2(x_2) = \int_{-\infty}^{\infty} f(x_1, x_2)\, dx_1.$$

Example 5.3 For the case given in Example 5.2, find the marginal probability density functions for X_1 and X_2.

Solution

Let us first try to visualize what the answers should look like, before going through the integration. To find $f_1(x_1)$, we accumulate all the probabilities in the x_2 direction. Look at Figure 5.2 and think of collapsing the wedge-shaped figure back onto the $[x_1, f_1(x_1, x_2)]$ plane. Then more probability mass will build up toward the zero point of the x_1-axis than toward the unity point. In other words, the function $f_1(x_1)$ should be high at zero and low at one.

FIGURE 5.3 Probability density function $f_1(x_1)$ for Example 5.3.

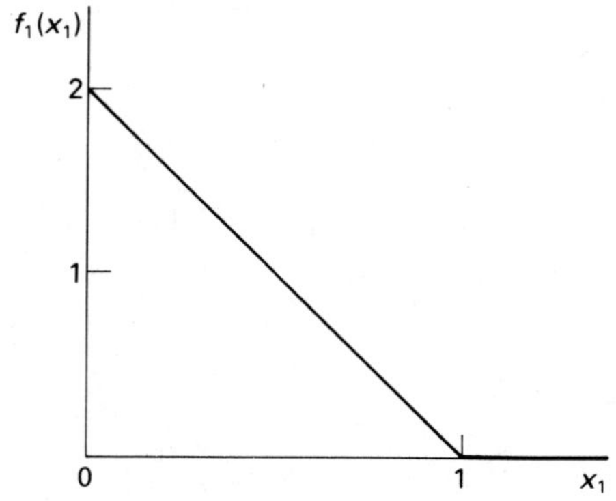

Formally,

$$\begin{aligned} f_1(x_1) &= \int_{-\infty}^{\infty} f(x_1, x_2)\,dx_2 \\ &= \int_0^1 2(1 - x_1)\,dx_2 \\ &= 2(1 - x_1) \qquad 0 < x \le 1. \end{aligned}$$

The function graphs as in Figure 5.3. Note that our conjecture is correct. Thinking of how $f_2(x_2)$ should look geometrically, suppose the wedge of Figure 5.2 is forced back onto the $[x_2, f(x_1, x_2)]$ plane. Then the probability mass should accumulate equally all along the $(0, 1)$ interval on the x_2-axis. Mathematically,

$$\begin{aligned} f_2(x_2) &= \int_{-\infty}^{\infty} f(x_1, x_2)\,dx_1 \\ &= \int_0^1 2(1 - x_1)\,dx_1 \\ &= [-(1 - x_1)]_0^1 \qquad 0 < x_2 \le 1 \end{aligned}$$

and again our conjecture is verified, as seen in Figure 5.4.

FIGURE 5.4 Probability density function $f_2(x_2)$ for Example 5.3.

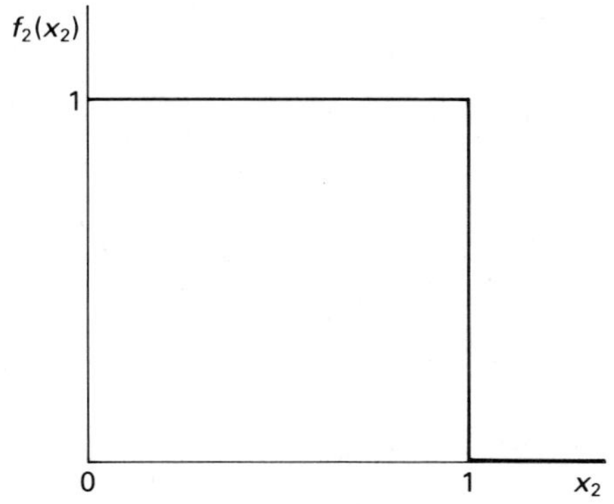

The bivariate continuous case is summarized in Definition 5.2.

Definition 5.2

Let X_1 and X_2 be continuous random variables. The **joint probability ~~distribution~~** density **function** of X_1 and X_2, if it exists, is given by a nonnegative function $f(x_1, x_2)$ such that

$$P(a_1 \le X_1 \le a_2, b_1 \le X_2 \le b_2) = \int_{b_1}^{b_2} \int_{a_1}^{a_2} f(x_1, x_2)\, dx_1\, dx_2.$$

The **marginal probability density functions** of X_1 and X_2, respectively, are given by

$$f_1(x_1) = \int_{-\infty}^{\infty} f(x_1, x_2)\, dx_2$$

and

$$f_2(x_2) = \int_{-\infty}^{\infty} f(x_1, x_2)\, dx_1.$$

As another (and somewhat more complicated) example, consider the following.

Example 5.4 Gasoline is to be stocked in a bulk tank once each week and then sold to customers. Let X_1 denote the proportion of the tank that is stocked in a particular week, and let X_2 denote the proportion of the tank that is sold in that same week. Due to limited supplies, X_1 is not fixed in advance but varies from week to week. Suppose a study of many weeks shows the joint relative

FIGURE 5.5 The joint density function for Example 5.4.

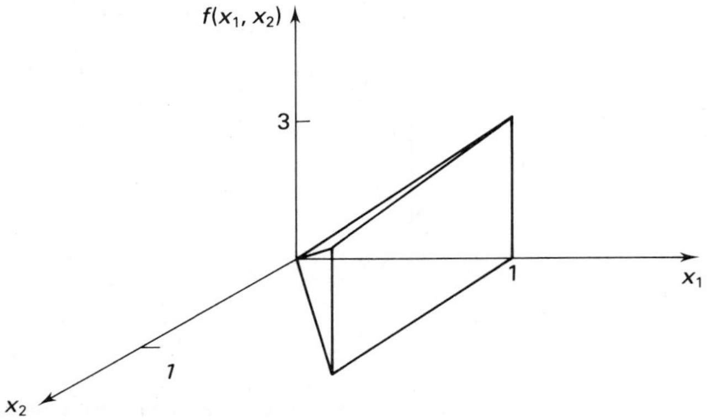

frequency behavior of X_1 and X_2 to be such that the following density function provides an adequate model:

$$f(x_1, x_2) = 3x_1 \qquad 0 \le x_2 \le x_1 \le 1$$
$$= 0 \qquad \text{elsewhere.}$$

Note that X_2 must always be less than or equal to X_1. This density function is graphed in Figure 5.5. Find the probability that X_2 will be between .2 and .4 for a given week.

Solution

The question refers to the marginal behavior of X_2. Thus, it is necessary to find

$$f_2(x_2) = \int_{-\infty}^{\infty} f(x_1, x_2)\, dx_1$$
$$= \int_{x_2}^{1} 3x_1\, dx = \frac{3}{2} x_1^2 \Big|_{x_2}^{1}$$
$$= \frac{3}{2}(1 - x_2^2) \qquad 0 < x_2 \le 1.$$

It follows directly that

$$P(.2 \le X_2 \le .4) = \int_{.2}^{.4} \frac{3}{2}(1 - x_2^2)\, dx_2$$
$$= \frac{3}{2}\left(x_2 - \frac{x_2^3}{3}\right)\Bigg|_{.2}^{.4}$$
$$= \frac{3}{2}\left\{\left[.4 - \frac{(.4)^3}{3}\right] - \left[.2 - \frac{(.2)^3}{3}\right]\right\}$$
$$= .272.$$

Note that the marginal density of X_2 graphs as a function that is high at $x_2 = 0$ and then tends to zero as x_2 tends to one. Does this agree with your intuition after looking at Figure 5.5?

5.2

CONDITIONAL PROBABILITY DISTRIBUTIONS

Recall that the bivariate discrete case, the conditional probabilities for X_1 for a given X_2 were found by fixing attention on the particular row in which $X_2 = x_2$, and then looking at the relative probabilities within that row. That is, the individual cell probabilities were divided by the marginal total for that row in order to obtain conditional probabilities.

In the bivariate continuous case, the form of the probability density function representing the conditional behavior of X_1 for a given value of X_2 is found by slicing through the joint density in the x_1 direction at the particular value of X_2. The function then has to be weighted by the marginal density function for X_2 at that point. We will look at a specific example before giving the general definition of conditional density functions.

Example 5.5 Referring to the joint density function of Example 5.4, find the conditional probability that X_2 is less than .2 given that X_1 was known to be .5.

Solution

Slicing through $f(x_1, x_2)$ in the x_2 direction at $x_1 = .5$ yields

$$f(.5, x_2) = 3(.5) = 1.5 \qquad 0 \leq x_2 \leq .5.$$

Thus, the conditional behavior of X_2 for a given X_1 of .5 is constant over the interval (0, .5). The marginal value of $f(x_1)$ at $x_1 = .5$ is obtained as follows:

$$\begin{aligned} f_1(x_1) &= \int_{-\infty}^{\infty} f(x_1, x_2)\,dx_2 \\ &= \int_0^{x_1} 3x_1\,dx_2 \\ &= 3x_1^2 \qquad 0 \leq x_1 \leq 1 \\ f_1(.5) &= 3(.5)^2 = .75. \end{aligned}$$

Upon dividing, we see that the conditional behavior of X_2 for a given X_1 of .5 is represented by the function

$$\begin{aligned} f(x_2 \mid x_1 = .5) &= \frac{f(.5, x_2)}{f_1(.5)} \\ &= \frac{1.5}{.75} = 2 \qquad 0 < x_2 < .5. \end{aligned}$$

This function of x_2 has all the properties of a probability density function and is the conditional density function for X_2 at $X_1 = .5$. Then,

$$\begin{aligned} P(X_2 < .2 \mid X_1 = .5) &= \int_0^{.2} f(x_2 \mid x_1 = .5)\,dx_2 \\ &= \int_0^{.2} 2\,dx_2 = .2(2) = .4. \end{aligned}$$

That is, among all weeks in which the tank was half full immediately after stocking, sales amounted to less than 20 % of the tank 40 % of the time.

We see that these manipulations to obtain conditional density functions in the continuous case are analogous to those to obtain conditional probabilities in the discrete case, except that integrals are used in place of sums. The formal definition of a conditional density function follows.

Definition 5.3

Let X_1 and X_2 be jointly continuous random variables with joint probability density function $f(x_1, x_2)$ and marginal densities $f_1(x_1)$ and $f_2(x_2)$, respectively. Then, the **conditional probability density function** of X_1 given $X_2 = x_2$ is given by

$$f(x_1 \mid x_2) = \begin{cases} \dfrac{f(x_1, x_2)}{f_2(x_2)} & f_2(x_2) > 0 \\ 0 & \text{elsewhere} \end{cases}$$

and the conditional probability density function of X_2 given $X_1 = x_1$ is given by

$$f(x_2 \mid x_1) = \begin{cases} \dfrac{f(x_1, x_2)}{f_1(x_1)} & f_1(x_1) > 0 \\ 0 & \text{elsewhere.} \end{cases}$$

Example 5.6 A soft-drink machine has a random supply Y_2 at the beginning of a given day and dispenses a random amount Y_1 during the day (with measurements in gallons). It is not resupplied during the day, hence $Y_1 \le Y_2$. It has been observed that Y_1 and Y_2 have joint density

$$f(y_1, y_2) = \begin{cases} 1/2 & 0 \le y_1 \le y_2,\ 0 \le y_2 \le 2 \\ 0 & \text{elsewhere.} \end{cases}$$

That is, the points (y_1, y_2) are uniformly distributed over the triangle with the given boundaries. Find the conditional density of Y_1 given $Y_2 = y_2$. Evaluate the probability that less than 1/2 gallon is sold, given that the machine contains 1 gallon at the start of the day.

Solution

The marginal density of Y_2 is given by

$$f_2(y_2) = \int_{-\infty}^{\infty} f(y_1, y_2)\, dy_1$$

$$= \begin{cases} \displaystyle\int_0^{y_2} (1/2)\, dy_1 = (1/2)y_2 & 0 \le y_2 \le 2 \\ 0 & \text{elsewhere.} \end{cases}$$

By Definition 5.3,

$$f_1(y_1 \mid y_2) = \frac{f(y_1, y_2)}{f_2(y_2)}$$

$$= \begin{cases} \dfrac{1/2}{(1/2)y_2} = \dfrac{1}{y_2} & 0 \le y_1 \le y_2 \le 2 \\ 0 & \text{elsewhere.} \end{cases}$$

The probability of interest is

$$P(Y_1 < 1/2 \mid Y_2 = 1) = \int_{-\infty}^{1/2} f(y_1 \mid y_2 = 1)\, dy_1$$

$$= \int_0^{1/2} (1)\, dy_1 = \frac{1}{2}.$$

Note that, if the machine had contained 2 gallons at the start of the day, then

$$P(Y_1 \le 1/2 \mid Y_2 = 2) = \int_0^{1/2} (1/2)\, dy_1$$

$$= \frac{1}{4}.$$

Thus, the amount sold is highly dependent upon the amount in supply.

5.3 INDEPENDENT RANDOM VARIABLES

Before defining independent random variables, we will recall once again that two events A and B are independent if $P(AB) = P(A)P(B)$. Somewhat analogously, two discrete random variables are independent if

$$P(X_1 = x_1, X_2 = x_2) = P(X_1 = x_1)P(X_2 = x_2)$$

for all real numbers x_1 and x_2. A similar idea carries over to the continuous case.

Definition 5.4

Discrete random variables X_1 and X_2 are said to be **independent** if

$$P(X_1 = x_1, X_2 = x_2) = P(X_1 = x_1)P(X_2 = x_2)$$

for all real numbers x_1 and x_2.

Continuous random variables X_1 and X_2 are said to be **independent** if

$$f(x_1, x_2) = f_1(x_1)f(x_2)$$

for all real numbers x_1 and x_2.

The concepts of joint probability density functions and independence extend immediately to n random variables, where n is any finite positive integer. The n random variables $X_1, X_2, \ldots, X_n$ are said to be independent if their joint density function, $f(x_1, \ldots, x_n)$, is given by

$$f(x_1, \ldots, x_n) = f_1(x_1)f_2(x_2) \cdots f_n(x_n)$$

for all real numbers $x_1, x_2, \ldots, x_n$.

Example 5.7 Show that the random variables having the joint distribution of Table 5.1 are not independent.

Solution

It is necessary to check only one entry in the table. We see that $P(X_1 = 0, X_2 = 0) = .6$, whereas $P(X_1 = 0) = .8$ and $P(X_2 = 0) = .8$. Because

$$P(X_1 = 0, X_2 = 0) \neq P(X_1 = 0)P(X_2 = 0)$$

the random variables cannot be independent.

Example 5.8 Show that the random variables in Example 5.2, pages 207–208, are independent.

Solution

Here,

$$\begin{aligned} f(x_1, x_2) &= 2(1 - x_1) \qquad 0 \leq x_1 \leq 1, 0 \leq x_2 \leq 1 \\ &= 0 \qquad \text{elsewhere.} \end{aligned}$$

We saw in Example 5.3 that

$$f_1(x_1) = 2(1 - x_1) \qquad 0 \leq x_1 \leq 1$$

and

$$f_2(x_2) = 1 \qquad 0 \leq x_2 \leq 1.$$

Thus, $f(x_1, x_2) = f_1(x_1)f_2(x_2)$, for all real numbers x_1 and x_2, and X_1 and X_2 are independent random variables.

EXERCISES

5.1 Two construction contracts are to be randomly assigned to one or more of three firms. Numbering the firms, I, II, and III, let X_1 be the number of contracts assigned to firm I and X_2 the number assigned to firm II. (A firm may receive more than one contract.)

(a) Find the joint probability distribution for X_1 and X_2.

(b) Find the marginal probability distribution for X_1.

(c) Find $P(X_1 = 1 \mid X_2 = 1)$.

5.2 A radioactive particle is randomly located in a square area with sides one unit in length. Let X_1 and X_2 denote the coordinates of the particle. Since the particle is equally likely to fall in any subarea of a fixed size, a reasonable model for (X_1, X_2) is given by

$$f(x_1, x_2) = \begin{cases} 1 & 0 \le x_1 \le 1, 0 \le x_2 \le 1 \\ 0 & \text{elsewhere.} \end{cases}$$

(a) Sketch the probability density surface.

(b) Find $P(X_1 \le .2, X_2 \le .4)$.

(c) Find $P(.1 \le X_1 \le .3, X_2 > .4)$.

5.3 A group of nine executives of a certain firm contain four who are married, three who never married, and two who are divorced. Three of the executives are to be selected for promotion. Let X_1 denote the number of married executives and X_2 the number of single executives among the three selected for promotion. Assuming that the three are randomly selected from the nine available, find the joint probability distribution for X_1 and X_2.

5.4 An environmental engineer measures the amount (by weight) of particulate pollution in air samples (of a certain volume) collected over the smokestack of a coal-fueled power plant. Let X_1 denote the amount of pollutant per sample when a certain cleaning device on the stack is not operating, and X_2 the amount of pollutant per sample when the cleaning device is operating, under similar environmental conditions. It is observed that X_1 is always greater than $2X_2$, and the relative frequency of (X_1, X_2) can be modeled by

$$f(x_1, x_2) = \begin{cases} k & 0 \le x_1 \le 2, 0 \le x_2 \le 1, 2x_2 \le x_1 \\ 0 & \text{elsewhere.} \end{cases}$$

(That is, X_1 and X_2 are randomly distributed over the region inside the triangle bounded by $x_1 = 2$, $x_2 = 0$, and $2x_2 = x_1$.)

(a) Find the value of k that makes this a probability density function.

(b) Find $P(X_1 \ge 3X_2)$. (That is, find the probability that the cleaning device will reduce the amount of pollutant by one-third or more.)

5.5 Referring to Exercise 5.2,

(a) find the marginal density function for X_1.

(b) find $P(X_1 \le .5)$.

(c) are X_1 and X_2 independent?

5.6 Referring to Exercise 5.4,

(a) find the marginal density function of X_2.

(b) find $P(X_2 \le .4)$.

(c) are X_1 and X_2 independent?

(d) find $P(X_2 \le 1/4 \mid X_1 = 1)$.

5.7 Let X_1 and X_2 denote the proportion of two different chemicals found in a sample mixture of chemicals used as an insecticide. Suppose X_1 and X_2 have joint probability density given by

$$f(x_1, x_2) = \begin{cases} 2 & 0 \le x_1 \le 1, 0 \le x_2 \le 1, 0 \le x_1 + x_2 \le 1 \\ 0 & \text{elsewhere.} \end{cases}$$

(Note that $X_1 + X_2$ must be unity at most, because the random variables denote proportions within the same sample.)

(a) Find $P(X_1 \le 3/4, X_2 \le 3/4)$.

(b) Find $P(X_1 \le 1/2, X_2 \le 1/2)$.

(c) Find $P(X_1 \le 1/2 \mid X_2 \le 1/2)$.

5.8 Referring to Exercise 5.7,

(a) find the marginal density functions for X_1 and X_2.

(b) are X_1 and X_2 independent?

(c) find $P(X_1 > 1/2 \mid X_2 = 1/4)$.

5.9 Let X_1 and X_2 denote the proportions of time, out of one work week, that employees I and II, respectively, actually spend on performing their assigned tasks. The joint relative frequency behavior of X_1 and X_2 is modeled by the probability density function

$$f(x_1, x_2) = \begin{cases} x_1 + x_2 & 0 \le x_1 \le 1, 0 \le x_2 \le 1 \\ 0 & \text{elsewhere.} \end{cases}$$

(a) Find $P(X_1 < 1/2, X_2 > 1/4)$.

(b) Find $P(X_1 + X_2 \le 1)$.

(c) Are X_1 and X_2 independent?

5.10 Referring to Exercise 5.9, find the probability that employee I spends more than 75% of the week on his assigned task, given that employee II spends exactly 50% of the work week on her assigned task.

5.11 An electronic surveillance system has one of each of two different types of components in joint operations. Letting X_1 and X_2 denote the random life lengths of the components of type I and type II, respectively, the joint probability density function is given by

$$f(x_1, x_2) = \begin{cases} (1/8)x_1 e^{-(x_1+x_2)/2} & x_1 > 0, x_2 > 0 \\ 0 & \text{elsewhere} \end{cases}$$

(measurements are in hundreds of hours).

(a) Are X_1 and X_2 independent?

(b) Find $P(X_1 > 1, X_2 > 1)$.

5.12 A bus arrives at a bus stop at a randomly selected time within a 1-hour period. A passenger arrives at the bus stop at a randomly selected time within the same hour. The passenger will wait for the bus up to 1/4 of an hour. What is the probability that the passenger will catch the bus? [Hint: Let X_1 denote the bus arrival time and X_2 the passenger arrival time. If these arrivals are independent, then

$$f(x_1, x_2) = \begin{cases} 1 & 0 \le x_1 \le 1, 0 \le x_2 \le 1 \\ 0 & \text{elsewhere.} \end{cases}$$

Now find $P(X_2 \le X_1 \le X_2 + 1/4)$.]

5.13 Two friends are to meet at a library. Each arrives at an independently selected time within a fixed 1-hour period. Each agrees to wait no more than 10 minutes for the other. Find the probability that they meet.

5.14 Two quality control inspectors each interrupt a production line at randomly, but independently, selected times within a given day (of 8 hours). Find the probability that the two interruptions will be more than 4 hours apart.

5.15 Two telephone calls come into a switchboard at random times in a fixed 1-hour period. If the calls are independent of each other,

(a) find the probability that both are made in the first half hour.

(b) find the probability that the two calls are within 5 minutes of each other.

5.16 A bombing target is in the center of a circle with radius 1 mile. A bomb falls at a randomly selected point inside that circle. If the bomb destroys everything within 1/2 mile of its landing point, what is the probability that it will destroy the target?

5.4

EXPECTED VALUES OF FUNCTIONS OF RANDOM VARIABLES

As we have seen, we often encounter problems that involve more than one random variable. We may be interested in the life lengths of five different electronic components within the same system or three different strength test measurements on the same section of cable. We now discuss how to find expected values of functions of more than one random variable. Definition 5.5 gives the basic result for finding expected values in the bivariate case. The definition, of course, can be generalized to more variables.

Definition 5.5

Suppose the discrete random variables (X_1, X_2) have joint probability function given by $p(x_1, x_2)$. If $g(X_1, X_2)$ is any real-valued function of (X_1, X_2), then

$$E[g(X_1, X_2)] = \sum_{x_1} \sum_{x_2} g(x_1, x_2) p(x_1, x_2).$$

The sum is over all values of (x_1, x_2) for which $p(x_1, x_2) > 0$. If (X_1, X_2) are continuous random variables with probability density function $f(x_1, x_2)$, then

$$E[g(X_1, X_2)] = \int_{-\infty}^{\infty}\int_{-\infty}^{\infty} g(x_1, x_2)f(x_1, x_2)\,dx_1\,dx_2.$$

Note that if X_1 and X_2 are independent, then

$$E[g(X_1)h(X_2)] = E[g(X_1)]E[h(x_2)].$$

A function of two variables that is commonly of interest in probabilistic and statistical problems is the covariance.

Definition 5.6

The **covariance** between two random variables X_1 and X_2 is given by

$$\text{cov}(X_1, X_2) = E[(X_1 - \mu_1)(X_2 - \mu_2)]$$

where

$$\mu_1 = E(X_1) \text{ and } \mu_2 = E(X_2).$$

The covariance helps us assess the relationship between two variables in the following sense. If X_2 tends to be large at the same time X_1 is large, and if X_2 tends to be small when X_1 is small, then X_1 and X_2 will have a positive covariance. If, on the other hand, X_2 tends to be small when X_1 is large and large when X_1 is small, then the variables will have a negative covariance. The next theorem gives an easier computational form for the covariance.

Theorem 5.1 If X_1 has a mean of μ_1 and X_2 has a mean of μ_2, then

$$\text{cov}(X_1, X_2) = E(X_1X_2) - \mu_1\mu_2.$$

If X_1 and X_2 are independent random variables, then $E(X_1 X_2) = E(X_1)E(X_2)$. Using Theorem 5.1, it is then clear that independence between X_1 and X_2 implies that $\text{cov}(X_1, X_2) = 0$. The converse is not necessarily true; that is, zero covariance does not imply that the variables are independent.

Example 5.8 A firm that sells word processing systems keeps track of the number of customers who call on any one day and the number of orders placed on any one day. Let X_1 denote the number of calls, X_2 the number of orders placed, and $p(x_1, x_2)$ the joint probability function for (X_1, X_2). Records indicate that

$$p(0, 0) = .04 \qquad p(2, 0) = .20$$
$$p(1, 0) = .16 \qquad p(2, 1) = .30$$
$$p(1, 1) = .10 \qquad p(2, 2) = .20.$$

That is, for any given day the probability of, say, two calls and one order is .30. Find $\text{cov}(X_1, X_2)$.

Solution

Theorem 5.1 suggests that we first find $E(X_1 X_2)$, which is

$$\begin{aligned} E(X_1 X_2) &= \sum_{x_1}\sum_{x_2} x_1 x_2 p(x_1, x_2) \\ &= (0 \times 0)p(0, 0) + (1 \times 0)p(1, 0) + (1 \times 1)p(1, 1) \\ &\quad + (2 \times 0)p(2, 0) + (2 \times 1)p(2, 1) + (2 \times 2)p(2, 2) \\ &= 0(.4) + 0(.16) + 1(.10) + 0(.20) + 2(.30) + 4(.20) \\ &= 1.50. \end{aligned}$$

Now we must find $E(X_1) = \mu_1$ and $E(X_2) = \mu_2$. The marginal distributions of X_1 and X_2 are given on the following charts:

x_1	$p(x_1)$
0	.04
1	.26
2	.70

y_2	$p(x_2)$
0	.40
1	.40
2	.20

It follows that

$$\mu_1 = 1(.26) + 2(.70) = 1.66$$

and

$$\mu_2 = 1(.40) + 2(.20) = .80.$$

Thus,

$$\begin{aligned} \text{cov}(X_1, X_2) &= E(X_1 X_2) - \mu_1\mu_2 \\ &= 1.50 - 1.66(.80) \\ &= 1.50 - 1.328 = .172. \end{aligned}$$

Does the positive covariance agree with your intuition?

When a problem involves n random variables $X_1, X_2, \ldots, X_n$, we are often interested in studying linear combinations of those variables. For example, if the random variables measure the quarterly incomes for n plants in a corporation, we may want to look at their sum or average. If X_1 represents the monthly cost of servicing defective plants before a new quality control system was installed, and X_2 denotes that cost after the system was in operation, then we may want to study $X_1 - X_2$. Theorem 5.2 gives a general result on the mean and variance of a linear combination of random variables.

Theorem 5.2 Let $Y_1, \ldots, Y_n$ and $X_1, \ldots, X_m$ be random variables with $E(Y_i) = \mu_i$ and $E(X_i) = \xi_i$. Define

$$U_1 = \sum_{i=1}^{n} a_i Y_i, \qquad U_2 = \sum_{j=1}^{m} b_j X_j$$

for constants $a_1, \ldots, a_n, b_1, \ldots, b_m$. Then

(a) $E(U_1) = \sum_{i=1}^{n} a_i \mu_i$.

(b) $V(U_1) = \sum_{i=1}^{n} a_1^2 V(Y_i) + 2 \sum\sum_{i<j} a_i a_j \operatorname{cov}(Y_i, Y_j)$,

where the double sum is over all pairs (i, j) with $i < j$ and

(c) $\operatorname{cov}(U_1, U_2) = \sum_{i=1}^{n} \sum_{j=1}^{m} a_i b_j \operatorname{cov}(Y_i, X_j)$.

Proof

Part (a) follows directly from the definition of expected value and properties of sums or integrals. To prove part (b), we appeal to the definition of variance and write

$$\begin{aligned}
V(U_1) &= E[U_1 - E(U_1)]^2 \\
&= E\left[\sum_{i=1}^{n} a_i Y_i - \sum_{i=1}^{n} a_i \mu_i\right]^2 \\
&= E\left[\sum_{i=1}^{n} a_i (Y_i - \mu_i)\right]^2 \\
&= E\left[\sum_{i=1}^{n} a_i^2 (Y_i - \mu_i)^2 + \sum\sum_{i \neq j} a_i a_j (Y_i - \mu_i)(Y_j - \mu_j)\right] \\
&= \sum_{i=1}^{n} a_i^2 E(y_i - \mu_i)^2 + \sum\sum_{i \neq j} a_i a_j E[(Y_i - \mu_i)(Y_j - \mu_j)].
\end{aligned}$$

By definition of variance and covariance, we then have

$$V(U_1) = \sum_{i=1}^{n} a_i^2 V(Y_i) + \sum\sum_{i \neq j} a_i a_j \operatorname{cov}(Y_i, Y_j).$$

Note that $\operatorname{cov}(Y_i, Y_j) = \operatorname{cov}(Y_j, Y_i)$, and hence we can write

$$V(U_1) = \sum_{i=1}^{n} a_i^2 V(Y_i) + 2 \sum\sum_{i<j} \operatorname{cov}(Y_i, Y_j).$$

Part (c) is obtained by similar steps. We have

$$\begin{aligned}
\operatorname{cov}(U_1, U_2) &= E\{[U_1 - E(U_1)][U_2 - E(U_2)]\} \\
&= E\left[\left(\sum_{i=1}^{n} a_i Y_i - \sum_{i=1}^{n} a_i \mu_i\right)\left(\sum_{j=1}^{m} b_j X_j - \sum_{j=1}^{m} b_j \xi_j\right)\right] \\
&= E\left\{\left[\sum_{i=1}^{n} a_i (Y_i - \mu_i)\right]\left[\sum_{j=1}^{m} b_j (X_j - \xi_j)\right]\right\} \\
&= E\left[\sum_{i=1}^{n}\sum_{j=1}^{m} a_i b_j (Y_i - \mu_i)(X_j - \xi_j)\right] \\
&= \sum_{i=1}^{n}\sum_{j=1}^{m} a_i b_j E(Y_i - \mu_i)(X_j - \xi_j) \\
&= \sum_{i=1}^{n}\sum_{j=1}^{m} a_i b_j \operatorname{cov}(Y_i, X_j).
\end{aligned}$$

On observing that $\operatorname{cov}(Y_i, Y_i) = V(Y_i)$, we can see that part (b) is a special case of part (c). ■

Example 5.9 Let $Y_1, Y_2, \ldots, Y_n$ be independent random variables with $E(Y_i) = \mu$ and $V(Y_i) = \sigma^2$. (These variables may denote the outcomes on n independent trials of an experiment.) Defining

$$\bar{Y} = \frac{1}{n}\sum_{i=1}^{n} Y_i,$$

show that $E(Y) = \mu$ and $V(\bar{Y}) = \sigma^2/n$.

Solution

Note that $\bar{Y}$ is a linear function with all constants, a_i equal to $1/n$. That is,

$$\bar{Y} = \left(\frac{1}{n}\right)Y_1 + \cdots + \left(\frac{1}{n}\right)Y_n.$$

By Theorem 5.2, part (a),

$$\begin{aligned}
E(\bar{Y}) &= \sum_{i=1}^{n} a_i \mu = \mu \sum_{i=1}^{n} a_i \\
&= \mu \sum_{i=1}^{n} \frac{1}{n} = \frac{n\mu}{n} = \mu.
\end{aligned}$$

By Theorem 5.2, part (b),

$$V(\bar{Y}) = \sum_{i=1}^{n} a_i^2 V(Y_i) + 2\sum\sum_{i<j} a_{ij} \operatorname{cov}(Y_i, Y_j),$$

but the covariance terms are all zero since the random variables are independent. Thus,

$$V(\bar{Y}) = \sum_{i=1}^{n} \left(\frac{1}{n}\right)^2 \sigma^2 = \frac{n\sigma^2}{n^2} = \frac{\sigma^2}{n}.$$

Example 5.10 With X_1 denoting the amount of gasoline stocked in a bulk tank at the beginning of a week and X_2 the amount sold during the week, $Y = X_1 - X_2$ represents the amount left over at the end of the week. Find the mean and variance of Y if the joint density function of (X_1, X_2) is given by

$$f(x_1, x_2) = \begin{cases} 3x_1 & 0 \le x_2 \le x_1 \le 1 \\ 0 & \text{elsewhere.} \end{cases}$$

Solution

We must first find the means and variances of X_1 and X_2. The marginal density of X_1 is found to be

$$f_1(x_1) = \begin{cases} 3x_1^2 & 0 \le x_1 \le 1 \\ 0 & \text{elsewhere.} \end{cases}$$

Thus,

$$E(X_1) = \int_0^1 x_1(3x_1^2)\, dx$$

$$= 3\left[\frac{x_1^4}{4}\right]_0^1 = \frac{3}{4}.$$

The marginal density of X_2 is found to be

$$f_2(x_2) = \begin{cases} \frac{3}{2}(1 - x_2^2) & 0 \le x_2 \le 1 \\ 0 & \text{elsewhere.} \end{cases}$$

Thus,

$$E(X_2) = \int_0^1 (x_2)\frac{3}{2}(1 - x_2^2)\, dx_2$$

$$= \frac{3}{2}\int_0^1 (x_2 - x_2^3)\, dx_2$$

$$= \frac{3}{2}\left\{\left[\frac{x_2^2}{2}\right]_0^1 - \left[\frac{x_2^4}{4}\right]_0^1\right\}$$

$$= \frac{3}{2}\left\{\frac{1}{2} - \frac{1}{4}\right\} = \frac{3}{8}.$$

By similar arguments, it follows that

$$E(X_1^2) = \frac{3}{5}$$

$$V(X_1) = \frac{3}{5} - \left(\frac{3}{4}\right)^2 = .375$$

$$E(X_2^2) = \frac{1}{5}$$

and

$$V(X_2) = \frac{1}{5} - \left(\frac{3}{8}\right)^2 = .0594.$$

The next step is to find $\text{cov}(X_1, X_2)$. Now

$$\begin{aligned} E(X_1 X_2) &= \int_0^1 \int_0^{x_1} (x_1 x_2) 3x_1 \, dx_2 \, dx_1 \\ &= 3\int_0^1 \int_0^{x_1} x_1^2 x_2 \, dx_2 \, dx_1 \\ &= 3\int_0^1 x_1^2 \left[\frac{x_2^2}{2}\right]_0^{x_1} dx_1 \\ &= \frac{3}{2}\int_0^1 x_1^4 \, dx_1 \\ &= \frac{3}{2}\left[\frac{x_1^5}{5}\right]_0^1 = \frac{3}{10} \end{aligned}$$

and

$$\begin{aligned} \text{cov}(X_1, X_2) &= E(X_1 X_2) - \mu_1\mu_2 \\ &= \frac{3}{10} - \left(\frac{3}{4}\right)\left(\frac{3}{8}\right) = .0188. \end{aligned}$$

From Theorem 5.2,

$$\begin{aligned} E(Y) &= E(X_1) - E(X_2) \\ &= \frac{3}{4} - \frac{3}{8} = \frac{3}{8} = .375 \end{aligned}$$

and

$$\begin{aligned} V(Y) &= V(X_1) + V(X_2) + 2(1)(-1)\,\text{cov}(X_1, X_2) \\ &= .0375 + .0594 - 2(.0188) \\ &= .0593. \end{aligned}$$

When the random variables in use are independent, the variance of a linear function simplifies since the covariances are zero.

Example 5.11 A firm purchases two types of industrial chemicals. The amount of type I chemical purchased per week, X_1, has $E(X_1) = 40$ gallons with $V(X_1) = 4$. The amount of type II chemical purchased, X_2, has $E(X_2) = 65$ gallons with $V(X_2) = 8$. Type I costs \$3 per gallon whereas type II costs \$5 per gallon. Find the mean and variance of the total weekly amount spent for these types of chemicals, assuming X_1 and X_2 are independent.

Solution

The dollar amount spent per week is given by

$$Y = 3X_1 + 5X_2.$$

From Theorem 5.2.

$$\begin{aligned} E(Y) &= 3E(X_1) + 5E(X_2) \\ &= 3(40) + 5(65) \\ &= 445 \end{aligned}$$

and

$$\begin{aligned} V(Y) &= (3)^2V(X_1) + (5)^2V(X_2) \\ &= 9(4) + 25(8) \\ &= 236. \end{aligned}$$

The firm can expect to spend \$445 per week on chemicals.

Example 5.12 Suppose an urn contains r white balls and $(N - r)$ black balls. A random sample of n balls is drawn without replacement and Y, the number of white balls in the sample, is observed. From Chapter 3 we know that Y has a hypergeometric probability distribution. Find the mean and variance of Y.

Solution

We first observe some characteristics of sampling without replacement. Suppose that the sampling is done sequentially and we observe outcomes for $X_1, X_2, \ldots, X_n$, where

$$X_i = \begin{cases} 1 & \text{if the } i\text{th draw results in a white ball} \\ 0 & \text{otherwise.} \end{cases}$$

Unquestionably, $P(X_1 = 1) = r/N$. But it is also true that $P(X_2 = 1) = r/N$

because

$$\begin{aligned} P(X_2 = 1) &= P(X_1 = 1, X_2 = 1) + P(X_1 = 0, X_2 = 1) \\ &= P(X_1 = 1)P(X_2 = 1 \mid X_1 = 1) \\ &\quad + P(X_1 = 0)P(X_2 = 1 \mid X_1 = 0) \\ &= \frac{r}{N}\frac{r-1}{N-1} + \frac{N-r}{N}\frac{r}{N-1} \\ &= \frac{r(N-1)}{N(N-1)} = \frac{r}{N}. \end{aligned}$$

The same is true for X_k; that is,

$$P(X_k = 1) = \frac{r}{N}, \qquad k = 1, \ldots, n.$$

Thus, the probability of drawing a white ball on any draw, given no knowledge of the outcomes on previous draws, is r/N.

In a similar way, it can be shown that

$$P(X_j = 1, X_k = 1) = \frac{r(r-1)}{N(N-1)}, \qquad j \neq k.$$

Now observe that $Y = \sum_{i=1}^{n} X_i$, and hence,

$$E(Y) = \sum_{i=1}^{n} E(X_i) = n\left(\frac{r}{N}\right).$$

To find $V(Y)$ we need $V(X_i)$ and $\text{cov}(X_i, X_j)$. Because X_i is 1 with probability r/N and 0 with probability $1 - (r/N)$, it follows that

$$V(X_i) = \frac{r}{N}\left(1 - \frac{r}{N}\right).$$

Also,

$$\begin{aligned} \text{cov}(X_i, X_j) &= E(X_i X_j) - E(X_i)E(X_j) \\ &= \frac{r(r-1)}{N(N-1)} - \left(\frac{r}{N}\right)^2 \\ &= -\frac{r}{N}\left(1 - \frac{r}{N}\right)\frac{1}{N-1} \end{aligned}$$

because $X_iX_j = 1$ if and only if $X_i = 1$ and $X_j = 1$. From Theorem 5.2, we have that

$$\begin{aligned} V(Y) &= \sum_{i=1}^{n} V(X_i) + 2\sum_{i<j}\sum \operatorname{cov}(X_i, X_j), \\ &= n\frac{r}{N}\left(1 - \frac{r}{N}\right) + 2\sum_{i<j}\sum\left[-\frac{r}{N}\left(1 - \frac{r}{N}\right)\frac{1}{N-1}\right] \\ &= n\frac{r}{N}\left(1 - \frac{r}{N}\right) - n(n-1)\frac{r}{N}\left(1 - \frac{r}{N}\right)\frac{1}{N-1} \end{aligned}$$

because there are $n(n - 1)/2$ terms in the double summation. A little algebra yields

$$V(Y) = n\frac{r}{N}\left(1 - \frac{r}{N}\right)\frac{N-n}{N-1}.$$

An appreciation for the usefulness of Theorem 5.2 can be gained if one tries to find the expected value and variance for the hypergeometric random variable by proceeding directly from the definition of an expectation. The necessary summations are exceedingly difficult to obtain.

EXERCISES

5.17 Table 5.1 on page 204 shows the joint probability distribution of the number of defectives observed on the first draw (X_1) and the second draw (X_2) from a box containing four good bulbs and one defective bulb.

(a) Find $E(X_1)$, $V(X_1)$, $E(X_2)$, and $V(X_2)$.

(b) Find $\operatorname{cov}(X_1, X_2)$.

(c) The variable $Y = X_1 + X_2$ denotes the total number of defectives observed on the two draws. Find $E(Y)$ and $V(Y)$.

5.18 In a study of particulate pollution in air samples over a smokestack, X_1 represents the amount of pollutant per sample when a cleaning device is not operating and X_2 the amount when the cleaning device is operating. Assume that (X_1, X_2) has joint probability density function

$$f(x_1, x_2) = \begin{cases} 1 & 0 \le x_1 \le 2,\ 0 \le x_2 \le 1,\ 2x_2 \le x_1 \\ 0 & \text{elsewhere.} \end{cases}$$

The random variable $Y = X_1 - X_2$ represents the amount by which the weight of pollutant can be reduced using the cleaning device.

(a) Find $E(Y)$ and $V(Y)$.

(b) Find an interval in which values of Y should lie at least 75% of the time.

5.19 The proportions X_1 and X_2 of two chemicals found in samples of an insecticide have joint probability density function

$$f(x_1, x_2) = \begin{cases} 2 & 0 \le x_1 \le 1, 0 \le x_2 \le 1, 0 \le x_1 + x_2 \le 1 \\ 0 & \text{elsewhere.} \end{cases}$$

The random variable $Y = X_1 + X_2$ denotes the proportion of the insecticides due to the combination of both chemicals.

(a) Find $E(Y)$ and $V(Y)$.

(b) Find an interval in which values of Y should lie for at least 50 % of the samples of insecticide.

5.20 For a sheet-metal stamping machine in a certain factory, the time between failures, X_1, has a mean time between failure (MTBF) of 56 hours and a variance of 16. The repair time, X_2, has a mean time to repair (MTTR) of 5 hours and a variance of 4.

(a) If X_1 and X_2 are independent, find the expected value and variance of $Y = X_1 + X_2$, which represents one operation-repair cycle.

(b) Would you expect an operation-repair cycle to last more than 75 hours? Why?

5.21 A particular fast-food outlet is interested in the joint behavior of the random variables Y_1, defined as the total time between a customer's arrival at the store and leaving the service window, and Y_2, the time that the customer waits in line before reaching the service window. Because Y_1 contains the time a customer waits in line, we must have $Y_1 \ge Y_2$. The relative frequency distribution of observed values of Y_1 and Y_2 can be modeled by the probability density function

$$f(y_1, y_2) = \begin{cases} e^{-y_1} & 0 \le y_2 \le y_1 < \infty \\ 0 & \text{elsewhere.} \end{cases}$$

(a) Find $P(Y_1 < 2, Y_2 > 1)$.

(b) Find $P(Y_1 \ge 2Y_2)$.

(c) Find $P(Y_1 - Y_2 \ge 1)$. [Note: $Y_1 - Y_2$ denotes the time spent at the service window.]

(d) Find the marginal density functions for Y_1 and Y_2.

5.22 Referring to Exercise 5.21, if the total waiting time for service is know to be more than 2 minutes, find the probability that the customer waited less than 1 minute to be served.

5.23 Referring to Exercise 5.21, the random variable $Y_1 - Y_2$ represents the time spent at the service window.

(a) Find $E(Y_1 - Y_2)$.

(b) Find $V(Y_1 - Y_2)$.

(c) Is it highly likely that a customer would spend more than 2 minutes at the service window?

5.24 Referring to Exercise 5.21, suppose a customer spends a length of time y_1 at the store.

Find the probability that this customer spends less than half of that time at the service window.

5.25 Prove Theorem 5.1.

5.5

THE MULTINOMIAL DISTRIBUTION

Suppose that an experiment consists of n independent trials, much like the binomial case, but that each trial can result in any one of k possible outcomes. For example, a customer checking out of a grocery store may choose any one of k checkout counters. Now suppose the probability that a particular trial results in outcome i is denoted by p_i, $i = 1, \ldots, k$, and p_i remains constant from trial to trial. Let Y_i, $i = 1, \ldots, k$, denote the number of the n trials resulting in outcome i. In developing a formula for $P(Y_1 = y_1, \ldots, Y_k = y_k)$, we first call attention to the fact that, because of independence of trials, the probability of having y_1 outcomes of type 1 through y_k outcomes of type k in a particular order will be

$$p_1^{y_1} p_2^{y_2} \cdots p_k^{y_k}.$$

The number of such orderings is the number of ways to partition the n trials into y_1 type 1 outcomes, y_2 type 2 outcomes, and so on through y_k type k outcomes, or

$$\frac{n!}{y_1!\, y_2! \cdots y_k!}$$

where

$$\sum_{i=1}^{k} y_i = n.$$

Hence,

$$P(Y_1 = y_1, \ldots, Y_k = y_k) = \frac{n!}{y_1!\, y_2! \cdots y_k!} p_1^{y_1} p_2^{y_2} \cdots p_k^{y_k}.$$

This is called the *multinomial probability distribution*. Note that if $k = 2$, we are back into the binomial case, because Y_2 is then equal to $n - Y_1$. The following example illustrates the computations.

Example 5.13 Items under inspection are subject to two types of defects. About 70% of the items in a large lot are judged to be free of defects, whereas 20% have only a type A defect and 10% have only a type B defect (none has both types of defect). If six of these items are randomly selected from the lot, find the probability that three have no defects, one has a type A defect, and two have type B defects.

Solution

If we can assume that the outcomes are independent from trial to trial (item to item in our sample), which they usually would be in a large lot, then the multinomial distribution provides a useful model. Letting Y_1, Y_2, and Y_3 denote the number of trials resulting in zero, type A, and type B defectives, respectively, we have $p_1 = .7$, $p_2 = .2$, and $p_3 = .1$. It follows that

$$P(Y_1 = 3, Y_2 = 1, Y_3 = 2) = \frac{6!}{3!\,1!\,2!}(.7)^3(.2)(.1)^2$$

$$= .041.$$

Example 5.14 Find $E(Y_i)$ and $V(Y_i)$ for the multinomial probability distribution.

Solution

We are concerned with the marginal distribution of Y_i, the number of trials falling in cell i. Imagine all the cells, excluding cell i, combined into a single large cell. Hence, every trial will result in cell i or not in cell i with probabilities p_i and $1 - p_i$, respectively; and Y_i possesses a binomial marginal probability distribution. Consequently,

$$E(Y_i) = np_i,$$
$$V(Y_i) = np_iq_i$$

where

$$q_i = 1 - p_i.$$

Note: The same results can be obtained by setting up the expectations and evaluating. For example,

$$E(Y_1) = \sum_{y_1}\sum_{y_2}\cdots\sum_{y_k} y_1 \frac{n!}{y_1!\,y_2!\cdots y_k!}p_1^{y_1}p_2^{y_2}\cdots p_k^{y_k}.$$

Because we have already derived the expected value and variance of Y_i, we leave the tedious summation of this expectation to the interested reader.

Example 5.15 If $Y_1, \ldots, Y_k$ have the multinomial distribution given in Example 5.14, find $\mathrm{cov}(Y_s, Y_t)$, $s \neq t$.

Solution

Thinking of the multinomial experiment as a sequence of n independent trials, we define

$$U_i = \begin{cases} 1 & \text{if trial } i \text{ results in class } s \\ 0 & \text{otherwise} \end{cases}$$

and

$$W_i = \begin{cases} 1 & \text{if trial } i \text{ results in class } t \\ 0 & \text{otherwise.} \end{cases}$$

Then,

$$Y_s = \sum_{i=1}^{n} U_i$$

and

$$Y_t = \sum_{j=1}^{n} W_j.$$

To evaluate $\text{cov}(Y_s, Y_t)$ we need the following results:

$$\begin{aligned} E(U_i) &= p_s, \\ E(W_j) &= p_t, \\ \text{cov}(U_i, W_j) &= 0 \end{aligned}$$

if $i \neq j$ since the trials are independent, and

$$\begin{aligned} \text{cov}(U_i, W_i) &= E(U_i W_i) - E(U_i)E(W_i) \\ &= 0 - p_s p_t \end{aligned}$$

because $U_i W_i$ always equals zero. From Theorem 5.2, we then have

$$\begin{aligned} \text{cov}(Y_s, Y_t) &= \sum_{i=1}^{n} \sum_{j=1}^{n} \text{cov}(U_i, W_j) \\ &= \sum_{i=1}^{n} \text{cov}(U_i, W_i) + \sum_{i=j}\sum \text{cov}(U_i, W_j), \\ &= \sum_{i=1}^{n} (-p_s p_t) + 0 \\ &= -np_s p_t. \end{aligned}$$

Note that the covariance is negative, which is to be expected because a large number of outcomes in cell s would force the number in cell t to be small.

Multinomial Experiment

1. The experiment consists of n identical trials.
2. The outcome of each trial falls into one of k classes or cells.
3. The probability that the outcome of a single trial will fall in a particular cell, say cell i, is p_i $(i = 1, 2, \ldots, k)$, and remains the same from trial to trial. Note that $p_1 + p_2 + p_3 + \cdots + p_k = 1$.
4. The trials are independent.
5. The random variables of interest are $Y_1, Y_2, \ldots, Y_k$, where Y_i $(i = 1, 2, \ldots, k)$ is equal to the number of trials in which the outcome falls in cell i. Note that $Y_1 + Y_2 + Y_3 + \cdots + Y_k = n$.

The Multinomial Distribution

$$P(Y_1 = y_1, \ldots, Y_K = y_k) = \frac{n!}{y_1!\, y_2! \cdots y_k!} p_1^{y_1} p_2^{y_2} \cdots p_k^{y_k}$$

where $\sum_{i=1}^{k} y_i = n$ and $\sum_{i=1}^{k} p_i = 1$

$$E(Y_i) = np_i \qquad V(Y_i) = np_i(1 - p_i) \qquad i = 1, \ldots, k$$

$$\operatorname{cov}(Y_i, Y_j) = -np_ip_j \qquad i \neq j$$

EXERCISES

5.26 The National Fire Incident Reporting Service says that, among residential fires, approximately 73 % are in family homes, 20 % are in apartments, and the other 7 % are in other types of dwellings. If four fires are reported independently in one day, find the probability that two are in family homes, one is in an apartment, and one is in another type of dwelling.

5.27 The typical cost of damages for a fire in a family home is \$20,000, whereas the typical cost in an apartment fire is \$10,000, and in other dwellings only \$2000. Using the information in Exercise 5.26, find the expected total damage cost for the four independently reported fires.

5.28 In the inspection of commercial aircraft, wing cracks are reported as nonexistent, detectable, or critical. The history of a certain fleet shows that 70 % of the planes inspected have no wing cracks, 25 % have detectable wing cracks, and 5 % have critical wing cracks. For the next five planes inspected, find the probability that

(a) one has a critical crack, two have detectable cracks, and two have no cracks.

(b) at least one critical crack is observed.

5.29 With the recent emphasis on solar energy, solar radiation has been carefully monitored at various sites in Florida. For typical July days in Tampa, 30 % have total radiation of at most 5 calories, 60 % have total radiation of at most 6 calories, and 10 % have total radiation of at most 8 calories. A solar collector for a hot water system is to be run for 6 days. Find the probability that 3 days produce no more than 5 calories, 1 day produces between 5 and 6 calories, and 2 days produce between 6 and 8 calories. What assumptions do your make for your answer to be correct?

5.30 The U.S. Bureau of Labor Statistics reports that, as of 1981, approximately 21 % of the adult population under age 65 is between 18 and 24 years of age, 28 % is between 25 and 34, 19 % is between 35 and 44, and 32 % is between 45 and 64. An automobile manufacturer wants to obtain opinions on a new design from five randomly chosen adults from the group. Of the five so selected, find the approximate probability that two are between 18 and 24, two are between 25 and 44, and one is between 45 and 64.

5.31 Customers leaving a subway station can exit through any one of three gates. Assuming that any one customer is equally likely to select any one of the three gates, find the probability that, among four customers,

(a) two select gate A, one selects gate B, and one selects gate C.

(b) all four select the same gate.

(c) all three gates are used.

5.32 Among a large number of applicants for a certain position, 60 % have only a high school education, 30 % have some college training, and 10 % have completed a college degree. If five applicants are selected to be interviewed, find the probability that at least one will have completed a college degree. What assumptions are necessary for your answer to be valid?

5.33 In a large lot of manufactured items, 10 % contain exactly one defect, and 5 % contain more than one defect. Ten items are randomly selected from this lot for sale, and the repair costs total

$$Y_1 + 3Y_2$$

where Y_1 denotes the number among the ten having one defect, and Y_2 the number with two or more defects. Find the expected value of the repair costs. Find the variance of the repair costs.

5.34 Referring to Exercise 5.33, if Y denotes the number of items containing at least one defect, among the ten sampled items, find the probability that

(a) Y is exactly 2.

(b) Y is at least 1.

5.35 Vehicles arriving at an intersection can turn right or left or continue straight ahead. In a study of traffic patterns at this intersection over a long period of time, engineers have noted that 40 % of the vehicles turn left, 25 % turn right, and the remainder continue straight ahead.

(a) For the next five cars entering this intersection, find the probability that one turns left, one turns right, and three continue straight ahead.
(b) For the next five cars entering the intersection, find the probability that at least one turns right.
(c) If 100 cars enter the intersection in a day, find the expected values and variance of the number turning left. What assumptions are necessary for your answer to be valid?

5.6

MORE ON THE MOMENT-GENERATING FUNCTION

The use of moment-generating functions to identify distributions of random variables is particularly useful when working with sums of independent random variables. For example, suppose X_1 and X_2 are independent, exponential random variables, each with mean θ, and $Y = X_1 + X_2$. Now, the moment-generating function of Y is given by

$$\begin{aligned} M_Y(t) &= E(e^{tY}) = E(e^{t(X_1+X_2)}) \\ &= E(e^{tX_1} \times e^{tX_2}) \\ &= E(e^{tX_1})E(e^{tX_2}) \\ &= M_{X_1}(t)M_{X_2}(t), \end{aligned}$$

because X_1 and X_2 are independent. From Chapter 4,

$$M_{X_1}(t) = M_{X_2}(t) = (1 - \theta t)^{-1}$$

and so

$$M_Y(t) = (1 - \theta)^{-2}.$$

Upon recognizing the form of this moment-generating function, we can immediately conclude that Y has a gamma distribution with $\alpha = 2$ and $\beta = \theta$. This result can be generalized to the sum of independent gamma random variables with common scale parameter, β.

Example 5.16 Let X_1 denote the number of vehicles passing a particular point on the eastbound lane of a highway in 1 hour. Suppose the Poisson distribution with mean λ_1 is a reasonable model for X_1. Now, let X_2 denote the number of vehicles passing a point on the westbound lane of the same highway in 1 hour. Suppose X_2 has a Poisson distribution with mean λ_2. Of interest is $Y = X_1 + X_2$, the total traffic count in both lanes, per hour. Find the probability distribution for Y if X_1 and X_2 are assumed to be independent.

Solution

It is known from Chapter 4 that

$$M_{X_1}(t) = e^{\lambda_1(e^t - 1)}$$

and

$$M_{X_2}(t) = e^{\lambda_2(e^t - 1)}$$

By the property of moment-generating functions seen above,

$$\begin{aligned} M_Y(t) &= M_{X_1}(t)M_{X_2}(t) \\ &= e^{\lambda_1(e^t-1)}e^{\lambda_2(e^t-1)} \\ &= e^{(\lambda_1+\lambda_2)(e^t-1)} \end{aligned}$$

Now the moment-generating function for Y has the form of a Poisson moment-generating function, with the mean $(\lambda_1 + \lambda_2)$. Thus, by the uniqueness property, Y must have a Poisson distribution with the mean $\lambda_1 + \lambda_2$. The ability to add independent Poisson random variables and still retain the Poisson properties is important in many applications.

EXERCISES

5.36 Find the moment-generating function for the negative binomial random variable. Use it to derive the mean and variance of that distribution.

5.37 There are two entrances to a parking lot. Cars arrive at entrance I according to a Poisson distribution with an average of three per hour, and at entrance II according to a Poisson distribution with an average of four per hour. Find the probability that exactly three cars arrive at the parking lot in a given hour.

5.38 Let X_1 and X_2 denote independent, normally distributed random variables, not necessarily having the same mean or variance. Show that, for any constants a and b, $Y = aX_1 + bX_2$ is normally distributed.

5.39 Resistors of a certain type have resistances that are normally distributed with a mean of 100 ohms and a standard deviation of 10 ohms. Two such resistors are connected in series, which causes the total resistance in the circuit to be the sum of the individual resistances. Find the probability that the total resistance

(a) exceeds 220 ohms.
(b) is less than 190 ohms.

5.40 A certain type of elevator has a maximum weight capacity X_1, which is normally distributed with a mean and standard deviation of 5000 and 300 pounds, respectively. For a certain building equipped with this type of elevator, the elevator loading, X_2, is a normally distributed random variable with a mean and standard deviation of 4000 and 400 pounds, respectively. For any given time that the elevator is in use, find the probability that it will be overloaded, assuming X_1 and X_2 are independent.

5.7

CONDITIONAL EXPECTATIONS

Section 5.2 contains a discussion of conditional probability functions and conditional density functions, which we shall now relate to conditional expectations. Conditional expectations are defined in the same manner as univariate expectations except that the conditional density is used in place of the marginal density function.

Definition 5.7

If X_1 and X_2 are any two random variables, the **conditional expectation** of X_1 given that $X_2 = x_2$ is defined to be

$$E(X_1 \mid X_2 = x_2) = \int_{-\infty}^{\infty} x_1 f(x_1 \mid x_2)\, dx_1$$

if X_1 and X_2 are jointly continuous and

$$E(X_1 \mid X_2 = x_2) = \sum_{y_1} x_1 p(x_1 \mid x_2)$$

if X_1 and X_2 are jointly discrete.

Example 5.17 Refer to Example 5.6, pages 213–214, with X_1 denoting amount of drink sold and X_2 denoting amount in supply, and

$$f(x_1, x_2) = \begin{cases} 1/2 & 0 \le x_1 \le x_2,\ 0 \le x_2 \le 2 \\ 0 & \text{elsewhere.} \end{cases}$$

Find the conditional expectation of amount of sales, X_1, given that $X_2 = 1$.

Solution

In Example 5.6, we found that

$$f(x_1 \mid x_2) = \begin{cases} 1/x_2 & 0 \le x_1 \le x_2 \le 2 \\ 0 & \text{elsewhere.} \end{cases}$$

Thus, from Definition 5.7,

$$\begin{aligned} E(X_1 \mid X_2 = 1) &= \int_{-\infty}^{\infty} x_1 f(x_1 \mid x_2)\, dx_1 \\ &= \int_0^1 x_1(1)\, dx_1 \\ &= \frac{x_1^2}{2}\Big|_0^1 = \frac{1}{2}. \end{aligned}$$

That is, if the soft-drink machine contains 1 gallon at the start of the day, the expected sales for that day is 1/2 gallon.

The conditional expectation of X_1 given $X_2 = x_2$ is a function of x_2. If we now let X_2 range over all its possible values, we can think of the conditional expectation as a function of the random variable X_2, and hence, we can find the expected value of the conditional expectation. The result of this type of iterated expectation is given in Theorem 5.3.

Theorem 5.3 Let X_1 and X_2 denote random variables. Then

$$E(X_1) = E[E(X_1 \mid X_2)],$$

where, on the right-hand side, the inside expectation is with respect to the conditional distribution of X_1 given X_2, and the outside expectation is with respect to the distribution of X_2.

Proof

Let X_1 and X_2 have joint density function $f(x_1, x_2)$ and marginal densities $f_1(x_1)$ and $f_2(x_2)$, respectively. Then

$$\begin{aligned} E(X_1) &= \int_{-\infty}^{\infty} x_1 f_1(x_2)\, dx \\ &= \int_{-\infty}^{\infty}\int_{-\infty}^{\infty} x_1 f(x_1, x_2)\, dx_1,\, dx_2 \\ &= \int_{-\infty}^{\infty}\int_{-\infty}^{\infty} x_1 f_1(x_1 \mid x_2) f_2(x_2)\, dx_1\, dx_2 \\ &= \int_{-\infty}^{\infty}\left[\int_{-\infty}^{\infty} x_1 f_1(x_1 \mid x_2)\, dx_1\right] f_2(x_2)\, dx_2 \\ &= \int_{-\infty}^{\infty} E(X_1 \mid X_2 = x_2) f_2(x_2)\, dx_2 \\ &= E[E(X_1 \mid X_2)]. \end{aligned}$$

The proof is similar for the discrete case. ∎

Example 5.18 A quality-control plan for an assembly line involves sampling $n = 10$ finished items per day and counting Y, the number of defectives. If p denotes the probability of observing a defective, then Y has a binomial distribution, when the number of items produced by the line is large. However, p varies from day to day and is assumed to have a uniform distribution on the interval 0 to 1/4. Find the expected value of Y for any given day.

Solution

From Theorem 5.3, we know that

$$E(Y) = E[E(Y \mid p)].$$

For a given p, Y has binomial distribution, and hence,

$$E(Y \mid p) = np.$$

Thus,

$$\begin{aligned} E(Y) &= E(np) = nE(p) \\ &= n\int_0^{1/4} 4p \, dp \\ &= n\left(\frac{1}{8}\right) \end{aligned}$$

and, for $n = 10$,

$$E(Y) = \frac{10}{8} = \frac{5}{4}.$$

This inspection policy should average 5/4 defective per day, in the long run. The calculations could be checked by actually finding the unconditional distribution of Y and computing $E(Y)$ directly.

5.8

COMPOUNDING AND ITS APPLICATIONS

The univariate probability distributions of Chapters 3 and 4 depend on one or more parameters; once the parameters are known, the distributions are specified completely. However, these parameters frequently are unknown and, as in Example 5.18, sometimes may be regarded as random quantities. Assigning distributions to these parameters and then finding the marginal distribution of the original random variable is known as *compounding*. This process has theoretical as well as practical uses, as we illustrate next.

Example 5.19 Suppose that Y denotes the number of bacteria per cubic centimeter in a certain liquid and that, for a given location, Y has a Poisson distribution with mean λ. Also, assume λ varies from location to location and, for a location chosen at random, λ has a gamma distribution with parameters α and β, where α is a positive integer. Find the probability distribution for the bacteria count, Y, at a randomly selected location.

Solution

Because λ is random, the Poisson assumption applies to the conditional distribution of Y for fixed λ. Thus,

$$p(y \mid \lambda) = \frac{\lambda^y e^{-\lambda}}{y!}, \qquad y = 0, 1, 2, \ldots.$$

Also,

$$\begin{aligned} f(\lambda) &= \frac{1}{\Gamma(\alpha)\beta^\alpha}\lambda^{\alpha-1}e^{-\lambda/\beta} && \lambda > 0 \\ &= 0 && \text{elsewhere.} \end{aligned}$$

Then the joint distribution of λ and Y is given by

$$\begin{aligned} g(y, \lambda) &= p(y \mid \lambda)f(\lambda) \\ &= \frac{1}{y!\,\Gamma(\alpha)\beta^\alpha}\lambda^{y+\alpha-1}e^{-\lambda[1+(1/\beta)]}. \end{aligned}$$

The marginal distribution of Y is found by integrating over λ and yields

$$\begin{aligned} p(y) &= \frac{1}{y!\,\Gamma(\alpha)\beta^\alpha}\int_0^\infty \lambda^{y+\alpha-1}e^{\lambda[1+(1/\beta)]}\,dx \\ &= \frac{1}{y!\,\Gamma(\alpha)\beta^\alpha}\Gamma(y+\alpha)\left(1+\frac{1}{\beta}\right)^{-(y+\alpha)}. \end{aligned}$$

Since α is an integer,

$$\begin{aligned} p(y) &= \frac{(y+\alpha-1)!}{(\alpha-1)!\,y!}\left(\frac{1}{\beta}\right)^\alpha\left(\frac{\beta}{1+\beta}\right)^{y+\alpha} \\ &= \binom{y+\alpha-1}{\alpha-1}\left(\frac{1}{1+\beta}\right)^\alpha\left(\frac{\beta}{1+\beta}\right)^y. \end{aligned}$$

If we let $y + \alpha = n$ and $1/(1+\beta) = p$, $p(y)$ has the form of a negative binomial distribution. Hence, the negative binomial distribution is a reasonable model for counts in which the mean count may be random.

Example 5.20 Suppose that a customer arrives at a checkout counter in a store just as the counter is opening. A random number of customers, N, will be ahead of him since some customers may arrive early. Suppose that this number has the probability distribution

$$p(n) = P(N = n) = pq^n, \qquad n = 0, 1, 2, \ldots$$

where $0 < p < 1$ and $q = 1 - p$ (this is a form of the geometric distribution).
Customer service times are assumed to be independent and identically distributed exponential random variables with a mean of θ. Find the expected waiting time for this customer to complete his checkout.

Solution

For a given value of n, the waiting time, W, is the sum of $n + 1$ independent exponential random variables and, thus, has a gamma distribution with $\alpha = n + 1$ and $\beta = \theta$. That is,

$$f(w \mid n) = \frac{1}{\Gamma(n + 1)\theta^{n+1}} w^n e^{-w/\theta}.$$

Hence,

$$f(w, n) = \frac{p}{\Gamma(n + 1)\theta^{n+1}} (qw)^n e^{-w/\theta}$$

and

$$\begin{aligned} f(w) &= \frac{p}{\theta} e^{-w/\theta} \sum_{n=0}^{\infty} \left(\frac{qw}{\theta}\right)^n \frac{1}{n!} \\ &= \frac{p}{\theta} e^{-w/\theta} e^{qw/\theta} \\ &= \frac{p}{\theta} e^{-(w/\theta)(1-q)} \\ &= \frac{p}{\theta} e^{-w(p/\theta)}. \end{aligned}$$

The waiting time, W, is still exponential, but its expected value is (θ/p).

5.9

BIRTH AND DEATH PROCESSES: BIOLOGICAL APPLICATIONS

In Section 3.11, we were interested in counting the number of occurrences of a single type of event, such as accidents, defects, or plants. Let us extend this idea to counts of two types of events, which we shall call *births* and *deaths*. For illustrative purposes, a birth may denote a literal birth in a population of organisms, and similarly for a death. More specifically, we shall think of a birth as a cell division that produces two new cells and of a death as the removal of a cell from the system.

Birth and death processes are of use in modeling the dynamics of population growth. In human populations such models may be used to influence decisions on housing programs, food management, natural resource management, and a host of related economic matters. In animal populations, birth and death models are used to predict the size of insect populations and to measure the effectiveness of nutrition or eradication programs.

Let λ denote the birth rate and θ the death rate of each cell in the population, and assume that the probability of birth or death for an individual cell is independent of the size and age of the population. If $Y(t)$ denotes the size of the population at time t, we write

$$P[Y(t) = n] = P_n(t).$$

More specifically, assume that the probability of a birth in a small interval of time, h, given the population size at the start of the interval is n, has the form $n\lambda h + o(h)$. Similarly, the probability of a death is given by $n\theta h + o(h)$. The probability of more than one birth or death in h is on order $o(h)$. For an individual cell, the case $n = 1$, this says that the probability of division in a small interval of time is $\lambda h + o(h)$.

A differential equation for $P_n(t)$ is developed as follows. If a population is of size n at time $(t + h)$, it must have been of size $(n - 1)$, n, or $(n + 1)$ at time t. Thus,

$$\begin{aligned} P_n(t + h) = \lambda(n - 1)hP_{n-1}(t) &+ [1 - n\lambda h - n\theta h]P_n(t) \\ &+ \theta(n + 1)hP_{n+1}(t) + o(h) \end{aligned}$$

or

$$\begin{aligned} \frac{1}{h}[P_n(t + h) - P_n(t)] &= \lambda(n - 1)P_{n-1}(t) - n(\lambda + \theta)P_n(t) \\ &\quad + \theta(n - 1)P_{n+1}(t) + \frac{o(h)}{h}. \end{aligned}$$

On taking limits as $h \to 0$, we have

$$\frac{dP_n(t)}{dt} = \lambda(n - 1)P_{n-1}(t) - n(\lambda + \theta)P_n(t) + \theta(n + 1)P_{n+1}(t).$$

A general solution to this equation can be obtained with some difficulty, but let us instead look at some special cases. If $\theta = 0$, which implies that no deaths are taking place in the time of interest, we have a pure birth process exemplified by the differential equation

$$\frac{dP_n(t)}{dt} = \lambda(n - 1)P_{n-1}(t) - n\lambda P_n(t).$$

A solution to this equation is

$$P_n(t) = \binom{n-1}{i-1} e^{-\lambda it}(1 - e^{-\lambda t})^{n-i}$$

where i is the size of the population at time $t = 0$; that is, $P_i(0) = 1$. Note that this solution is a negative binomial probability distribution. The pure birth process may be a reasonable model of a real population if the time interval is short or if deaths are neglected.

Example 5.21 Find the expected size of a population, governed by a pure birth process, at time t if the size was i at time 0.

Solution

It is known that the mean of a negative binomial distribution written as

$$\binom{n-1}{r-1} p^r (1 - p)^{n-r}$$

is given by r/p. Thus, the mean of the pure birth process is $i/e^{-\lambda t}$ or $ie^{\lambda t}$. If the birth rate is known, at least approximately, this provides a formula for estimating what the expected population size will be t time units in the future.

Returning to the case $\theta > 0$, we can show that

$$P_0(t) = \left[\frac{\theta e^{(\lambda-\theta)t} - \theta}{\lambda e^{(\lambda-\theta)t} - \theta}\right]^i$$

(see Baily, 1964).

On taking the limit of $P_0(t)$ as $t \to \infty$, we get the probability of ultimate extinction of the population, which turns out to be

$$\lim_{t\to\infty} P_0(t) = \begin{cases} (\theta/\lambda)^i & \lambda > \theta \\ 1 & \lambda < \theta \\ 1 & \lambda = \theta. \end{cases}$$

Thus, the population has a chance of persisting indefinitely only if the birth rate, λ, is larger than the death rate, θ; otherwise, it is certain to become extinct.

We can find $E[Y(t)]$ for the birth and death process without first finding the probability distribution, $P_n(t)$. The method is outlined in Exercise 5.72, and the solution is

$$E[Y(t)] = ie^{(\lambda - \theta)t}.$$

5.10

QUEUES: ENGINEERING APPLICATIONS

Queueing theory is concerned with probabilistic models governing the behavior of customers arriving at a certain station and demanding some kind of service. The word *customer* may refer to actual customers at a service counter or, more generally, it may refer to automobiles entering a highway or a service station, telephone calls coming into a switchboard, breakdowns of machines in a factory, or any similar phenomenon. Queues are classified according to an input distribution (the distribution of customer arrivals), a service distribution (the distribution of service time per customer), and a queue discipline, which specifies the number of servers and the manner of dispensing service (such as "first come, first served").

Queueing theory forms useful probabilistic models for a wide range of practical problems, including the design of stores and public buildings, the optimal arrangement of workers in a factory, and the planning of highway systems. Consider a system that involves one station dispensing service to customers on a first-come–first-served basis. This could be a store with a single checkout counter or a service station with a single gasoline pump. Suppose that customer arrivals constitute a Poisson process with intensity λ per hour and departures of customers from the station constitute an independent Poisson process with intensity θ per hour. (This implies that the service time is an exponential random variable with mean $1/\theta$.)

Hence, the probability of a customer arrival in a small interval of time, h, is $\lambda h + o(h)$, and the probability of a departure is $\theta h + o(h)$, with the probability of more than one arrival (or departure) being $o(h)$. Note that these probabilities are identical to those for births and deaths, Section 5.9, with the dependence on n suppressed. Thus, if $Y(t)$ denotes the number of customers in the system (being served and waiting to be served) at time t and if $P_n(t) = P[Y(t) = n]$, then, as in Section 5.9.

$$\frac{dP_n(t)}{dt} = \lambda P_{n-1}(t) - (\lambda + \theta)P_n(t) + \theta P_{n+1}(t).$$

It can be shown that for a large t this equation has a solution, P_n, which does not depend on t. Such a solution is called an *equilibrium distribution*. If the solution is free of t, it must satisfy

$$0 = \lambda P_{n-1} - (\lambda + \theta)P_n + \theta P_{n+1}.$$

A solution to this equation is given by

$$P_n = \left(1 - \frac{\lambda}{\theta}\right)\left(\frac{\lambda}{\theta}\right)^n, \qquad n = 0, 1, 2, \ldots,$$

provided that $\lambda < \theta$. P_n provides the probability of there being n customers in the system at any time, t, which is far removed from the start of the system.

Example 5.22 In a system operating as just indicated, find the expected number of customers, including the one being served, at the station at some time far removed from the start.

Solution

The equilibrium distribution is a version of the geometric distribution given in Chapter 3. If Y has the distribution given by P_n, then $X = Y + 1$ will have the distribution given by

$$P(X = m) = P(Y = m - 1) = \left(1 - \frac{\lambda}{\theta}\right)\left(\frac{\lambda}{\theta}\right)^{m-1}, \qquad m = 1, 2, \ldots.$$

From Chapter 3,

$$E(X) = 1\Big/\left(1 - \frac{\lambda}{\theta}\right) = \frac{\theta}{\theta - \lambda}.$$

Hence,

$$\begin{aligned} E(Y) &= E(X) - 1 \\ &= \frac{\theta}{\theta - \lambda} - 1 = \frac{\lambda}{\theta - \lambda} \end{aligned}$$

again assuming that $\lambda < \theta$.

SUPPLEMENTARY EXERCISES

5.41 Let X_1 and X_2 have the joint probability density function given by

$$f(x_1, x_2) = \begin{cases} Kx_1, x_2 & 0 \le x_1 \le 1, 0 \le x_2 \le 1 \\ 0 & \text{elsewhere.} \end{cases}$$

(a) Find the value of K that makes this a probability density function.
(b) Find the marginal densities of X_1 and X_2.
(c) Find the joint distributin function for X_1 and X_2.
(d) Find the probability $P(X_1 < 1/2, X_2 < 3/4)$.
(e) Find the probability $P(X_1 \le 1/2 \mid X_2 > 3/4)$.

5.42 Let X_1 and X_2 have the joint density function given by

$$f(x_1, x_2) = \begin{cases} 3x_1 & 0 \le x_2 \le x_1 \le 1 \\ 0 & \text{elsewhere.} \end{cases}$$

(a) Find the marginal density functions of X_1 and X_2.
(b) Find $P(X_1 \le 3/4, X_2 \le 1/2)$.
(c) Find $P(X_1 \le 1/2 \mid X_2 \ge 3/4)$.

5.43 A committee of three persons is to be randomly selected from a group consisting of four Republicans, three Democrats, and two Independents. Let X_1 denote the number of Republicans and X_2 the number of Democrats on the committee.

(a) Find the joint probability distribution of X_1 and X_2.
(b) Find the marginal distributions of X_1 and X_2.
(c) Find the probability $P(X_1 = \;\mid X_2 \ge 1)$.

5.44 For Exercise 5.41, find the conditional density of X_1 given $X_2 = x_2$. Are X_1 and X_2 independent?

5.45 For Exercise 5.42,

(a) find the conditional density of X_1 given $X_2 = x_2$.
(b) find the conditional density of X_2 given $X_1 = x_1$.
(c) show that X_1 and X_2 are dependent.
(d) find the probability $P(X_1 \le 3/4 \mid X_2 = 1/2)$.

5.46 Let x_1 denote the amount of a certain bulk item stocked by a supplier at the beginning of a week and suppose that X_1 has a uniform distribution over the interval $0 \le x_1 \le 1$. Let X_2 denote the amount of this item sold by the supplier during the week and suppose that X_2 has a uniform distribution over the interval $0 \le x_2 \le x_1$, where x_1 is a specific value of X_1.

(a) Find the joint density function of x_1 and x_2.
(b) If the supplier stocks an amount of 1/2, what is the probability that she sells an amount greater than 1/4?
(c) If it is known that the supplier sold an amount equal to 1/4, what is the probability that she had stocked an amount greater than 1/2?

5.47 Let (X_1, X_2) denote the coordinates of a point at random inside a unit circle with center at the origin. That is, X_1 and X_2 have joint density function given by

$$f(x_1, x_2) = \begin{cases} 1/\pi & x_1^2 + x_2^2 \le 1 \\ 0 & \text{elsewhere.} \end{cases}$$

(a) Find the marginal density function of X_1.
(b) Find $P(X_1 \le X_2)$.

5.48 Let X_1 and X_2 have the joint density function given by

$$f(x_1, x_2) = \begin{cases} x_1 + x_2 & 0 \le x_1 \le 1, 0 \le x_2 \le 1 \\ 0 & \text{elsewhere.} \end{cases}$$

(a) Find the marginal density functions of X_1 and X_2.

(b) Are X_1 and X_2 independent?
(c) Find the conditional density of X_1 given $X_2 = x_2$.

5.49 Let X_1 and X_2 have the joint density function given by

$$f(x_1, x_2) = \begin{cases} K & 0 \le x_1 \le 2,\ 0 \le x_2 \le 1,\ 2x_2 \le x_1 \\ 0 & \text{elsewhere.} \end{cases}$$

(a) Find the value of K that makes the function a probability density.
(b) Find the marginal densities of X_1 and X_2.
(c) Find the conditional density of X_1 given $X_2 = x_2$.
(d) Find the conditional density of X_2 given $X_1 = x_1$.
(e) Find $P(X_1 \le 1.5, X_2 \le .5)$.
(f) Find $P(X_2 \le .5 \mid X_1 \le 1.5)$.

5.50 Let X_1 and X_2 have a joint distribution that is uniform over the region shaded in the following diagram.
(a) Find the marginal density for X_2.
(b) Find the marginal density for X_1.
(c) Find $P[(X_1 - X_2) \ge 0]$.

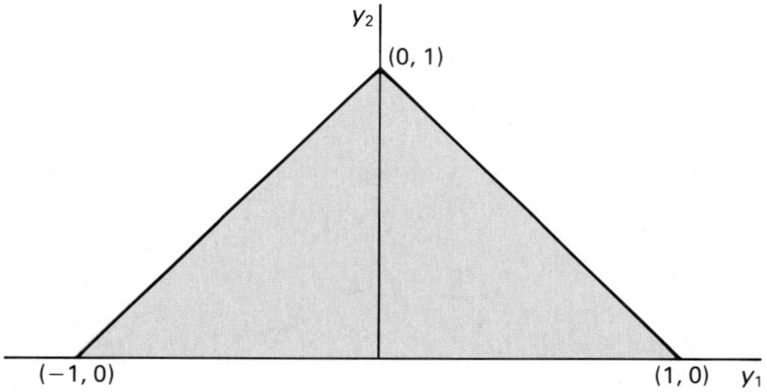

5.51 Referring to Exercise 5.41,
(a) find $E(X_1)$.
(b) find $V(X_1)$.
(c) find $\text{cov}(X_1, X_2)$.

5.52 Referring to Exercise 5.42, find $\text{cov}(X_1, X_2)$.

5.53 Referring to Exercise 5.43,
(a) find $\text{cov}(X_1, X_2)$.
(b) find $E(X_1 + X_2)$ and $V(X_1 + X_2)$ by finding the probability distribution of $X_1 + X_2$.
(c) find $E(X_1 + X_2)$ and $V(X_1 + X_2)$ by using Theorem 5.2.

5.54 Referring to Exercise 5.48,

(a) find $\text{cov}(X_1, X_2)$.
(b) find $E(3X_1 - 2X_2)$.
(c) find $V(3X_1 - 2X_2)$.

5.55 Referring to Exercise 5.49,

(a) find $E(X_1 + 2X_2)$.
(b) find $V(X_1 + 2X_2)$.

5.56 A quality-control plan calls for randomly selecting three items from the daily production (assumed large) of a certain machine and observing the number of defectives. However, the proportion, p, of defectives produced by the machine varies from day to day and is assumed to have a uniform distribution on the interval (0, 1). For a randomly chosen day, find the unconditional probability that exactly two defectives are observed in the sample.

5.57 The number of defects per yard, denoted by X, for a certain fabric is known to have a Poisson distribution with parameter λ. However, λ is not known and is assumed to be random with its probability density function given by

$$f(\lambda) = \begin{cases} e^{-\lambda} & \lambda \geq 0 \\ 0 & \text{elsewhere.} \end{cases}$$

Find the unconditional probability function for X.

5.58 The length of life, X, of a fuse has a probability density,

$$f(x) = \begin{cases} e^{-\lambda/\theta}/\theta & x > 0,\ \theta > 0 \\ 0 & \text{elsewhere.} \end{cases}$$

Three such fuses operate independently. Find the joint density of their lengths of life, X_1, X_2, and X_3.

5.59 A retail grocer figures that his daily gain from sales, X, is a normally distributed random variable with $\mu = 50$ and $\sigma^2 = 10$ (measurements in dollars). X could be negative if he is forced to dispose of perishable goods. Also, he figures daily overhead costs, Y, to have a gamma distribution with $\alpha = 4$ and $\beta = 2$. If X and Y are independent, find the expected value and variance of his net daily gain. Would you expect his net gain for tomorrow to go above \$70?

5.60 A coin has probability p of coming up heads when tossed. In n independent tosses of the coin, let $X_i = 1$ if the ith toss results in heads and $X_i = 0$ if the ith toss results in tails. Then Y, the number of heads in the n tosses, has a binomial distribution and can be represented as

$$Y = \sum_{i=1}^{n} X_i.$$

Find $E(Y)$ and $V(Y)$ using Theorem 5.2.

5.61 Referring to Exercises 5.42 and 5.45,

(a) find $E(X_2 \mid X_1 = x_1)$.

(b) use Theorem 5.3 to find $E(X_2)$.

(c) find $E(X_2)$ directly from the marginal density of X_2.

5.62 Referring to Exercise 5.57,

(a) find $E(X)$ by first finding the conditional expectation of X for given λ and then using Theorem 5.3.

(b) find $E(X)$ directly from the probability distribution of X.

5.63 Referring to Exercise 5.46, if the supplier stocks an amount equal to 3/4, what is the expected amount sold during the week?

5.64 Let X be a continuous random variable with distribution function $F(x)$ and density function $f(x)$. We can then write, for $x_1 \le x_2$,

$$P(X \le x_2 \mid X \ge x_1) = \frac{F(x_2) - F(x_1)}{1 - F(x_1)}.$$

As a function of x_2 for fixed x_1, the right-hand side of this expression is called the *conditional distribution function* of X given that $X \ge x_1$. On taking the derivative with respect to x_2, we see that the corresponding conditional density function is given by

$$\frac{f(x_2)}{1 - F(x_1)}, \qquad x_2 \ge x_1.$$

Suppose that a certain type of electronic component has life length, X, with the density function (life length measured in hours)

$$f(x) = \begin{cases} (1/200)e^{-x/200} & x \ge 0 \\ 0 & \text{elsewhere.} \end{cases}$$

Find the expected length of life for a component of this type that has already been in use for 100 hours.

5.65 Let X_1, X_2, and X_3 be random variables, either continuous or discrete. The joint moment-generating function of X_1, X_2, and X_3 is defined by

$$M(t_1, t_2, t_3) = E(e^{t_1X_1 + t_2X_2 + t_3X_3}).$$

(a) Show that $M(t, t, t)$ gives the moment-generating function of $X_1 + X_2 + X_3$.

(b) Show that $M(t, t, 0)$ gives the moment-generating function of $X_1 + X_2$.

(c) Show that

$$\left.\frac{\partial^{k_1+k_2+k_3} M(t_1, t_2, t_3)}{\partial t_1^{k_1} \partial t_2^{k_2} \partial t_3^{k_3}}\right|_{t_1=t_2=t_3=0} = E(X_1^{k_1} X_2^{k_2} X_3^{k_3}).$$

5.66 Let X_1, X_2, and X_3 have multinomial distribution with probability function

$$p(x_1, x_2, x_3) = \frac{n!}{x_1!\, x_2!\, x_3!} p_1^{x_1} p_2^{x_2} p_3^{x_3}, \qquad \sum_{i=1}^{n} x_i = n.$$

Employ the results of Exercise 5.65 to answer the following.

(a) Find the joint moment-generating function of X_1, X_2, and X_3.

(b) Use the joint moment-generating function to find $\text{cov}(X_1, X_2)$.

5.67 The negative binomial variable, X, is defined as the number of the trial on which the

rth success occurs in a sequence of independent trials with constant probability, p, of success on each trial. Let X_i denote a geometric random variable, defined as the number of the trial on which the first success occurs. Then, we can write

$$X = \sum_{i=1}^{n} X_i$$

for independently random variables $X_1, \ldots, X_r$. Use Theorem 5.2 to show that $E(X) = r/p$ and $V(X) = r(1 - p)/p^2$.

5.68 A box contains four balls, numbered 1 through 4. One ball is selected at random from this box. Let

$$X_1 = 1 \text{ if ball 1 or ball 2 is drawn,}$$
$$X_2 = 1 \text{ if ball 1 or ball 3 is drawn,}$$
$$X_3 = 1 \text{ if ball 1 or ball 4 is drawn,}$$

and the values of X_i are zero otherwise. Show that any two of the random variables X_1, X_2, and X_3 are independent, but the three together are not.

5.69 Let X_1 and X_2 be jointly distributed random variables with finite variances.

(a) Show that $[E(X_1 X_2)]^2 \le E(X_1^2)E(X_2^2)$. Hint: Observe that, for any real number t, $E[(tX_1 - X_2)^2] \ge 0$ or, equivalently,

$$t^2E(X_1^2) - 2t(X_1 X_2) + E(X_2^2) \ge 0.$$

This is a quadratic expression of the form $At^2 + Bt + C$, and because it is not negative, we must have $B^2 - 4AC \le 0$. The preceding inequality follows directly.

(b) Let ρ denote the correlation coefficient of X_1 and X_2. That is, $\rho = \dfrac{\text{cov}(X_1, X_2)}{\sqrt{V(X_1)V(X_2)}}$. Using the inequality of part (a), show that $\rho^2 \le 1$.

5.70 A box contains N_1 white balls, N_2 black balls, and N_3 red balls ($N_1 + N_2 + N_3 = N$). A random sample of n balls is selected from the box without replacement. Let X_1, X_2, and X_3 denote the number of white, black, and red balls, respectively, observed in the sample. Find the correlation coefficient for X_1 and X_2. (Let $p_i = N_i/N$, $i = 1, 2, 3$.)

5.71 Let $X_1, X_2, \ldots, X_n$ be independent random variables with $E(X_i) = \mu$ and $V(X_i) = \sigma^2$, $i = 1, \ldots, n$. Let

$$U_1 = \sum_{i=1}^{n} a_i X_i$$

and

$$U_2 = \sum_{i=1}^{n} b_i X_i,$$

where $a_1, \ldots, a_n$, $b_1, \ldots, b_n$ are constants. U_1 and U_2 are said to be orthogonal if $\text{cov}(U_1, U_2) = 0$. Show that U_1 and U_2 are orthogonal if and only if $\sum_{i=1}^{n} a_i b_i = 0$.

5.72 For the birth and death process of Section 5.9, $E[Y(t)] = ie^{(\lambda - \theta)t}$, where i is the population size at $t = 0$. Show this by observing that

$$m(t) = E[Y(t)] = \sum_{n=0}^{\infty} nP_n(t),$$

and

$$m'(t) = \frac{dm(t)}{dt} = \sum_{n=0}^{\infty} nP_n'(t).$$

Now, use the expression for $P_n'(t)$ given in Section 5.9 and evaluate the sum to obtain a differential equation relating $m'(t)$ to $m(t)$. The solution follows.

5.73 Referring to Section 5.10, show that the geometric equilibrium distribution, P_n, is a solution to the differential equation defining the equilibrium state.

5.74 In a single-server queue as defined in Section 5.10, let W denote the total waiting time, including her own service time, of a customer entering the queue a long time after the start of operations. (Assume that $\lambda < \theta$.) Show that W has an exponential distribution by writing

$$P[W \le w] = \sum_{n=0}^{\infty} P[W \le w \mid n \text{ customers in queue}]P_n$$

$$= \sum_{n=0}^{\infty} P[W \le w \mid n \text{ customers in queue}]\left(\frac{\lambda}{\theta}\right)^n\left(1 - \frac{\lambda}{\theta}\right)$$

and observing that the conditional distribution of W for fixed n is a gamma distribution with $\alpha = n + 1$ and $\beta = 1/\theta$. (The interchange of Σ and $\int$ is permitted here.)

5.75 The life length, X, of fuses of a certain type is modeled by the exponential distribution with

$$f(x) = \begin{cases} (1/3)e^{-x/3} & x > 0 \\ 0 & \text{elsewhere.} \end{cases}$$

The measurements are in hundreds of hours.

(a) If two such fuses have independent life lengths X_1 and X_2, find their joint probability density function.

(b) One fuse in (a) is in a primary system and the other is in a backup system, which comes into use only if the primary system fails. The total effective life length of the two fuses is then $X_1 + X_2$. Find $P(X_1 + X_2 \le 1)$.

5.76 Referring to Exercise 5.75, suppose three such fuses are operating independently in a system.

(a) Find the probability that exactly two of the three last longer than 500 hours.

(b) Find the probability that at least one of the three fails before 500 hours.

CHAPTER 6

FUNCTIONS OF RANDOM VARIABLES

6.1 INTRODUCTION

As we saw in Chapter 5, many situations we wish to study produce a set of random variables, $X_1, \ldots, X_n$, instead of a single random variable. Questions on the average life of components, the maximum price of a stock in a quarter, the time between two incoming telephone calls, or the total production costs across the plants in a manufacturing firm all involve the study of functions of random variables. This chapter considers the problem of finding the probability density function for a function of random variables with known probability distributions.

We have already seen some results along these lines in Chapter 5. Moment-generating functions were used to show that sums of exponential random variables have gamma distributions, and that linear functions of independent normal random variables are again normal. However, those were special cases, and now more general methods for finding distributions of functions of random variables will be introduced.

If X_1 and X_2 represent two different features of the phenomenon under study, it is sometimes convenient (and even necessary) to look at one random variable as a function of the other. For example, if the probability distribution for the velocity of a molecule in a uniform gas is known, the distribution of the kinetic energy can be found (see Exercise 6.40). Or, if the distribution of the radius of a sphere is known, the distribution of the volume can be found (see Exercise 6.39). Similarly, knowledge of the probability distribution for points in a plane allows us to find, in some cases, the distribution of the distance from a selected point to its nearest neighbor. Knowledge of the distribution of life lengths of components in a complex system allows us to find the probability distribution for the life length of the system as a whole. Examples are endless, but those mentioned should adequately illustrate the point that we are embarking on the study of a very important area of applied probability.

6.2

METHOD OF DISTRIBUTION FUNCTIONS

If X has a probability density function $f(x)$, and if U is some function of X, then we can find $F_U(u) = P(U \le u)$ directly by integrating $f(x)$ over the region for which $U \le u$. We can find the probability density function for U by differentiating $F_U(u)$. The method is illustrated in Example 6.1.

Example 6.1 The percentage of time, X, that a lathe is in use during a typical 40-hour work week is a random variable with probability density function given by

$$f(x) = \begin{cases} 3x^2 & 0 \le x \le 1 \\ 0 & \text{elsewhere.} \end{cases}$$

The actual number of *hours*, out of a 40-hour week, that the lathe is *not* in use, then, is

$$U = 40(1 - X).$$

Find the probability density function for U.

Solution

Because the distribution function for U looks at a region of the form $U \le u$, we must first find that region on the x scale. Now

$$U \le u \Rightarrow 40(1 - X) \le u$$

$$\Rightarrow X > 1 - \frac{u}{40},$$

so that

$$F_U(u) = P(U \le u) = P[40(1 - X) \le u]$$
$$= P\left(X > 1 - \frac{u}{40}\right)$$
$$\int_{1-u/40}^{1} f(x)\,\mathrm{d}x$$
$$\int_{1-u/40}^{1} 3x^2 = [x^3]_{1-u/40}^{1},$$

or

$$F_U(u) = 1 - \left(1 - \frac{u}{40}\right)^3, \qquad 0 \le u \le 40.$$

Now, the probability density function is found by differentiating the distribution function, so that

$$f_U(u) = \frac{\mathrm{d}F_U(u)}{\mathrm{d}u} = \frac{3}{40}\left(1 - \frac{u}{40}\right)^2, \qquad 0 \le u \le 40$$
$$= 0 \qquad \text{elsewhere.}$$

We could now use $f_U(u)$ to evaluate probabilities or find expected values related to the number of hours that the lathe is not in use. The reader should verify that $f_U(u)$ has all the properties of a probability density function.

The bivariate case is handled similarly, although it is often more difficult to transform the bivariate regions from statements about U, a function of (X_1, X_2), to statements about X_1 and X_2. Example 6.2 illustrates the point.

Example 6.2 Two friends plan to meet at the library during a given 1-hour period. Their arrival times are independent and randomly distributed across the 1-hour period. Each agrees to wait for 15 minutes, or until the end of the hour. If the friend hasn't appeared during that time, she will leave. What is the probability that the two friends will meet?

Solution

If X_1 denotes one person's arrival time in $(0, 1)$, the 1-hour period, and X_2 denotes the second person's arrival time, then (X_1, X_2) can be modeled as having a two-dimensional uniform distribution over the unit square. That is,

$$f(x_1, x_2) = 1 \qquad 0 \le x_1 \le 1, 0 \le x_2 \le 1$$
$$= 0 \qquad \text{elsewhere.}$$

The event that the two friends will meet depends upon the time, U, between their arrivals, where

$$U = |X_1 - X_2|.$$

We will solve the specific problem by finding the probability density for U. Now,

$$U \leq u \Rightarrow |X_1 - X_2| \leq u$$
$$\Rightarrow -u \leq X_1 - X_2 \leq u.$$

Figure 6.1 shows the square region over which (X_1, X_2) has positive probability, and the region defined by $U \leq u$. The probability that $U \leq u$ can be found by integrating the joint density function of (X_1, X_2) over the six-sided region shown in the center of Figure 6.1. This can be simplified by integrating over the triangles (A_1 and A_2) and subtracting from one, as we now see. We have

$$\begin{aligned}
F_U(u) = P(U \leq u) &= \iint_{|x_1 - x_2| \leq u} f(x_1, x_2)\,dx_1\,dx_2 \\
&= 1 - \iint_{A_1} f(x_1, x_2)\,dx_1\,dx_2 - \iint_{A_2} f(x_1, x_2)\,dx_1\,dx_2 \\
&= 1 - \int_u^1 \int_0^{x_2 - u} (1)\,dx_1\,dx_2 - \int_0^{1-u} \int_{x_2+u}^1 (1)\,dx_1\,dx_2 \\
&= 1 - \int_u^1 (x_2 - u)\,dx_2 - \int_0^{1-u} (1 - u - x_2)\,dx_2 \\
&= 1 - \frac{1}{2}[(x_2 - u)^2]_u^1 - \frac{1}{2}[-(1 - u - x_2)^2]_0^{1-u} \\
&= 1 - \frac{1}{2}(1 - u)^2 - \frac{1}{2}(1 - u)^2 \\
&= 1 - (1 - u)^2 \qquad 0 \leq u \leq 1.
\end{aligned}$$

FIGURE 6.1 Region $U \leq u$ for Example 6.2.

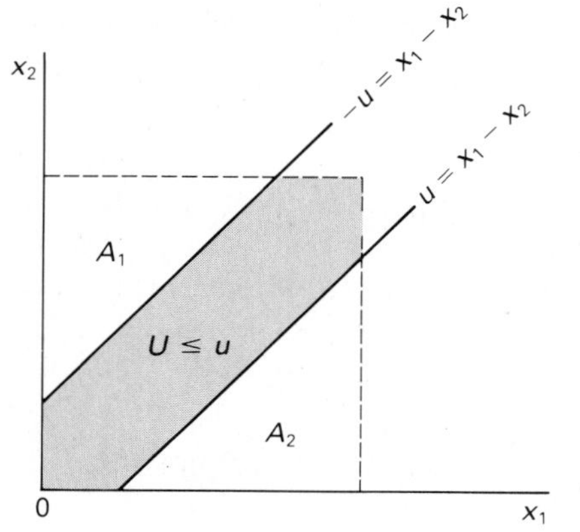

(Note that the two double integrals really evaluate the volumes of prisms with triangular bases.) Because the probability density function is found by differentiating the distribution function, we have

$$f_U(u) = \frac{dF_U(u)}{du} = \begin{cases} 2(1-u), & 0 \le u \le 1 \\ 0 & \text{elsewhere.} \end{cases}$$

The problem asks for the probability of a meeting if each friend waits up to 15 minutes. Because 15 minutes is 1/4 hour, this can be found by evaluating

$$\begin{aligned} P\left(U \le \frac{1}{4}\right) &= \int_0^{1/4} f_U(u)\, du = F_u\left(\frac{1}{4}\right) \\ &= 1 - \left(1 - \frac{1}{4}\right)^2 = 1 - \left(\frac{3}{4}\right)^2 \\ &= \frac{7}{16} = .4375. \end{aligned}$$

There is less than a 50–50 chance that the two friends will meet under this rule.

Applications of probability often call for the use of sums of random variables. A study of downtimes of computer systems might require knowledge of the sum of the downtimes over a day or a week. The total cost of a building project can be studied as the sum of the costs for the major components of the project. The size of an animal population can be modeled as the sum of the sizes of the colonies within the population. The list of examples is endless, and we now present an example that shows how to use distribution functions to find the probability distribution of a sum of independent random variables. This complements the work in Chapter 5 on using moment-generating functions to find distributions of sums.

Example 6.3 Suppose X_1 and X_2 are independent, exponentially distributed random variables, each with a mean of one. Find the probability density function for $U = X_1 + X_2$.

Solution

The joint density function of X_1 and X_2 is given by

$$\begin{aligned} f(x_1, x_2) &= f_1(x_1)f_2(x_2) \\ &= e^{-x_1}e^{-x_2} \qquad x_1 \ge 0, x_2 \ge 0 \\ &= 0 \qquad \text{elsewhere} \end{aligned}$$

and the distribution function of either X_1 or X_2 is given by

$$F(x) = 1 - e^{-x} \qquad x \ge 0.$$

To find $P(U \leq u)$ we must integrate over the region shown in Figure 6.2. Thus,

$$F_U(u) = P(U \leq u) = \int_0^u \int_0^{u-x_2} f_1(x_1)f_2(x_2)\,dx_1\,dx_2$$
$$= \int_0^u F(u - x_2)f_2(x_2)\,dx_2$$
$$= \int_0^u (1 - e^{-(u-x_2)})e^{-x_2}\,dx_2$$
$$= [1 - e^{-x_2}]_0^u - [x_2 e^{-u}]_0^u$$
$$= 1 - e^{-u} - ue^{-u} \qquad u \geq 0$$

and

$$f_U(u) = \frac{dF_U(u)}{du} = ue^{-u} \qquad u \geq 0$$
$$= 0 \qquad \text{elsewhere.}$$

Note that $f_U(u)$ is a gamma function, which is consistent with results for the distribution of sums of independent exponential random variables found using moment-generating functions.

FIGURE 6.2 Region $U \leq u$ for Example 6.3.

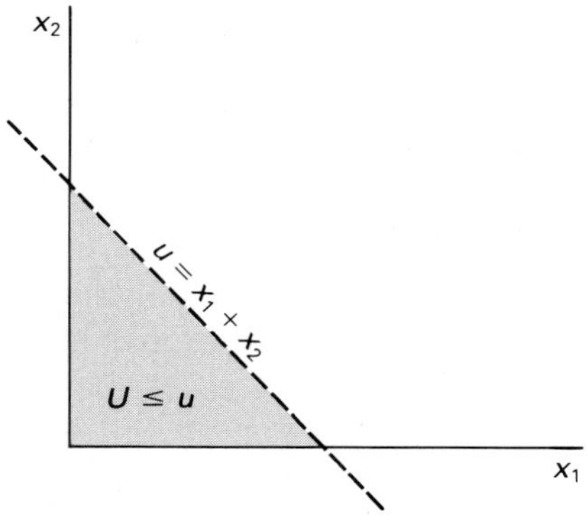

Summary of the Distribution Function Method

Let U be a function of the continuous random variables $X_1, X_2, \ldots, X_n$. Then,

1. find the region $U = u$ in the $(x_1, x_2, \ldots, x_n)$ space.
2. find the region $U \leq u$.

3. find $F_U(u) = P(U \leq u)$ by integrating $f(x_1, x_2, \ldots, x_n)$ over the region $U \leq u$.
4. find the density function $f_U(u)$ by differentiating $F_U(u)$. Thus, $f_U(u) = dF_U(u)/du$.

EXERCISES

6.1 Let X be a random variable with probability density function given by

$$f(x) = \begin{cases} 2(1 - x) & 0 \leq x \leq 1 \\ 0 & \text{elsewhere.} \end{cases}$$

Find the density function of

(a) $U_1 = 2X - 1$.
(b) $U_2 = 1 - 2X$.
(c) $U_3 = X^2$.

6.2 Let X be a random variable with density function given by

$$f(x) = \begin{cases} (3/2)x^2 & -1 \leq x \leq 1 \\ 0 & \text{elsewhere.} \end{cases}$$

Find the density function of

(a) $U_1 = 3X$.
(b) $U_2 = 3 - X$.
(c) $U_3 = X^2$.

6.3 A supplier of kerosene has a weekly demand X possessing a probability density function given by

$$f(x) = \begin{cases} x & 0 \leq x \leq 1 \\ 1 & 1 < x \leq 1.5 \\ 0 & \text{elsewhere} \end{cases}$$

with measurements in hundreds of gallons. The supplier's profit is given by $U = 10X - 4$.

(a) Find the probability densisty function for U.
(b) Use the answer in part (a) to find $E(U)$.
(c) Find $E(U)$ by the methods of Chapter 4.

6.4 The waiting time until delivery of a new component for an industrial operation, X, is uniformly distributed over the interval from 1 to 5 days. The cost of this delay is given by $U = 2X^2 + 3$. Find the probability density function for U.

6.5 The joint distribution of amount of pollutant emitted from a smokestack without a cleaning device (X_1) and with a cleaning device (X_2) is given by

$$f(x_1, x_2) = \begin{cases} k & 0 \leq x_1 \leq 2,\ 0 \leq x_2 \leq 1,\ 2x_2 \leq x_1 \\ 0 & \text{elsewhere.} \end{cases}$$

The reduction in amount due to the cleaning device is given by $U = X_1 - X_2$.

(a) Find the probability density function for U.

(b) Use the answer in part (a) to find $E(U)$. (Compare with the result of Exercise 5.4.)

6.6 The total time from arrival to completion of service at a fast-food outlet, X_1, and the time spent waiting in line before arriving at the service window, X_2, have a joint density function given by

$$f(x_1, x_2) = \begin{cases} e^{-x_1} & 0 \le x_2 \le x_1 < \infty \\ 0 & \text{elsewhere} \end{cases}$$

where $U = X_1 - X_2$ represents the time spent at the service window.

(a) Find the probability density function for U.

(b) Find $E(U)$ and $V(U)$ using the answer to part (a). (Compare with the result of Exercise 5.21.)

6.7 Suppose that a unit of mineral ore contains a proportion X_1 of metal A and a proportion X_2 of metal B. Experience has shown that the joint probability density function of (X_1, X_2) is uniform over the region $0 \le x_1 \le 1$, $0 \le x_2 \le 1$, $0 \le x_1 + x_2 \le 1$. Let $U = X_1 + X_2$, the proportion of metals A and B per unit.

(a) Find the probability density function for U.

(b) Find $E(U)$ by using part (a).

(c) Find $E(U)$ by using only the marginal densities of X_1 and X_2.

6.8 Suppose a continuous random variable, X, has distribution function $F(x)$. Show that $F(X)$ is uniformly distributed over the interval $(0, 1)$.

6.3

METHOD OF TRANSFORMATIONS

The transformation method for finding the probability distribution of a function of random variables is simply a generalization of the distribution function method, Section 6.2. Through the distribution function approach, we can arrive at a single method of writing down the density function of $U = h(X)$ provided that $h(x)$ is either decreasing or increasing. [By $h(x)$ increasing, we mean that if $x_1 < x_2$ then $h(x_1) < h(x_2)$ for any real numbers x_1 and x_2.]

Suppose that $h(x)$ is an increasing function of x and that $U = h(X)$, where X has density function $f_X(x)$. The symbol $h^{-1}(u)$ denotes the inverse function; that is, if $u = h(x)$, we can solve for x, obtaining $x = h^{-1}(u)$. The graph of an increasing function $h(x)$ appears in Figure 6.3, where we see that the set of points x such that $h(x) \le u_1$ is precisely the same as the set of points x such that $x \le h^{-1}(u_1)$. To find the density of $U = h(X)$ by the distribution function

FIGURE 6.3 Increasing function.

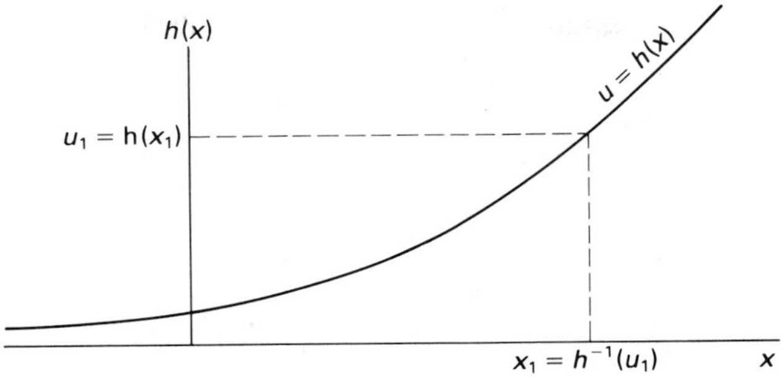

method, we write

$$\begin{aligned} F_U(u) &= P(U \le u) \\ &= P[h(X) \le u] \\ &= P[X \le h^{-1}(u)] \\ &= F_X[h^{-1}(u)], \end{aligned}$$

where $F_X(x)$ is the distribution function of X.

To find the density function of $U, f_U(u)$, we must differentiate $F_U(u)$. Because $x = h^{-1}(u)$,

$$F_U(u) = F_X[h^{-1}(u)] = F_X(x).$$

Then,

$$\begin{aligned} f_U(u) &= \frac{dF_U(u)}{du} \\ &= \frac{dF_X(x)}{dx}\frac{dx}{du} \\ &= f_X(x)\frac{dx}{du} \\ &= f_X[h^{-1}(u)]\frac{dx}{du}. \end{aligned}$$

Note that $dx/du = 1/(du/dx)$.

Example 6.4 Let X have the probability density function given by

$$f_X(x) = \begin{cases} 2x & 0 < x < 1 \\ 0 & \text{elsewhere.} \end{cases}$$

Find the density function of $U = 3X - 1$.

Solution

The function of interest here is $h(x) = 3x - 1$, which is increasing in x. If $u = 3x - 1$, then

$$x = h^{-1}(u) = \frac{u + 1}{3}$$

and

$$\frac{dx}{du} = \frac{1}{3}.$$

Thus,

$$\begin{aligned} f_U(u) &= f_X[h^{-1}(u)]\frac{dx}{du} = 2x\frac{dx}{du} \\ &= 2 \cdot \frac{u + 1}{3} \cdot \frac{1}{3} \\ &= \frac{2(u + 1)}{9} \qquad -1 < u < 2 \\ &= 0 \qquad \text{elsewhere.} \end{aligned}$$

The range over which $f_U(u)$ is positive is simply the interval $0 < x < 1$ transformed to the u-axis by the function $u = 3x - 1$.

If $h(x)$ is a decreasing function, as in Figure 6.4, then the set of points x such that $h(x) \leq u_1$ is the same as the set of points x such that $x \geq h^{-1}(u_1)$. It

FIGURE 6.4 Decreasing function.

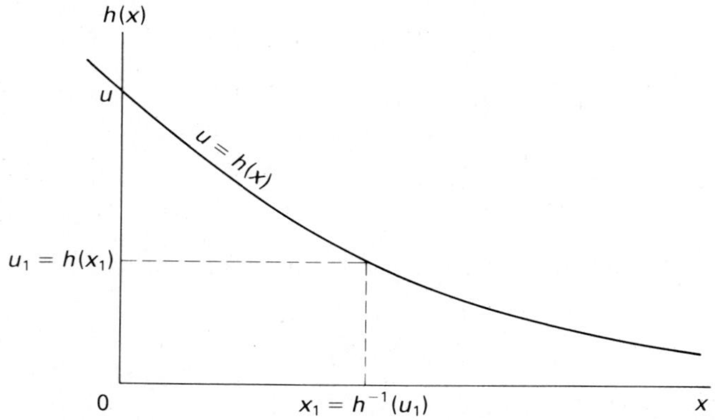

follows that, for $U = h(X)$,

$$\begin{aligned} F_U(u) &= P(U \le u) \\ &= P[h(X) \le u] \\ &= P[X \ge h^{-1}(u)] \\ &= 1 - F_X[h^{-1}(u)]. \end{aligned}$$

On taking derivatives with respect to u, we have

$$f_U(u) = -f_X[h^{-1}(u)]\frac{dx}{du}.$$

Since dx/du is negative for a decreasing function, this equation is equivalent to

$$f_U(u) = f_X[h^{-1}(u)]\left|\frac{dx}{du}\right|.$$

The results for h increasing and h decreasing can now be combined into the following statement.

> Let X have probability density function $f_X(x)$. If $h(x)$ is either increasing or decreasing in x, then $U = h(X)$ has denisty function given by
>
> $$f_U(u) = f_X[h^{-1}(u)]\left|\frac{dx}{du}\right|.$$

Example 6.5 Let X have the probability density function given by

$$f_X(x) = \begin{cases} 2x & 0 < x < 1 \\ 0 & \text{elsewhere.} \end{cases}$$

Find the density function of $U = -4X + 3$.

Solution

The function of interest here, $h(x) = -4x + 3$, is decreasing in x. If $u = -4x + 3$, then

$$x = h^{-1}(u) = \frac{3 - u}{4}$$

and

$$\frac{dx}{du} = -\frac{1}{4}.$$

Thus,

$$\begin{aligned}
f_U(u) &= f_X[h^{-1}(u)]\left|\frac{dx}{du}\right| = 2x\left|\frac{dx}{du}\right| \\
&= 2\cdot\frac{3-u}{4}\cdot\frac{1}{4} \\
&= \frac{3-u}{8} \qquad -1 < u < 3 \\
&= 0 \qquad \text{elsewhere.}
\end{aligned}$$

The transformation method as outlined here can be applied readily to some functions that are neither increasing nor decreasing. To illustrate, consider the case $U = h(X) = X^2$, where X is still continuous with distribution function $F_X(x)$ and density function $f_X(x)$. We then have

$$\begin{aligned}
F_U(u) &= P(U \le u) = P(X^2 \le u) \\
&= P(-\sqrt{u} \le X \le \sqrt{u}) \\
&= F_X(\sqrt{u}) - F_X(-\sqrt{u}).
\end{aligned}$$

On differentiating with respect to u, we see that

$$\begin{aligned}
f_U(u) &= f_X(\sqrt{u})\frac{1}{2\sqrt{u}} + f_X(-\sqrt{u})\frac{1}{2\sqrt{u}} \\
&= \frac{1}{2\sqrt{u}}[f_X(\sqrt{u}) + f_X(-\sqrt{u})].
\end{aligned}$$

Example 6.6 Let X have the probability density function given by

$$f_X(x) = \begin{cases} \dfrac{x+1}{2} & -1 \le x \le 1 \\ 0 & \text{elsewhere.} \end{cases}$$

Find the density function for $U = X^2$.

Solution

We know that

$$f_U(u) = \frac{1}{2\sqrt{u}}[f_X(\sqrt{u}) + f_X(-\sqrt{u})]$$

and, on substituting into this equation,

$$
\begin{aligned}
f_U(u) &= \frac{1}{2\sqrt{u}}\left(\frac{\sqrt{u}+1}{2} + \frac{-\sqrt{u}+1}{2}\right) \\
&= \frac{1}{2\sqrt{u}} \qquad 0 \le u \le 1 \\
&= 0 \qquad \text{elsewhere.}
\end{aligned}
$$

Note that, because X has positive density over the interval $-1 \le x \le 1$, $U = X^2$ has positive density over the interval $0 \le u \le 1$.

The transformation method can also be used in multivariate situations. The following example illustrates for the bivariate case.

Example 6.7 Let X_1 and X_2 have joint density function given by

$$
f(x_1, x_2) = \begin{cases} e^{-(x_1+x_2)} & 0 \le x_1, 0 \le x_2 \\ 0 & \text{elsewhere.} \end{cases}
$$

Find the density function for $U = X_1 + X_2$.

Solution

This problem must be solved in two stages; first, we find the joint density of X_1 and U, and second, we find the marginal density of U. The approach is to let one of the original variables, say X_1, be fixed at a value x_1. Then $U = x_1 + X_2$ and we can consider the one-dimensional transformation problem in which $U = h(X_2) = x_1 + X_2$. Letting $g(x_1, u)$ denote the joint density of X_1 and U, we have

$$
\begin{aligned}
g(x_1, u) &= f[x_1, h^{-1}(u)]\left|\frac{dx_2}{du}\right| \\
&= e^{-u}(1) \qquad 0 \le u, 0 \le x_1 \le u \\
&= 0 \qquad \text{otherwise.}
\end{aligned}
$$

(Note that $X_1 \le U$.) The marginal density of U is then given by

$$
\begin{aligned}
f_U(u) &= \int_{-\infty}^{\infty} g(x_1, u)\, dx_1 \\
&= \int_0^u e^{-u}\, dx_1 \\
&= ue^{-u} \qquad 0 \le u \\
&= 0 \qquad \text{otherwise.}
\end{aligned}
$$

Summary of the Transformation Method

Let U be an increasing or decreasing function of the random variable X; say, $U = h(X)$.

1. Find the inverse function, $X = h^{-1}(U)$.
2. Evaluate $\mathrm{d}x/\mathrm{d}u$.
3. Find $f_U(u)$ by

$$f_U(u) = f_X[h^{-1}(u)] \left|\frac{\mathrm{d}x}{\mathrm{d}u}\right|.$$

6.4

METHOD OF CONDITIONING

Conditional density functions frequently provide a convenient path for finding distributions of random variables. Suppose that X_1 and X_2 have a joint density function, and we want to find the density function for $U = h(X_1, X_2)$. We note that the density of U, $f_U(u)$, can be written

$$f_U(u) = \int_{-\infty}^{\infty} f(u, x_2)\,\mathrm{d}x_2.$$

Since the conditional density of U gives X_2 is given by

$$f(u \mid x_2) = \frac{f(u, x_2)}{f_2(x_2)},$$

it follows that

$$f_U(u) = \int_{-\infty}^{\infty} f(u \mid x_2) f_2(x_2)\,\mathrm{d}x_2.$$

Example 6.8 Let X_1 and X_2 be independent random variables each with density function

$$f_X(x) = \begin{cases} e^{-x} & x > 0 \\ 0 & \text{elsewhere.} \end{cases}$$

Find the density function for $U = X_1/X_2$.

Solution

We find the conditional density of U given $X_2 = x_2$, which is obtained by the transformation method. When X_2 is held fixed at the constant value x_2, U is simply X_1/x_2, where X_1 has the exponential density function. Thus, we want the density function of $U = X_1/x_2$. The inverse function is $X_1 = Ux_2$. By the method of Section 6.3,

$$f(u \mid x_2) = f_{X_1}(ux_2)\left|\frac{dx_1}{du}\right|$$
$$= e^{-ux_2}x_2 \qquad u > 0$$
$$= 0 \qquad \text{elsewhere.}$$

Now,

$$f_U(u) = \int_{-\infty}^{\infty} f(u \mid x_2)f(x_2)\,dx_2$$
$$= \int_0^{\infty} e^{-ux_2}x_2(e^{-x_2})\,dx_2$$
$$= \int_0^{\infty} x_2e^{-x_2(u+1)}\,dx_2$$
$$= (u+1)^{-2} \qquad u > 0$$
$$= 0 \qquad \text{elsewhere.}$$

The evaluation of this integral is facilitated by observing that the integrand is a gamma function.

Summary of the Conditioning Method

Let U be a function of the random variables X_1 and X_2.

1. Find the conditional density of U given $X_2 = x_2$. (This is usually found by transformations.)
2. Find $f_U(u)$ from the relation

$$f_U(u) = \int_{-\infty}^{\infty} f(u \mid x_2)f(x_2)\,dx_2.$$

EXERCISES

6.9 Referring to Exercise 6.1, find the answers by using the method of transformations.

6.10 Referring to Exercise 6.2, find the answers by using the method of transformations.

6.11 In a process of sintering two types of copper powder, the density function for X_1, the volume proportion of solid copper in a sample, was given by

$$f_1(x_1) = \begin{cases} 6x_1(1 - x_1) & 0 \le x_1 \le 1 \\ 0 & \text{elsewhere.} \end{cases}$$

The density function for X_2, the proportion of type A crystals among the solid copper, was given by

$$f_2(x_2) = \begin{cases} 3x_2^2 & 0 \le x_2 \le 1 \\ 0 & \text{elsewhere.} \end{cases}$$

The variable $U = X_1X_2$ gives the proportion of the sample volume due to type A crystals. Find the probability density function for U, assuming X_1 and X_2 are independent.

6.12 A density function sometimes used to model lengths of life of electronic components is the Rayleigh density, given by

$$f(x) = \begin{cases} \left(\frac{2x}{\theta}\right)e^{-x^2/\theta} & x > 1 \\ 0 & \text{elsewhere.} \end{cases}$$

(a) If X has a Rayleigh density, find the probability density function for $U = X^2$.
(b) Use the result of part (a) to find $E(X)$ and $V(X)$.

6.13 The Weibull density function is given by

$$f(x) = \begin{cases} \frac{1}{\alpha}mx^{m-1}e^{-x^m/\alpha} & x > 0 \\ 0 & \text{elsewhere} \end{cases}$$

where α and m are positive constants. If X has the Weibull density,
(a) find the density function of $U = X^m$.
(b) find $E(X^k)$ for any positive integer k.

6.5

ORDER STATISTICS

Many functions of random variables of interest in practice depend on the relative magnitudes of the observed variables. For instance, we may be interested in the fastest time in an automobile race or the heaviest mouse among those fed a certain diet. Thus, we often order observed random variables according to their magnitudes. The resulting ordered variables are called *order statistics*.

Formally, let $X_1, X_2, \ldots, X_n$ denote independent continuous random variables with distribution function $F(x)$ and density function $f(x)$. We shall denote the ordered random variables, X_i, by $X_{(1)}, X_{(2)}, \ldots, X_{(n)}$, where $X_{(1)} \le X_{(2)} \le \cdots \le X_{(n)}$. (Because the random variables are continuous, the equality signs can be ignored.) That is,

$$X_{(1)} = \min(X_1, \ldots, X_n),$$

the minimum value of X_i, and

$$X_{(n)} = \max(X_1, \ldots, X_n),$$

the maximum value of X_i.

The probability density functions for $X_{(1)}$ and $X_{(n)}$ are found easily. Looking at $X_{(n)}$ first, we see that

$$P[X_{(n)} \le x] = P(X_1 \le x, X_2 \le x, \ldots, X_n \le x),$$

because $[X_{(n)} \le x]$ implies that all the values of X_i must be less than or equal to x, and vice versa. However, $X_1, \ldots, X_n$ are independent; hence,

$$P[X_{(n)} \le x] = [F(x)]^n.$$

Letting $g_n(x)$ denote the density function of $X_{(n)}$, we see that, on taking derivatives on both sides,

$$g_n(x) = n[F(x)]^{n-1}f(x).$$

The density function of $X_{(1)}$, denoted by $g_1(x)$, can be found by a similar device. We have

$$\begin{aligned} P[X_{(1)} \le x] &= 1 - P[X_{(1)} > x] \\ &= 1 - P(X_1 > x, X_2 > x, \ldots, X_n > x) \\ &= 1 - [1 - F(x)]^n. \end{aligned}$$

Hence,

$$g_1(x) = n[1 - F(x)]^{n-1}f(x).$$

Let us consider the case $n = 2$ and find the joint density of $X_{(1)}$ and $X_{(2)}$. Now, the event $[X_{(1)} \le x_1, X_{(2)} \le x_2]$ means that either $(X_1 \le x_1, X_2 \le x_2)$ or $(X_2 \le x_1, X_1 \le x_2)$. [Note that $X_{(1)}$ could be either X_1 or X_2, whichever is smaller.] Thus, for $x_1 \le x_2$,

$$\begin{aligned} P[X_{(1)} \le x_1, X_{(2)} \le x_2] &= P[(X_1 \le x_1, X_2 \le x_2) \cup (X_2 \le x_1, X_1 \le x_2)] \\ &= P(X_1 \le x_1, X_2 \le x_2) + P(X_2 \le x_1, X_1 \le x_2) \\ &\quad - P(X_1 \le x_1, X_2 \le x_1) \end{aligned}$$

(by the additive laws of probabilities)

$$= 2F(x_1)F(x_2) - [F(x_1)]^2.$$

Letting $g_{12}(x_1, x_2)$ denote the joint density of $X_{(1)}$ and $X_{(2)}$, we see that, on differentiating first with respect to x_2 and then with respect to x_1,

$$g_{12}(x_1, x_2) = \begin{cases} 2f(x_1)f(x_2) & x_1 \le x_2 \\ 0 & \text{elsewhere.} \end{cases}$$

The same method can be used to find the joint density of $X_{(1)}, \ldots, X_{(n)}$, which turns out to be

$$g_{12\cdots n}(x_1, \ldots, x_n) = \begin{cases} n!f(x_1), \ldots, f(x_n) & x_1 \le x_2 \le \cdots \le x_n \\ 0 & \text{elsewhere.} \end{cases}$$

The marginal density function for any of the order statistics can be found from this joint density function, but we shall not pursue the matter in this text.

Example 6.9 Electronic components of a certain type have life length, X, with probability density given by

$$f(x) = \begin{cases} (1/100)e^{-x/100} & x > 0 \\ 0 & \text{elsewhere.} \end{cases}$$

(Life length is measured in hours.) Suppose that two such components operate independently and in series in a certain system; that is, the system fails when either component fails. Find the density function for U, the life length of the system.

Solution

Because the system fails at the failure of the first component, $U = \min(X_1, X_2)$, where X_1 and X_2 are independent random variables with the given density. Then because $F(x) = 1 - e^{-x/100}$, $x \geq 0$,

$$\begin{aligned} f_U(u) &= g_1(x) = n[1 - F(x)]^{n-1}f(x) \\ &= 2e^{-x/100}\left(\frac{1}{100}\right)e^{-x/100} \\ &= \left(\frac{1}{50}\right)e^{-x/50} \qquad x > 0 \\ &= 0 \qquad \text{elsewhere.} \end{aligned}$$

Thus, we see that the minimum of two exponentially distributed random variables has an exponential distribution.

Example 6.10 Suppose that in Example 6.9 the components operate in parallel; that is, the system does not fail until both components fail. Find the density functions for U, the life length of the system.

Solution

Now, $U = \max(X_1, X_2)$ and

$$\begin{aligned} f_U(u) &= g_2(x) = n[F(x)]^{n-1}f(x) \\ &= \begin{cases} 2(1 - e^{-x/100})\left(\frac{1}{100}\right)e^{-x/100} & x > 0 \\ 0 & \text{elsewhere} \end{cases} \\ &= \begin{cases} \frac{1}{50}(e^{-x/100} - e^{-x/50}) & x > 0 \\ 0 & \text{elsewhere.} \end{cases} \end{aligned}$$

We see that the maximum of two exponential random variables is not an exponential random variable.

EXERCISES

6.14 Let X_1 and X_2 be independent and uniformly distributed over the interval (0, 1). Find the probability density function of

(a) $U_1 = \min(X_1, X_2)$.
(b) $U_2 = \max(X_1, X_2)$.

6.15 The opening prices per share of two similar stocks, X_1 and X_2, are independent random variables, each with density function

$$f(x) = \begin{cases} (1/2)e^{-(1/2)(x-4)} & x \geq 4 \\ 0 & \text{elsewhere.} \end{cases}$$

On a given morning, Mr. A is going to buy shares of whichever stock is less expensive. Find the probability density function for the price per share that Mr. A will have to pay.

6.16 Suppose that the length of time it takes a worker to complete a certain task, X, has the probability density function

$$f(x) = \begin{cases} e^{-(x-\theta)} & x > \theta \\ 0 & \text{elsewhere} \end{cases}$$

where θ is a positive constant that represents the minimum time to task completion. Let $X_1, \ldots, X_n$ denote independent random variables from this distribution.

(a) Find the probability density function for $X_{(1)} = \min(X_1, \ldots, X_n)$.
(b) Find $E(X_{(1)})$.

6.6

PROBABILITY-GENERATING FUNCTIONS: APPLICATIONS TO RANDOM SUMS OF RANDOM VARIABLES

Probability-generating functions (see Section 3.10, pages 106–109) also are useful tools for determining the probability distribution of a function of random variables. We shall show their applicability by developing some properties of random sums of random variables.

Let $X_1, X_2, \ldots$ be a sequence of independent and identically distributed random variables, each with probability-generating function $R(s)$. Let N be an integer-valued random variable with probability-generating function $Q(s)$.

Define S_N as

$$S_N = X_1 + X_2 + \cdots + X_N.$$

The following are illustrative of the many applications of such random variables. N could represent the number of customers per day at a checkout counter and X_i the number of items purchased by the ith customer; then, S_N represents the total number of items sold per day. N could represent the number of seeds produced by a plant and X_i could be defined by

$$X_i = \begin{cases} 1 & \text{if the } i\text{th seed germinates} \\ 0 & \text{otherwise.} \end{cases}$$

Then S_N is the total number of germinating seeds produced by the plant.

We now assume that S_N has probability-generating function $P(s)$ and investigate its construction. By definition,

$$P(s) = E(s^{S_N}),$$

which can be written

$$\begin{aligned} P(s) &= E[E(s^{S_n} \mid N = n)], \\ &= \sum_{n=0}^{\infty} E(s^{S_n} \mid N = n)P(N = n). \end{aligned}$$

Note that

$$\begin{aligned} E(s^{S_n}) &= E(s^{X_1+X_2+\cdots+X_n}) \\ &= E(s^{X_1}s^{X_2}\cdots s^{X_n}) \\ &= E(s^{X_1})E(s^{X_2})\cdots(S^{X_n}) \\ &= [R(s)]^n. \end{aligned}$$

Also,

$$Q(s) = E(s^N) = \sum_{n=0}^{\infty} s^n P(N = n).$$

If N is independent of the values of X_i, then

$$\begin{aligned} P(s) &= \sum_{n=0}^{\infty} E(s^{S_n})P(N = n) \\ &= \sum_{n=0}^{\infty} [R(s)]^n P(N = n) \end{aligned}$$

or

$$P(s) = Q[R(s)].$$

The expression $Q[R(s)]$ is the probability-generating function $Q(s)$ evaluated at $R(s)$ instead of s.

Suppose that N has Poisson distribution with mean λ. Then,

$$Q(s) = e^{\lambda(s-1)} = e^{-\lambda+\lambda s}.$$

It follows that

$$\begin{aligned} P(s) &= Q[R(s)] \\ &= e^{-\lambda+\lambda R(s)}. \end{aligned}$$

This is often referred to as the probability-generating function of the compound Poisson distribution.

Example 6.11 Suppose that N, the number of animals caught in a trap per day, has a Poisson distribution with mean λ. The probability of any one animal being male is p. Find the probability-generating function and expected value of the number of males caught per day.

Solution

Let $X_i = 1$ if the ith animal caught is male and $X_i = 0$ otherwise. Then $S_N = X_1 + \cdots + X_N$ denotes the total number of males caught per day. Now,

$$R(s) = E(s^X) = q + ps.$$

where $q = 1 - p$, and $P(s)$, the probability-generating function of S_N, is given by

$$\begin{aligned} P(s) &= e^{-\lambda+\lambda R(s)} \\ &= e^{-\lambda+\lambda(q+ps)} \\ &= e^{-\lambda(1-q)+\lambda ps} \\ &= e^{-\lambda p+\lambda ps} \end{aligned}$$

and thus S_N has a Poisson distribution with mean λp.

6.7

ARRIVAL TIMES FOR THE POISSON PROCESS

Let $Y(t)$ denote a Poisson process (see Section 3.11, pages 110–113) with mean λ per unit time. It is sometimes of interest to study properties of the actual times at which events have occurred, given that $Y(t)$ is fixed. Suppose that $Y(t) = n$, and let $0 < U_{(1)} < U_{(2)} < \cdots < U_{(n)} < t$ be the actual times that events occur in the interval 0 to t. Let the joint density function of $U_{(1)}, \ldots, U_{(n)}$ be denoted by $g(u_1, \ldots, u_n)$. If we think of du_i as a very small interval, then $g(u_1, \ldots, u_n)\, du_1 \cdots du_n$ is equal to the probability that one event occurs in each of the intervals $(u_i, u_i + du_i)$ and none occur elsewhere, given that n events occur in $(0, t)$. Thus,

$$g(u_i, \ldots, u_n)\,du_1 \cdots du_2 = \frac{1}{\dfrac{(\lambda t)^n e^{-\lambda t}}{n!}} [\lambda\,du_1 e^{-\lambda du_1} \cdots \lambda\,du_n e^{-\lambda du_n} e^{-\lambda(t - du_1 - \cdots - du_n)}]$$

$$= \frac{n!}{t^n}\,du_1 \cdots du_n.$$

It follows that

$$g(u_i, \ldots, u_n) = \frac{n!}{t^n} \qquad u_1 < u_2 < \cdots < u_n$$

or, in other words, $U_{(1)}, \ldots, U_{(n)}$ behave as an ordered set of n independent observations from the uniform distribution on $(0, t)$. This implies that the unordered occurrence times, $U_1, \ldots, U_n$ are independent uniform random variables.

Example 6.12 Suppose that telephone calls coming into a switchboard follow a Poisson process with a mean of ten calls per minute. A particularly slow period of 2 minutes duration had only four calls. Find the probability that all four calls came in the first minute.

Solution

Here we are given that $Y(2) = 4$. We can then evaluate the density function of $U_{(4)}$ to be

$$g_4(u) = \left(\frac{4}{2^4}\right)u^3 \qquad 0 \le u \le 2.$$

This comes from the fact that $U_{(4)}$ is the largest order statistic in a sample $n = 4$ observations from a uniform distribution. Thus, the probability in question is

$$P[U_{(4)} \le 1] = \int_0^1 g_4(u)\,du = \frac{4}{2^4}\int_0^1 u^3\,du$$

$$= \frac{1}{2^4} = \frac{1}{16}.$$

6.8

INFINITE-SERVER QUEUE

We now consider a queue with an infinite number of servers. This queue could model a telephone system with a very large number of channels or the claims department of an insurance company in which all claims are processed immediately. Customers arrive at random times and keep a server busy for a random length of time. The number of servers is large enough, however, that a customer never has to wait for his service to begin. A random variable of much interest in this system is $X(t)$, the number of servers busy at time t.

Let the customer arrivals be a Poisson process with a mean of λ per unit time. The customer arriving at time U_n, measured from the start of the system at $t = 0$, keeps a server busy for a random service time Y_n. The service times, $Y_1, Y_2, \ldots$, are assumed to be independent and identically distributed, with distribution function $F(y)$. $X(t)$, the number of servers busy at time t, can then be written

$$X(t) = \sum_{n=1}^{N(t)} w(t, U_n, Y_n),$$

where $N(t)$, the number of arrivals in $(0, t)$, has a Poisson distribution and

$$w(t, U_n, Y_n) = \begin{cases} 1 & \text{if } 0 < U_n \le t \le U_n + Y_n \\ 0 & \text{otherwise.} \end{cases}$$

The condition $U_n \le t \le U_n + Y_n$ is precisely the condition that a customer arriving at time U_n is still being served at time t. The function $w(t, U_n, Y_n)$ just keeps track of those customers still under service at t.

Operations similar to those used in Section 6.6 for the compound Poisson distribution will yield that $X(t)$ has Poisson distribution with a mean given by

$$E[X(t)] = E[N(t)]E[w(t, U_n, Y_n)].$$

Because $E[N(t)] = \lambda t$, it remains to evaluate

$$E[w(t, U_n, Y_n)] = P(0 \le U_n \le t \le U_n + Y_n) = P(Y_n \ge t - U_n).$$

Recall that U_n is uniformly distributed on $(0, t)$, and thus,

$$\begin{aligned} P(Y_n \ge t - U_n) &= \int_0^t P(Y_n \ge t - u \mid U_n = u)\frac{du}{t} \\ &= \frac{1}{t}\int_0^t P(Y_n \ge t - u)\,du \\ &= \frac{1}{t}\int_0^t [1 - F(t - u)]\,du. \end{aligned}$$

Making the change of variable $t - u = s$, we have

$$P(Y_n \ge t - U_n) = \frac{1}{t}\int_0^t [1 - F(s)]\,ds$$

or

$$E[X(t)] = \lambda \int_0^t [1 - F(s)]\,ds.$$

Example 6.13 Suppose that the telephone switchboard of Example 6.12, with $\lambda = 10$ calls per minute, has a large number of channels. The calls are of random and of independent length, but average 4 minutes each. Approximate the probability that, after the switchboard has been in service for a long time, there will be no busy channels at some specified time t_0.

Solution

Note that $\int_0^\infty [1 - F(s)]\,ds = E(Y)$. Thus, for a large value t_0,

$$\begin{aligned} E[X(t_0)] &= \lambda \int_0^{t_0} [1 - F(s)]\,ds \\ &= \lambda E(Y) \\ &= 10(4) = 40. \end{aligned}$$

Since $X(t)$ has a Poisson distribution,

$$\begin{aligned} P[X(t_0) = 0] &= \exp\left\{-\lambda \int_0^{t_0} [1 - F(s)]\,ds\right\} \\ &= \exp(-40). \end{aligned}$$

EXERCISES

6.17 In the compound Poisson distribution of Section 6.6, let

$$P(X_i = n) = \frac{1}{n}\alpha q^n \qquad n = 1, 2, \ldots,$$

for some constants α and q, $0 < q < 1$. (X_i is said to have a *logarithmic series distribution,* which is frequently used in ecological models.) Show that the resulting compound Poisson distribution is related to the negative binomial distribution.

6.18 Show that the random variable S_N defined in Section 6.6 has mean $E(N)E(X)$ and variance $E(N)V(X) + V(N)E^2(X)$. Use probability-generating functions.

6.19 Let $X_1, X_2, \ldots$ denote a sequence of independent and identically distributed random variables with $P(X_i = 1) = p$ and $P(X_i = -1) = q = 1 - p$. If a value of 1 signifies success, then

$$S_n = X_1 + \cdots + X_n$$

is the excess of successes over failures for n trials. A success could represent, for example, the completion of a sale or a win at a gambling device. Let $p_n^{(y)}$ equal the probability that S_n reaches the value y for the first time on the nth trial. We outline the steps for obtaining the probability-generating function, $P(s)$, for $p_n^{(1)}$:

$$P(s) = \sum_{n=1}^{\infty} s^n p_n^{(1)}.$$

(a) Let $P^{(2)}(s)$ denote the probability-generating function for $p_n^{(2)}$; that is,

$$P^{(2)}(s) = \sum_{n=1}^{\infty} s^n p_n^{(2)}.$$

Since a first accumulation of S_n to 2 represents two successive first accumulations to 1, reason that

$$P^{(2)}(s) = P^2(s).$$

(b) For $n > 1$, $S_n = 1$ for the first time on trial n implies that S_n must have been at 2 on trial $n - 1$, and that a failure must have occurred on trial n. Since $p_n^{(1)} = qp_{n-1}^{(2)}$, show that

$$P(s) = ps + qsP^2(s).$$

(c) Solve the quadratic equation in part (b) to show that

$$P(s) = \frac{1}{2qs}[1 - (1 - 4pqs^2)^{1/2}].$$

(d) Show that

$$P(1) = \frac{1}{2q}(1 - |p - q|)$$

so that the probability that S_n ever becomes positive equals 1 or p/q, whichever is smaller.

(e) Show that for the case $p = q = 1/2$ the expected number of trials until the first passage of S_n through 1 is infinite.

6.20 If $Y(t)$ and $X(t)$ are independent Poisson processes, show that $Y(t) + X(t)$ is also a Poisson process.

6.21 Claims arrive at an insurance company according to a Poisson process with a mean of twenty per day. The claims are placed in the processing system immediately upon arrival, but the service time per claim is an exponential random variable with a mean of 10 days. On a certain day, the company was completely caught up on its claims service (no claims were left to be processed). Find the probability that 5 days later the company had at least two claims being serviced.

6.22 Assume that telephone calls arrive at a switchboard according to a Poisson process with an average rate of λ per minute. One time period t minutes in length is known to have included exactly two calls. Find the probability that the length of time between these calls exceeds d minutes, for some constant $d \leq t$.

SUPPLEMENTARY EXERCISES

6.23 Let X_1 and X_2 be independent and uniformly distributed over the interval $(0, 1)$. Find the probability density function of

(a) $U_1 = X_1^2$.
(b) $U_2 = X_1/X_2$.
(c) $U_3 = -\ln(X_1X_2)$.
(d) $U_4 = X_1X_2$.

6.24 Let X_1 and X_2 be independent Poisson random variables with mean λ_1 and λ_2, respectively.

(a) Find the probability function of $X_1 + X_2$.

(b) Find the conditional probability function of X_1, given $X_1 + X_2 = m$.

6.25 Referring to Exercise 6.1,

(a) find the expected values of U_1, U_2, and U_3 directly (without using the density functions of U_1, U_2, and U_3).

(b) find the expected values of U_1, U_2, and U_3 by using the derived density functions for these random variables.

6.26 A parachutist wants to land at a target, T, but finds that he is equally likely to land at any point on a straight line (A, B) of which T is the midpoint. Find the probability density function of the distance between his landing point and the target. (Hint: Denote A by -1, B by $+1$, and T by 0. Then, the parachutist's landing point has a coordinate, X, which is uniformly distributed between -1 and $+1$. The distance between X and T is $|X|$.)

6.27 Let X_1 denote the amount of a bulk item stocked by a supplier at the beginning of a day, and let X_2 denote the amount of that item sold during the day. Suppose that X_1 and X_2 have joint density function

$$f(x_1, x_2) = \begin{cases} 2 & 0 \le x_2 \le x_1 \le 1 \\ 0 & \text{elsewhere.} \end{cases}$$

Of interest to this supplier is the random variable $U = X_1 - X_2$, which denotes the amount left at the end of the day.

(a) Find the probability density function for U.

(b) Find $E(U)$.

(c) Find $V(U)$.

6.28 An efficiency expert takes two independent measurements, X_1 and X_2, on the length of time it takes workers to complete a certain task. Each measurement is assumed to have the density function given by

$$f(x) = \begin{cases} (1/4)xe^{-x/2} & x > 0 \\ 0 & \text{elsewhere.} \end{cases}$$

Find the density function for the average

$$U = (1/2)(X_1 + X_2).$$

6.29 The length of time that a certain machine operates without failure is denoted by X_1 and the length of time to repair a failure is denoted by X_2. After repair, the machine is assumed to operate like a new machine. X_1 and X_2 are independent and each has the density function

$$f(x) = \begin{cases} e^{-x} & x > 0 \\ 0 & \text{elsewhere.} \end{cases}$$

Find the probability density function for

$$U = \frac{X_1}{X_1 + X_2},$$

the proportion of time that the machine is in operation during any one operation-repair cycle.

6.30 Two sentries are sent to patrol a road 1 mile long. The sentries are sent to points chosen independently and at random along the road. Find the probability that the sentries will be less than 1/2 mile apart when they reach their assigned posts.

6.31 Let X_1 and X_2 be independent standard normal random variables. Find the probability density function of $U = X_1/X_2$.

6.32 Let X be uniformly distributed over the interval $(-1, 3)$. Find the probability density function of $U = X^2$.

6.33 If X denotes the life length of a component and $F(x)$ is the distribution function of X, then $P(X > x) = 1 - F(x)$ is called the *reliability* of the component. Suppose that a system consists of four components with identical reliability functions, $1 - F(x)$, operating as indicated:

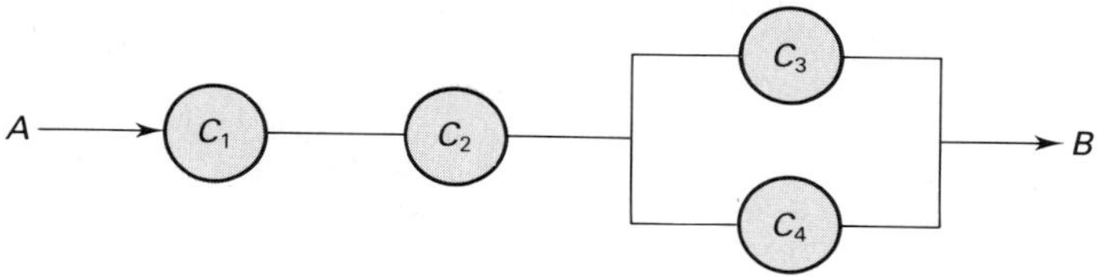

The system operates correctly if an unbroken chain of components is in operation between A and B. If the four components operate independently, find the reliability of the system, in terms of $F(x)$.

6.34 Let $X_1, X_2, \ldots, X_n$ denote a random sample from the uniform distribution, $f(x) = 1, 0 \le x \le 1$. Find the probability density function for the range, $R = X_{(n)} - X_{(1)}$.

6.35 Suppose that U and V are independent random variables with U having a standard normal distribution and V having a χ^2 distribution with n degrees of freedom. Define T by

$$T = \frac{U}{\sqrt{V/n}}.$$

T has a t (or Student's t) distribution, and the density can be obtained as follows:

(a) If V is fixed at v, then T is given by U/c, where $c = \sqrt{v/n}$. Use this idea to find the conditional density of T for fixed $V = v$.

(b) Find the joint density of T and V, $f(t, v)$, by using

$$f(t, v) = f(t \mid v)f(v).$$

(c) Integrate over v to show that

$$f(t) = \frac{\Gamma[(n+1)/2]}{\sqrt{\pi n}\,\Gamma(n/2)}(1 + t^2/n)^{-(n+1)/2} \qquad -\infty < t < \infty.$$

6.36 Suppose that V and W are independent χ^2 random variables with n_1 and n_2 degrees of freedom, respectively. Then F, defined by

$$F = \frac{V/n_1}{W/n_2}$$

is said to have an F distribution with n_1 and n_2 degrees of freedom, and the density function can be obtained as follows:

(a) If W is fixed at w, then $F = V/c$, where $c = wn_1/n_2$. Thus, find the conditional density of F for fixed $W = w$.

(b) Find the joint density of F and W.

(c) Integrate over w to show that the probability density function of F, $g(f)$, is given by

$$g(f) = \frac{\Gamma[(n_1 + n_2)/2](n_1/n_2)^{n_1/2}}{\Gamma(n_1/2)\Gamma(n_2/2)} (f)^{(n_1/2)-1}\left(1 + \frac{n_1 f}{n_2}\right)^{-(n_1+n_2)/2}$$

$$0 < f < \infty.$$

6.37 An object is to be dropped at a point, 0, in the plane, but lands at the point (X, Y) instead, where X and Y denote distances on a coordinate system centered at 0. If X and Y are independent, normally distributed random variables, each with a mean of zero and variance σ^2, find the distribution of the distance between the landing point, (X, Y), and 0. (The resulting distribution is called the *Rayleigh distribution.*)

6.38 Suppose that n electronic components, each having an exponentially distributed life length with a mean of θ, are put into operation at the same time. The components operate independently and are observed until r have failed ($r \leq n$). Let W_j denote the length of time until the jth failure ($W_1 \leq W_2 \leq \cdots \leq W_r$). Let $T_j = W_j - W_{j-1}$ for $j \geq 2$ and $T_1 = W_1$.

(a) Show that T_j, for $j = 1, \ldots, r$, has an exponential distribution with mean $\theta/(n - j + 1)$.

(b) Show that $U_r = \sum_{j=1}^{r} W_j + (n - r)W_r = \sum_{j=1}^{r} (n - j + 1)T_j$; and, hence, that $E(U_r) = r\theta$. [This suggests that $(1/r)U_r$ can be used as an estimator of θ.]

6.39 A machine produces spherical containers with the radii varying according to the probability density function

$$f(r) = \begin{cases} 2r & 0 \leq r \leq 1 \\ 0 & \text{elsewhere.} \end{cases}$$

Find the probability density function for the volume of the containers.

6.40 Let V denote the velocity of a molecule having mass m in a uniform gas at equilibrium. The probability density function of V is known to be

$$f(v) = \frac{4}{\sqrt{\pi}} b^{3/2} v^2 e^{-bv^2} \qquad v > 0$$

where $b = m/2KT$, T denotes the absolute temperature of the gas, and $K =$ Boltzmann's constant. Find the probability density function for the kinetic energy, E, given by $E = (1/2)mV^2$.

6.41 Suppose that members of a certain animal species are randomly distributed over a planar area, so that the number found in a randomly selected quadrat of unit area has a Poisson distribution with mean λ. Over a specified period of time, each animal has a probability θ of dying, and deaths are assumed to be independent from animal to animal. Ignoring births and other population changes, what is the distribution of animals at the end of the time period?

6.42 Let $Y(A)$ denote a Poisson process in a plane, as given in Section 3.11 (pages 110–113), and let R denote the distance from a random point to the nearest point of realization of the process.

(a) Show that $U = R^2$ is an exponential random variable with mean $(1/\lambda\pi)$.

(b) Suppose that n independent values of R can be obtained by sampling n points. Let W denote the sum of the squares of these values. Show that W has a gamma distribution.

(c) Show that $(n - 1)/\pi W$ is an unbiased estimator of λ; that is, $E[(n - 1)/\pi W] = \lambda$.

6.43 Let V denote the volume of a three-dimensional figure, such as a sphere. The axioms of the Poisson process as applied to three-dimensional point processes, such as stars in the heavens or bacteria in water, yield the fact that $Y(V)$, the number of points in a figure of volume V, has a Poisson distribution with mean λV. Also, $Y(V_1)$ and $Y(V_2)$ are independent if V_1 and V_2 are not overlapping volumes.

(a) If a point is chosen at random in three-dimensional space, show that the distance, R, to the nearest point in the realization of a Poisson process has the density function

$$f(r) = \begin{cases} 4\lambda\pi r^2 e^{-(4/3)\lambda\pi r^3} & r \geq 0 \\ 0 & \text{elsewhere.} \end{cases}$$

(b) If R is as in part (a), show that $U = R^3$ has an exponential distribution.

CHAPTER 7

SOME APPROXIMATIONS TO PROBABILITY DISTRIBUTIONS: LIMIT THEOREMS

7.1 INTRODUCTION

As noted in Chapter 6, we frequently are interested in functions of random variables, such as their average or sum. Unfortunately, the application of the methods in Chapter 6 for finding probability distributions for functions of random variables may lead to intractable mathematical problems. Hence, we need some simple methods for approximating the probability distributions of functions of random variables.

In Chapter 7, we discuss some properties of functions of random variables when the number of variables, n, gets large (approaches infinity). We shall see, for example, that the distribution for certain functions of random variables easily can be approximated for large n even though the exact distribution for fixed n may be difficult to obtain. Even more important, the approximations are sometimes good for samples of modest size and, in some instances, for samples as small as $n = 5$ or 6. In Section 7.2, we present extremely useful theorems, called *limit theorems,* which give properties of random variables as n tends to infinity.

7.2

CONVERGENCE IN PROBABILITY

Suppose that a coin has probability p, $0 \le p \le 1$, of coming up heads on a single toss and that we toss the coin n times. What can be said about the fraction of heads observed in the n tosses? Intuition tells us that the sampled fraction of heads provides an estimate of p, and we would expect the estimate to fall closer to p for larger sample sizes; that is, as the quantity of information in the sample is increased. Although our supposition and intuition may be correct for many problems of estimation, it is *not* always true that larger sample sizes lead to better estimates. Hence, this example gives rise to a question that occurs in all estimation problems: What can be said about the random distance between an estimate and its target parameter?

Notationally, let X denote the number of heads observed in the n tosses. Then $E(X) = np$ and $V(X) = np(1 - p)$. One way to measure the closeness of X/n to p is to examine the probability that the distance, $|(X/n) - p|$, will be less than a preassigned real number ε. This probability,

$$P\left(\left|\frac{X}{n} - p\right| \le \varepsilon\right),$$

should be close to unity for large n if our intuition is correct. The following definition formalizes this convergence concept.

> ***Definition 7.1***
>
> The sequence of random variables, $X_1, X_2, \ldots, X_n$, is said to **converge in probability** to the constant c if, for every positive number ε,
>
> $$\lim_{n\to\infty} P(|X_n - c| \le \varepsilon) = 1.$$

The following theorem often provides a mechanism for proving convergence in probability.

Theorem 7.1 Let $X_1, \ldots, X_n$ be independent and identically distributed random variables, with $E(X_i) = \mu$ and $V(X_i) = \sigma^2 < \infty$. Let $\bar{X}_n = (1/n) \sum_{i=1}^{n} X_i$. Then, for any positive real number ε,

$$\lim_{n\to\infty} P(|\bar{X}_n - \mu| \geq \varepsilon) = 0$$

or

$$\lim_{n\to\infty} P(|\bar{X}_n - \mu| < \varepsilon) = 1.$$

That is, $\bar{X}_n$ converges in probability to μ.

Proof

Note that $E(\bar{X}_n) = \mu$ and $V(\bar{X}_n) = \sigma^2/n$. To prove the theorem, we appeal to Tchebysheff's theorem (see Section 3.2, pages 60–73, or 4.2, pages 138–141), which states that

$$P(|X - \mu| \geq k\sigma) \leq \frac{1}{k^2},$$

where $E(X) = \mu$ and $V(X) = \sigma^2$. In the context of our theorem, X is to be replaced by $\bar{X}_n$ and σ^2 by σ^2/n. It then follows that

$$P\left(|\bar{X}_n - \mu| \geq k\frac{\sigma}{\sqrt{n}}\right) \leq \frac{1}{k^2}.$$

Note that k can be any real number, so we shall choose

$$k = \frac{\varepsilon}{\sigma}\sqrt{n}.$$

Then,

$$P\left(|\bar{X}_n - \mu| \geq \frac{\varepsilon\sqrt{n}}{\sigma}\frac{\sigma}{\sqrt{n}}\right) \leq \frac{\sigma^2}{\varepsilon n}$$

or

$$P(|\bar{X}_n - \mu| \geq \varepsilon) \leq \frac{\sigma^2}{\varepsilon^2 n}.$$

Now let us take the limit of this expression as n tends to infinity. Recall that σ^2 is finite and ε is a positive real number. On taking the limit as n tends to infinity, we have

$$\lim_{n\to\infty} P(|\bar{X}_n - \mu| \geq \varepsilon) = 0.$$

That $\lim_{n\to\infty} P(|\bar{X}_n - \mu| < \varepsilon) = 1$ follows directly because

$$P(|\bar{X}_n - \mu| \geq \varepsilon) = 1 - P(|\bar{X}_n - \mu| < \varepsilon).$$

We now apply Theorem 7.1 to our coin-tossing example. ■

Example 7.1 Let X be a binomial random variable with probability of success p and number of trials n. Show that X/n converges in probability to p.

Solution

We have seen that we can write X as $\sum_{i=1}^{n} X_i$, where $X_i = 1$ if the ith trial results in success, and $X_i = 0$ otherwise. Then

$$\frac{X}{n} = \frac{1}{n}\sum_{i=1}^{n} X_i.$$

Also, $E(X_i) = p$ and $V(X_i) = p(1 - p)$. The conditions of Theorem 7.1 are then fulfilled with $\mu = p$ and $\sigma^2 = p(1 - p)$, and we conclude that

$$\lim_{n\to\infty} P\left(\left|\frac{X}{n} - p\right| \geq \varepsilon\right) = 0$$

for any positive ε.

Theorem 7.1, sometimes called the (*weak*) *law of large numbers,* is the theoretical justification for the averaging process employed by many experimenters to obtain precision in measurements. For example, an experimenter may take the average of five measurements of the weight of an animal to obtain a more precise estimate of an animal's weight. His feeling, a feeling borne out by Theorem 7.1, is that the average of a number of independently selected weights should be quite close to the true weight, with high probability.

Like the law of large numbers, the theory of convergence in probability has many applications. Theorem 7.2, which we present without proof, points out some properties of the concept of convergence in probability.

Theorem 7.2 Suppose X_n converges in probability to μ_1 and Y_n converges in probability to μ_2. Then,

1. $X_n + Y_n$ converges in probability to $\mu_1 + \mu_2$.
2. $X_n Y_n$ converges in probability to $\mu_1 \mu_2$.
3. X_n/Y_n converges in probability to μ_1/μ_2, provided that $\mu_2 \neq 0$.
4. $\sqrt{X_n}$ converges in probability to $\sqrt{\mu_1}$, provided that $P(X_n \geq 0) = 1$.

Example 7.2 Suppose that $X_1, X_2, \ldots, X_n$ are independent and identically distributed random variables with $E(X_i) = \mu$, $E(X_i^2) = \mu_2'$, $E(X_i^3) = \mu_3'$, and $E(X_i^4) = \mu_4'$

all assumed finite. Let S'^2 denote the sample variance given by

$$S'^2 = \frac{1}{n}\sum_{i=1}^{n}(X_i - \bar{X})^2.$$

Show that S'^2 converges in probability to $V(X_i)$.

Solution

First, note that we can write

$$S'^2 = \frac{1}{n}\sum_{i=1}^{n} X_i^2 - \bar{X}^2,$$

where

$$\bar{X} = \frac{1}{n}\sum_{i=1}^{n} X_i.$$

To show that S'^2 converges in probability to $V(X_i)$, we apply both Theorems 7.1 and 7.2. Look at the terms in S'^2. The quantity $(1/n)\sum_{i=1}^{n} X_i^2$ is the average of n independent and identically distributed variables of the form X_i^2, with $E(X_i^2) = \mu_2'$ and $V(X_i^2) = \mu_4' - (\mu_2')^2$. Because $V(X_i^2)$ is assumed to be finite, Theorem 7.1 tells us that $(1/n)\sum_{i=1}^{n} X_i^2$ converges in probability to μ_2'. Now consider the limit of $\bar{X}^2$ as n approaches infinity. Theorem 7.1 tells us that $\bar{X}$ converges in probability to μ, and it follows from Theorem 7.2, part (b), that $\bar{X}^2$ converges in probability to μ^2. Having shown that $(1/n)\sum_{i=1}^{n} X_i^2$ and $\bar{X}^2$ converge in probability to μ_2' and μ^2, respectively, it follows from Theorem 7.2 that

$$S'^2 = \frac{1}{n}\sum_{i=1}^{n} X_i^2 - \bar{X}^2$$

converges in probability to $\mu_2' - \mu^2 = V(X_i)$.

This example shows that, for large samples, the sample variance should be close to the population variance with high probability.

7.3 CONVERGENCE IN DISTRIBUTION

In Section 7.2, we dealt only with the convergence of certain random variables to constants and said nothing about the form of the probability distributions. In this section, we look at what happens to the probability distributions of certain types of random variables as n tends to infinity. We need the following definition before proceeding.

Definition 7.2

Let Y_n be a random variable with distribution function $F_n(y)$. Let Y be a random variable with distribution function $F(y)$. If

$$\lim_{n\to\infty} F_n(y) = F(y)$$

at every point y for which $F(y)$ is continuous, then Y_n is said to **converge in distribution** to Y. $F(y)$ is called the *limiting distribution function of* Y_n.

We illustrate Definition 7.2 with the following example.

Example 7.3 Let $X_1, \ldots, X_n$ be independent uniform random variables over the interval $(0, \theta)$ for a positive constant θ. In addition, let $Y_n = \max(X_1, \ldots, X_n)$. Find the limiting distribution of Y_n.

Solution

The distribution function for the uniform random variable X_i is

$$F_X(y) = P(X_i \le y) = \begin{cases} 0 & y \le 0 \\ \dfrac{y}{\theta} & 0 < y < \theta \\ 1 & y \ge \theta. \end{cases}$$

In Section 6.5, we found that the distribution function for Y_n is

$$G(y) = P(Y_n \le y) = [F_X(y)]^n,$$

where $F_X(y)$ is the distribution function for each X_i. Then

$$\lim_{n\to\infty} G(y) = \begin{cases} 0 & y \le 0 \\ \lim\limits_{n\to\infty} \left(\dfrac{y}{\theta}\right)^n = 0 & 0 < y < \theta \\ 1 & y \ge \theta. \end{cases}$$

Thus, Y_n converges in distribution to a random variable that has a probability of 1 at the point θ and a probability of 0 elsewhere.

It is often easier to find limiting distributions by working with moment-generating functions. The following theorem gives the relationship between convergence of distribution functions and convergence of moment-generating functions.

Theorem 7.3 Let Y_n and Y be random variables with moment-generating functions $M_n(t)$ and $M(t)$, respectively. If

$$\lim_{n\to\infty} M_n(t) = M(t)$$

for all real t, then Y_n converges in distribution to Y.

Proof

Proof of Theorem 7.3 is beyond the scope of this text. ■

Example 7.4 Let X_n be a binomial random variable with n trials and probability p of success on each trial. If n tends to infinity and p to zero, with np remaining fixed, show that X_n converges in distribution to a Poisson random variable.

Solution

This problem was solved in Chapter 3 when we derived the Poisson probability distribution. We now solve it using moment-generating functions and Theorem 7.3.

We know that the moment-generating function of X_n, $M_n(t)$, is given by

$$M_n(t) = (q + pe^t)^n,$$

where $q = 1 - p$. This can be rewritten as

$$M_n(t) = [1 + p(e^t - 1)]^n.$$

Letting $np = \lambda$ and substituting into $M_n(t)$, we obtain

$$M_n(t) = \left[1 + \frac{\lambda}{n}(e^t - 1)\right]^n.$$

Now let us take the limit of this expression as n approaches infinity.

From the calculus you recall that

$$\lim_{n\to\infty} \left(1 + \frac{k}{n}\right)^n = e^k.$$

Letting $k = \lambda(e^t - 1)$, we have

$$\lim_{n\to\infty} M_n(t) = \exp[\lambda(e^t - 1)].$$

We recognize the right-hand expression as the moment-generating function for the Poisson random variable. Hence, it follows from Theorem 7.3 that X_n converges in distribution to a Poisson random variable.

Example 7.5 In the interest of pollution control, an experimenter wants to count the number of bacteria per small volume of water. The sample size in this problem is really the volume of water in which the count is made. We do not have an n as in previous problems. For purposes of approximating the probability distribution of counts, we can think of the volume, and hence the average count per volume, as the quantity that is getting large.

Let X denote the bacteria count per cubic centimeter of water and assume that X has a Poisson probability distribution with mean λ. We want to approximate the probability distribution of X for large values of λ; we do this by showing that

$$Y = \frac{X - \lambda}{\sqrt{\lambda}}$$

converges in distribution to a standard normal random variable as λ tends to infinity.

Specifically, if the allowable pollution in a water supply is a count of 110 per cubic centimeter, approximate the probability that X will be at most 110, assuming that $\lambda = 100$.

Solution

We proceed by taking the limit of the moment-generating function of Y as $\lambda \to \infty$, and then use Theorem 7.3. The moment-generating function of X, $M_X(t)$, is given by

$$M_X(t) = e^{\lambda(e^t - 1)},$$

and hence the moment-generating function for Y, $M_Y(t)$, is

$$\begin{aligned} M_Y(t) &= e^{-\sqrt{\lambda}t} M_X\left(\frac{t}{\sqrt{\lambda}}\right) \\ &= e^{-\sqrt{\lambda}t} \exp[\lambda(e^{t/\sqrt{\lambda}} - 1)]. \end{aligned}$$

The term $e^{t/\sqrt{\lambda}} - 1$ can be written as

$$e^{t/\sqrt{\lambda}} - 1 = \frac{t}{\sqrt{\lambda}} + \frac{t^2}{2\lambda} + \frac{t^3}{6\lambda^{3/2}} + \cdots$$

and, on adding exponents,

$$\begin{aligned} M_Y(t) &= \exp\left[-\sqrt{\lambda}t + \lambda\left(\frac{t}{\sqrt{\lambda}} + \frac{t^2}{2\lambda} + \frac{t^3}{6\lambda^{3/2}} + \cdots\right)\right] \\ &= \exp\left(\frac{t^2}{2} + \frac{t^3}{6\sqrt{\lambda}} + \cdots\right). \end{aligned}$$

In the exponent of $M_Y(t)$, the first term, $t^2/2$, is free of λ and the remaining terms all have a λ to some positive power in the denominator. Therefore, as

$\lambda \to \infty$, all the terms after the first will tend to zero sufficiently fast to allow

$$\lim_{\lambda \to \infty} M_Y(t) = e^{t^2/2},$$

and the right-hand expression is the moment-generating function for a standard normal random variable. We now want to approximate $P(X \leq 110)$. Note that

$$P(X \leq 110) = P\left(\frac{X - \lambda}{\sqrt{\lambda}} \leq \frac{110 - \lambda}{\sqrt{\lambda}}\right).$$

We have shown that $Y = (X - \lambda)/\sqrt{\lambda}$ is approximately a standard normal random variable for large λ. Hence, for $\lambda = 100$, we have

$$P\left(Y \leq \frac{110 - 100}{10}\right) = P(Y \leq 1)$$
$$= .8413,$$

from Table 4 in the Appendix.

The normal approximation to Poisson probabilities works reasonably well for $\lambda \geq 25$.

EXERCISES

7.1 Let $Y_1, \ldots, Y_n$ be independent random variables, each with probability density function

$$f(y) = \begin{cases} 3y^2 & 0 \leq y \leq 1 \\ 0 & \text{elsewhere.} \end{cases}$$

Show that

$$\bar{Y} = \frac{1}{n}\sum_{i=1}^{n} Y_i$$

converges in probability to a constant as $n \to \infty$, and find the constant.

7.2 Let $Y_1, \ldots, Y_n$ be independent gamma-type random variables with a density function given by

$$f(y) = \begin{cases} \dfrac{1}{b^a \Gamma(a)} y^{a-1} e^{-y/b} & y \geq 0 \\ 0 & \text{elsewhere.} \end{cases}$$

Show that the mean, $\bar{Y}$, converges in probability to a constant, and find the constant.

7.3 Let $Y_1, \ldots, Y_n$ be independent random variables, each uniformly distributed over the interval $(0, \theta)$.

(a) Show that the mean, $\bar{Y}$, converges in probability to a constant as $n \to \infty$, and find the constant.

(b) Show that $\max(Y_1, \ldots, Y_n)$ converges in probability to θ as $n \to \infty$.

7.4 Let $Y_1, \ldots, Y_n$ be independent random variables, each possessing the density function

$$f(y) = \begin{cases} \dfrac{2}{y^2} & y \geq 2 \\ 0 & \text{elsewhere.} \end{cases}$$

Does the law of large numbers apply to $\bar{Y}$ in this case? If so, find the limit in probability of $\bar{Y}$.

7.5 If the probability that a person suffers a bad reaction from an injection of a certain serum is .001, use the Poisson distribution to approximate the probability that, of 1000 persons, 2 or more will suffer a bad reaction.

7.6 The number of accidents per year at a given intersection, Y, is assumed to have a Poisson distribution. Over the past few years, there has been an average of 36 accidents per year at this intersection. If the number of accidents per year is at least 45, an intersection can qualify to be rebuilt under an emergency program set up by the state. Approximate the probability that the intersection in question will come under the emergency program at the end of next year.

7.4

THE CENTRAL LIMIT THEOREM

Example 7.5 gives a random variable that converges in distribution to the standard normal random variable. That this phenomenon is shared by a large class of random variables is shown by the following theorem.

Theorem 7.4 *The Central Limit Theorem.*
Let $X_1, \ldots, X_n$ be independent and identically distributed random variables with $E(X_i) = \mu$ and $V(x_i) = \sigma^2 < \infty$. Define Y_n as

$$Y_n = \sqrt{n}\,\frac{\bar{X} - \mu}{\sigma},$$

where

$$\bar{X} = \frac{1}{n}\sum_{i=1}^{n} X_i.$$

Then Y_n converges in distribution to a standard normal random variable.

Proof

We sketch a proof for the case in which the moment-generating function for X_i exists. (This is not the most general proof, because moment-generating functions do not always exist.)

Define a random variable Z_i by

$$Z_i = \frac{X_i - \mu}{\sigma}.$$

Note that $E(Z_i) = 0$ and $V(Z_i) = 1$. The moment-generating function of Z_i, $M_Z(t)$, can then be written as

$$M_Z(t) = 1 + \frac{t^2}{2} + \frac{t^3}{3!}E(Z_i^3) + \cdots.$$

Now,

$$Y_n = \sqrt{n}\frac{\bar{X} - \mu}{\sigma} = \frac{1}{\sqrt{n}}\frac{\sum_{i=1}^{n} X_i - n\mu}{\sigma}$$

$$= \frac{1}{\sqrt{n}}\sum_{i=1}^{n} Z_i$$

and the moment-generating function of Y_n, $M_n(t)$, can be written as

$$M_n(t) = \left[M_Z\left(\frac{t}{\sqrt{n}}\right)\right]^n.$$

Recall that the moment-generating function of the sum of independent random variables is the product of their individual moment-generating functions. Hence,

$$M_n(t) = \left[M_Z\left(\frac{t}{\sqrt{n}}\right)\right]^n$$

$$= \left(1 + \frac{t^2}{2n} + \frac{t^3}{3!\,n^{3/2}}k + \cdots\right)^n,$$

where $k = E(Z_i^3)$.

Now take the limit of $M_n(t)$ as $n \to \infty$. One way to evaluate the limit is to consider $\log M_n(t)$, where

$$\log M_n(t) = n \log\left[1 + \left(\frac{t^2}{2n} + \frac{t^3k}{6n^{3/2}} + \cdots\right)\right].$$

A standard series expansion for $\log(1 + x)$ is

$$\log(1 + x) = x - \frac{x^2}{2} + \frac{x^3}{3} - \frac{x^4}{4} + \cdots.$$

Let

$$x = \left(\frac{t^2}{2n} + \frac{t^{3k}}{6n^{3/2}} + \cdots\right).$$

Then,

$$\log M_n(t) = n\log(1+x) = n\left(x - \frac{x^2}{2} + \cdots\right)$$

$$= n\left[\left(\frac{t^2}{2n} + \frac{t^{3k}}{6n^{3/2}} + \cdots\right) - \frac{1}{2}\left(\frac{t^2}{2n} + \frac{t^{3k}}{6n^{3/2}} + \cdots\right)^2 + \cdots\right],$$

where the succeeding terms in the expansion involve x^3, x^4, and so on. Multiplying through by n, we see that the first term, $t^2/2$, does not involve n, whereas all other terms will have n to a positive power in the denominator. Thus, it can be shown that

$$\lim_{n\to\infty} \log M_n(t) = t^2/2$$

or

$$\lim_{n\to\infty} M_n(t) = e^{t^2/2},$$

the moment-generating function for a standard normal random variable. Applying Theorem 7.3, we conclude that Y_n converges in distribution to a standard normal random variable. ∎

Another way to say that a random variable converges in distribution to a standard normal is to say that it is asymptotically normal. We note in passing that Theorem 7.4 is not the most general form of the central limit theorem. Similar theorems exist for some cases in which the values of X_i are not identically distributed and in which they are dependent.

The probability distribution that arises from looking at many independent values of $\bar{X}$, for a fixed sample size, selected from the same population is called the *sampling distribution* of $\bar{X}$. The practical importance of the central limit theorem is that, for large n, the sampling distribution of $\bar{X}$ can be closely approximated by a normal distribution. More precisely,

$$P(\bar{X} \le b) = P\left(\frac{\bar{X}-\mu}{\sigma/\sqrt{n}} \le \frac{b-\mu}{\sigma/\sqrt{n}}\right)$$

$$\approx P\left(Z \le \frac{b-\mu}{\sigma/\sqrt{n}}\right)$$

where Z is a standard normal random variable.

We can observe an approximate sampling distribution of $\bar{X}$ by looking at the following results from a computer simulation. Samples of size n were drawn from a population having the probability density function

$$f(x) = \begin{cases} \frac{1}{10}e^{-x/10} & x > 0 \\ 0 & \text{elsewhere.} \end{cases}$$

The sample mean was computed for each sample. The relative frequency histogram of these mean values for 1000 samples of size $n = 5$ is shown in Figure 7.1. Figures 7.2 and 7.3 show similar results for 1000 samples of size $n = 25$ and $n = 100$, respectively. Although all the relative frequency histograms have a sort of bell shape, notice that the tendency toward a symmetric normal curve is better for the larger n. A smooth curve drawn through the bar graph of Figure 7.3 would be nearly identical to a normal density function with a mean of 10 and a variance of $(10)^2/100 = 1$.

The central limit theorem provides a very useful result for statistical inference, for we now know not only that $\bar{X}$ has mean μ and variance σ^2/n if the population has mean μ and variance σ^2, but also that the probability distribution for $\bar{X}$ is approximately normal. For example, suppose we wish to find an interval (a, b) such that

$$P(a \leq \bar{X} \leq b) = .95.$$

This probability is equivalent to

$$P\left(\frac{a - \mu}{\sigma/\sqrt{n}} \leq \frac{\bar{X} - \mu}{\sigma/\sqrt{n}} \leq \frac{b - \mu}{\sigma/\sqrt{n}}\right) = .95$$

for constants μ and σ. Because $(\bar{X} - \mu)/(\sigma/\sqrt{n})$ has approximately a standard

FIGURE 7.1 Relative frequency histogram for $\bar{x}$ from 1000 samples of size $n = 5$.

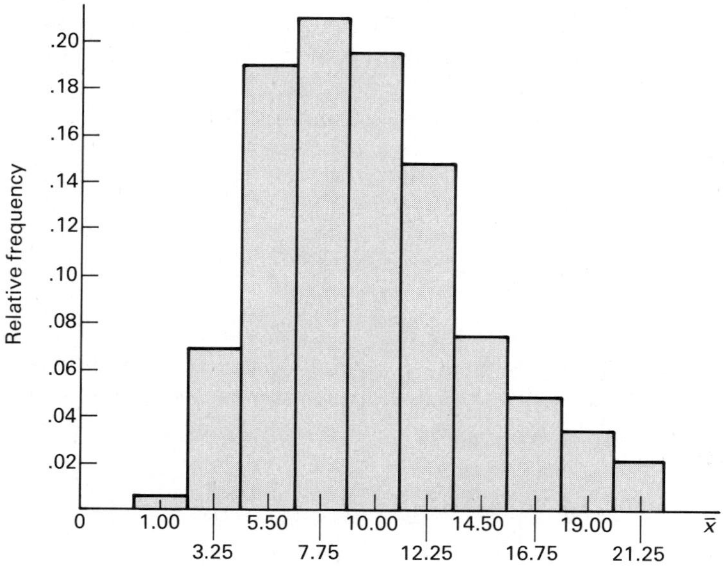

FIGURE 7.2 Relative frequency histogram for $\bar{x}$ from 1000 samples of size $n = 25$.

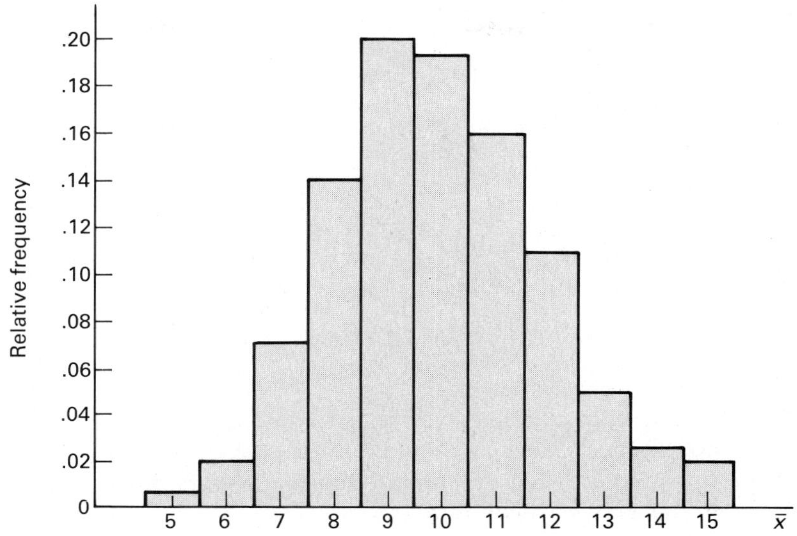

FIGURE 7.3 Relative frequency histogram for $\bar{x}$ from 1000 samples of size $n = 100$.

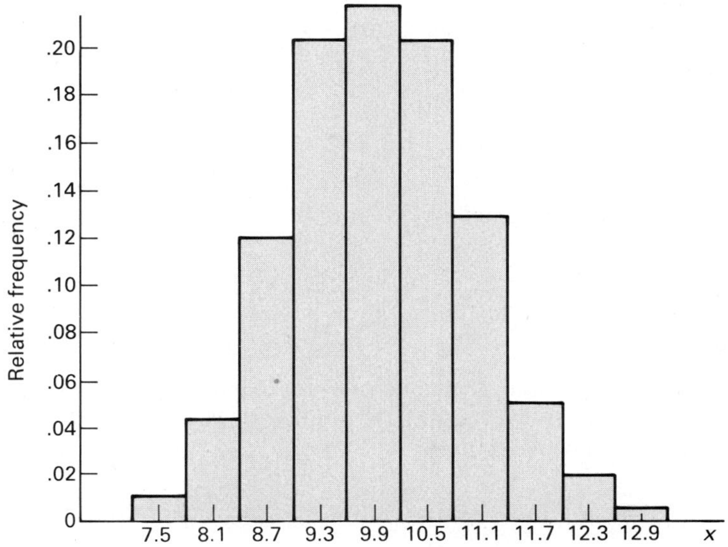

normal distribution, this equality can be approximated by

$$P\left(\frac{a-\mu}{\sigma/\sqrt{n}} \le Z \le \frac{b-\mu}{\sigma/\sqrt{n}}\right) = .95$$

where Z has a standard normal distribution. From Table 4 in the Appendix, we know that

$$P(-1.96 \le Z \le 1.96) = .95$$

and, hence,

$$\frac{a-\mu}{\sigma/\sqrt{n}} = -1.96 \qquad \frac{b-\mu}{\sigma/\sqrt{n}} = 1.96$$

or

$$a = \mu - 1.96\sigma/\sqrt{n} \qquad b = \mu + 1.96\sigma/\sqrt{n}.$$

Example 7.6 The fracture strengths of a certain type of glass average 14 (thousands of pounds per square inch) and have a standard deviation of 2.

(a) What is the probability that the average fracture strength for 100 pieces of this glass exceeds 14.5?

(b) Find an interval that includes the average fracture strength for 100 pieces of this glass with probability .95.

Solution

(a) The average strength $\bar{X}$ has approximately normal distribution with $\mu = 14$ and standard deviation

$$\frac{\sigma}{\sqrt{n}} = \frac{2}{\sqrt{100}} = .2.$$

Thus,

$$P(\bar{X} > 14.5) = P\left(\frac{\bar{X}-\mu}{\sigma/\sqrt{n}} > \frac{14.5-\mu}{\sigma/\sqrt{n}}\right)$$

is approximately equal to

$$P\left(Z > \frac{14.5-14}{.2}\right) = P\left(Z > \frac{.5}{.2}\right)$$
$$= P(Z > 2.5) = .5 - .4938 = .0062$$

from Table 4. The probability of seeing an average value (for $n = 100$) more than .5 units above the population mean, in this case, is very small.

(b) We have seen that

$$P\left(\mu - 1.96\frac{\sigma}{\sqrt{n}} \le \bar{X} \le \mu + 1.96\frac{\sigma}{\sqrt{n}}\right) = .95$$

for a normally distributed $\bar{X}$. In this problem,

$$\mu - 1.96\frac{\sigma}{\sqrt{n}} = 14 - 1.96\frac{2}{\sqrt{100}} = 13.6$$

and

$$\mu + 1.96\frac{\sigma}{\sqrt{n}} = 14 + 1.96\frac{2}{\sqrt{100}} = 14.4.$$

Approximately 95 % of sample mean fracture strengths, for samples of size 100, should lie between 13.6 and 14.4.

Example 7.7 A certain machine used to fill bottles with liquid has been observed over a long period of time, and the variance in the amounts of fill is found to be approximately $\sigma^2 = 1$ ounce. However, the mean ounces of fill, μ, depends on an adjustment that may change from day to day or operator to operator. If $n = 25$ observations on ounces of fill dispensed are to be taken on a given day (all with the same machine setting), find the probability that the sample mean will be within .3 ounce of the true population mean for that setting.

Solution

We will assume $n = 25$ is large enough for the sample mean $\bar{X}$ to have approximately a normal distribution. Then,

$$\begin{aligned} P(|\bar{X} - \mu| \le .3) &= P[-.3 \le (\bar{X} - \mu) \le .3] \\ &= P\left[-\frac{.3}{\sigma/\sqrt{n}} \le \frac{X - \mu}{\sigma/\sqrt{n}} \le \frac{.3}{\sigma/\sqrt{n}}\right] \\ &= P\left[-.3\sqrt{25} \le \frac{\bar{X} - \mu}{\sigma/\sqrt{n}} \le .3\sqrt{25}\right] \\ &= P\left[-1.5 \le \frac{\bar{X} - \mu}{\sigma/\sqrt{n}} \le 1.5\right]. \end{aligned}$$

Since $(\bar{X} - \mu)/(\sigma\sqrt{n})$ has approximately a standard normal distribution, the above probability is approximately

$$P[-1.5 \le Z \le 1.5] = .8664.$$

Example 7.8 Achievement test scores from all high school seniors in a certain state have a mean and variance of 60 and 64, respectively. A specific high school class of $n = 100$ students had a mean score of 58. Is there evidence to suggest that this high school is inferior? (Calculate the probability that the sample mean is at most 58 when $n = 100$.)

Solution

Let $\bar{X}$ denote the mean of a random sample of $n = 100$ scores from a population with $\mu = 60$ and $\sigma^2 = 64$. We want to approximate $P(\bar{X} \le 58)$. We know from Theorem 7.4 that $\sqrt{n}(\bar{X} - \mu)/\sigma$ is approximately a standard normal variable, which we denote by Z. Hence,

$$\begin{aligned} P(\bar{X} \le 58) &\approx P\left(Z \le \frac{58 - 60}{\sqrt{64/100}}\right) \\ &= P(Z \le -2.5) \\ &= .0062, \end{aligned}$$

from Table 4 in the Appendix. Because this probability is so small, it is unlikely that the specific class of interest can be regarded as a random sample from a population with $\mu = 60$ and $\sigma^2 = 64$. There is evidence to suggest that this class could be set aside as inferior.

We have seen in Chapter 3 that a binomially distributed random variable Y can be written as a sum of independent Bernoulli random variables X_i. That is,

$$Y = \sum_{i=1}^{n} X_i$$

where $X_i = 1$ with probability p and $X_i = 0$ with probability $1 - p$, $i = 1, \ldots, n$. Y can represent the number of successes in a sample of n trials, or measurements, such as the number of thermistors conforming to standards in a sample of n thermistors.

Now the *fraction* of successes in the n trials is

$$\frac{Y}{n} = \frac{1}{n}\sum_{i=1}^{n} X_i = \bar{X}$$

so Y/n is a sample mean. In particular, for large n, Y/n has approximately a normal distribution with a mean of

$$E(X_i) = p$$

and a variance of

$$\begin{aligned} V(Y/n) &= \frac{1}{n^2}\sum_{i=1}^{n} V(X_i) \\ &= \frac{1}{n^2}\sum_{i=1}^{n} p(1 - p) = \frac{p(1 - p)}{n}. \end{aligned}$$

The normality follows from the central limit theorem. Because $Y = n\bar{X}$, Y has approximately a normal distribution with a mean of np and a variance of $np(1 - p)$. Because binomial probabilities are cumbersome to calculate for a large n, we make extensive use of this normal approximation to the binomial distribution.

Figure 7.4 shows the histogram of a binomial distribution for $n = 20$ and $p = .6$. The heights of the bars represent the respective binomial probabilities. For this distribution the mean is $np = 200(.6) = 12$, and the variance is $np(1 - p) = 20(.6)(.4) = 4.8$. Superimposed on the binomial distribution is a normal distribution with mean $\mu = 12$ and variance $\sigma^2 = 4.8$. Notice how the normal curve closely approximates the binomial histogram.

For the situation displayed in Figure 7.4, suppose we wish to find $P(Y \leq 10)$. By the exact binomial probabilities found in Table 2 of the

FIGURE 7.4 A binomial distribution $n = 20$, $p = 0.6$; and a normal distribution, $\mu = 12$, $\sigma^2 = 4.8$.

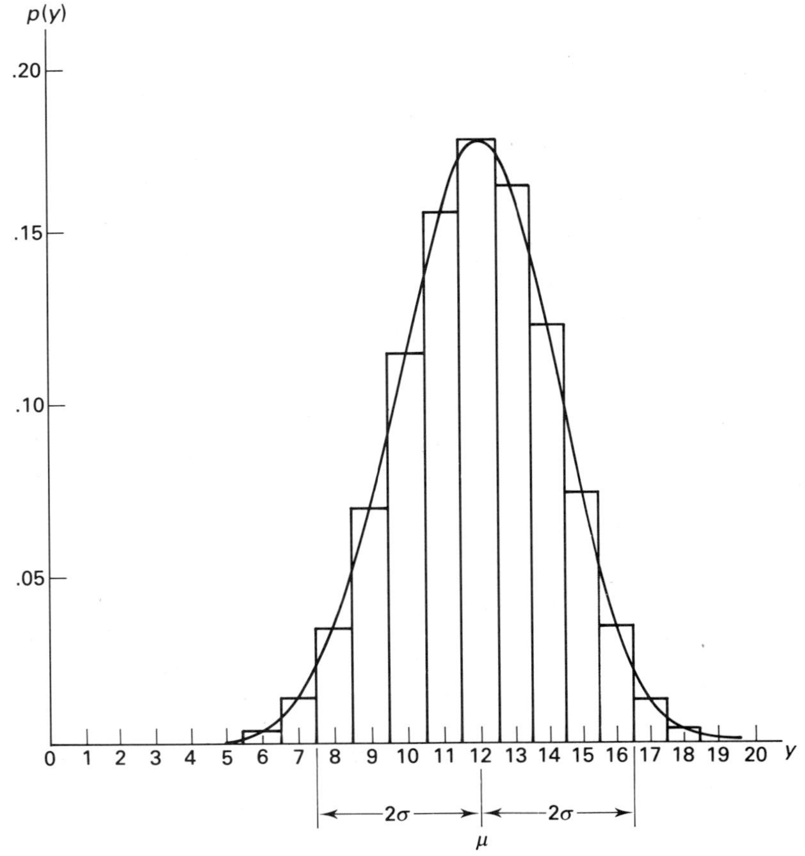

Appendix,

$$P(Y \leq 10) = .245.$$

The value is the sum of the heights of the bars from $y = 0$ up to and including $y = 10$.

Looking at the normal curve in Figure 7.4, we can see that the areas in the bars at $y = 10$ and below are best approximated by the area under the curve to the left of *10.5*. The extra .5 is added so that the total bar at $y = 10$ is included in the area under consideration. Thus, if W represents a normally distributed random variable with $\mu = 12$ and $\sigma^2 = 4.8$ ($\sigma = 2.2$), then

$$\begin{aligned} P(Y \leq 10) &\approx P(W \leq 10.5) \\ &= P\left(\frac{W - \mu}{\sigma} \leq \frac{10.5 - 12}{2.2}\right) = P(Z \leq -.68) \\ &= .5 - .2517 = .2483 \end{aligned}$$

from Table 4 in the Appendix. We see that the normal approximation of .248 is close to the exact binomial probability of .245. The approximation would even be better if n were larger.

The normal approximation to the binomial distribution works well for even moderately large n, as long as p is not close to zero or one. A useful rule of thumb is to make sure n is large enough so that $p \pm 2\sqrt{p(1 - p)/n}$ lies within the interval $(0, 1)$ before the normal approximation is used. Otherwise, the binomial distribution may be so asymmetric that the symmetric normal distribution cannot provide a good approximation.

Example 7.9 Silicon wafers coming into a microchip plant are inspected for conformance to specifications. From a large lot of wafers, $n = 100$ are inspected. If the number of nonconformances, Y, is no more than 12, the lot is accepted. Find the approximate probability of acceptance if the proportion of nonconformances in the lot is $p = .2$.

Solution

The number of nonconformances, Y, has a binomial distribution if the lot, indeed, is large. Before using the normal approximation, we should check to see that

$$\begin{aligned} p \pm 2\sqrt{\frac{p(1 - p)}{n}} &= .2 \pm 2\sqrt{\frac{(.2)(.8)}{100}} \\ &= .2 \pm .08 \end{aligned}$$

is entirely within the interval $(0, 1)$, which it is. Thus, the normal approximation should work well.

Now the probability of accepting the lot is

$$P(Y \leq 12) \approx P(W \leq 12.5)$$

where W is a normally distributed random variable with $\mu = np = 20$ and $\sigma = \sqrt{np(1-p)} = 4$. It folows that

$$P(W \leq 12.5) = P\left(\frac{W-\mu}{\sigma} \leq \frac{12.5-20}{4}\right)$$
$$= P(Z \leq -1.88) = .5 - .4699 = .0301.$$

There is only a small probability of accepting any lot that has 20 % nonconforming wafers.

Example 7.10 Candidate A believes that she can win a city election if she can poll at least 55 % of the votes in precinct I. She also believes that about 50 % of the city's voters favor her. If $n = 100$ voters show up to vote at precinct I, what is the probability that candidate A receives at least 55 % of the votes?

Solution

Let X denote the number of voters at precinct I who vote for candidate A. We must approximate $P(X/n \geq .55)$, when p, the probability that a randomly selected voter favors candidate A, is .5. If we think of the $n = 100$ voters at precinct I as a random sample from the city, then X has a binomial distribution with $p = .5$.

Applying Theorem 7.4, it follows that

$$P(X/n \geq .55) \approx P\left[Z \geq \frac{.545 - .5}{\sqrt{.5(.5)/100}}\right]$$
$$= P(Z \geq .9) = .1841,$$

from Table 4 in the Appendix.

We can make use of the central limit theorem to generate normally distributed random variables. Recall that it is stated as follows:

> Let $X_1, X_2, \ldots, X_n$ be independent and identically distributed random variables with $E(X_i) = \mu$ and $V(X_i) = \sigma^2$. Let $Y_n = \sqrt{n}(\bar{X} - \mu)/\sigma$, where $\bar{X} = (\sum X_i)/n$. Y_n converges in distribution to a standard normal random variable as $n \to \infty$.

We can generate normal random variables as follows:

1. Generate n x's from a specified distribution with known μ and σ^2. Compute the sample mean, $\bar{x}_i$.

2. Evaluate $z_i = \sqrt{n}(\bar{x}_i - \mu)/\sigma$. The z_i will be a standard normal random variable with a mean of 0 and variance of 1.
3. The normal random variable, y_i, will be given by $y_i = \mu_y + z_i\sigma_y$.
4. Repeat steps 1–3 m times, where m is the number of normal random variables desired.

The distribution most often used in step 1 is the uniform distribution on (0, 1). When using this distribution to evaluate z_i, $\mu = .5$ and $\sigma = 1/\sqrt{12}$. To make the algorithm more efficient, let $n = 12$.

EXERCISES

7.7 Shear strength measurements for spot welds of a certain type have been found to have a standard deviation of approximately 10 psi. If 100 test welds are to be measured, find the approximate probability that the sample mean will be within 1 psi of the true population mean.

7.7 If shear strength measurements have a standard deviation of 10 psi, how many test welds should be used in the sample if the sample mean is to be within 1 psi of the population mean with probability approximately .95?

7.9 The soil acidity is measured by a quantity called the pH, which may range from 0 to 14 for soils ranging from high alkalinity to high acidity. Many soils have an average pH in the neutral 5 to 8 range. A scientist wants to estimate the average pH for a large field from n randomly selected core samples, and measuring the pH in each sample. If the scientist selects $n = 40$ samples, find the approximate probability that the sample mean of the 40 pH measurements will be within .2 unit of the true average pH for the field.

7.10 Suppose the scientist of Exercise 7.9 would like the sample mean to be within .1 of the true mean with probability .90. How many core samples should he take?

7.11 Resistors of a certain type have resistances that average 200 ohms with a standard deviation of 10 ohms. Twenty-five of these resistors are to be used in a circuit.

(a) Find the probability that the average resistance of the 25 resistors is between 199 and 202 ohms.

(b) Find the probability that the *total* resistance of the 25 resistors does not exceed 5100 ohms.

$$\left[\text{Hint: Note that } P\left(\sum_{i=1}^{n} X_i > a\right) = P(n\bar{X} > a) = P(\bar{X} > a/n).\right]$$

(c) What assumptions are necessary for the answers in parts (a) and (b) to be good approximations?

7.12 One-hour carbon monoxide concentrations in air samples from a large city average 12 ppm, with a standard deviation of 9 ppm. Find the probability that the average concentration in 100 samples selected randomly will exceed 14 ppm.

7.13 Unaltered bitumens, as commonly found in lead–zinc deposits, have atomic hydrogen/carbon ratios that average 1.4 with a standard deviation of .05. Find the probability that 25 samples of bitumen have an average H/C ratio below 1.3.

7.14 The downtime per day for a certain computing facility averages 4.0 hours with a standard deviation of .8 hour.

(a) Find the probability that the average daily downtime for a period of 30 days is between 1 and 5 hours.

(b) Find the probability that the *total* downtime for the 30 days is less than 115 hours.

(c) What assumptions are necessary for the answers in parts (a) and (b) to be valid approximations?

7.15 The strength of a thread is a random variable with mean .5 lb and standard deviation of .2 lb. Assume the strength of a rope is the sum of the strengths of the threads in the rope.

(a) Find the probability that a rope consisting of 100 threads with hold 45 lb.

(b) How many threads are needed for a rope that will hold 50 lb with 99% assurance?

7.16 Many bulk products, such as iron ore, coal, and raw sugar, are sampled for quality by a method that requires many small samples to be taken periodically as the material moves along a conveyor belt. The small samples are then aggregated and mixed to form one composite sample. Let Y_i denote the volume of the ith small sample from a particular lot, and suppose $Y_1, \ldots, Y_n$ constitutes a random sample with each Y_i having a mean of μ and a variance of σ^2. The average volume of the samples, μ, can be set by adjusting the size of the sampling device. Suppose the variance of sampling volumes, σ^2, is known to be approximately 4 for a particular situation (measurements are to be in cubic inches). The total volume of the composite sample is required to exceed 200 cubic inches with probability approximately .95 when $n = 50$ small samples are selected. Find a setting for μ that will satisfy the sampling requirements.

7.17 The service times for customers coming through a checkout counter in a retail store are independent random variables with a mean of 1.5 minutes and a variance of 1.0. Approximate the probability that 100 customers can be serviced in less than 2 hours of total service time.

7.18 Referring to Exercise 7.17, find the number of customers n such that the probability of servicing all n customers in less than 2 hours is approximately .1.

7.19 Suppose that $X_1, \ldots, X_{n_1}$ and $Y_1, \ldots, Y_{n_2}$ constitute independent random samples from populations with means μ_1 and μ_2 and variances σ_1^2 and σ_2^2, respectively. Then the central limit theorem can be extended to show that $\bar{X} - \bar{Y}$ is approximately normally distributed, for large n_1 and n_2, with mean $\mu_1 - \mu_2$ and variance $(\sigma_1^2/n_1 + \sigma_2^2/n_2)$.

Water flow through soil depends, among other things, on the porosity (volume proportion due to voids) of the soil. To compare two types of sandy soil, $n_1 = 50$ measurements are to be taken on the porosity of soil A, and $n_2 = 100$ measurements are to be taken on soil B. Assume that $\sigma_1^2 = .01$ and $\sigma_2^2 = .02$. Find the approximate probability that the difference between the sample means will be within .05 unit of the true difference between the population means, $\mu_1 - \mu_2$.

7.20 Referring to Exercise 7.19, suppose samples are to be selected with $n_1 = n_2 = n$. Find the value of n that will allow the difference between the sample means to be within .04 unit of $\mu_1 - \mu_2$ with probability approximately .90.

7.21 An experiment is designed to test whether operator A or B gets the job of operating a new machine. Each operator is timed on 50 independent trials involving the performance of a certain task on the machine. If the sample means for the 50 trials differ by more than 1 second, the operator with the smaller mean gets the job. Otherwise, the experiment is considered a tie. If the standard deviations of times for both operators are assumed to be 2 seconds, what is the probability that operator A gets the job even though both operators have equal ability?

7.22 The median age of residents of the United States is 31 years. If a survey of 100 randomly selected United States residents is taken, find the approximate probability that at least 60 of them will be under 31 years of age.

7.23 A lot acceptance sampling plan for large lots calls for sampling fifty items and accepting the lot if the number of nonconformances is no more than five. Find the approximate probability of acceptance if the true proportion of nonconformances in the lot is

(a) 10%.

(b) 20%.

(c) 30%.

7.24 Of the customers entering a showroom for stereo equipment, only 30% make purchases. If forty customers enter the showroom tomorrow, find the approximate probability that at least fifteen make purchases.

7.25 The quality of computer disks is measured by the number of missing pulses. For a certain brand of disk 80% are generally found to contain no missing pulses. If 100 such disks are inspected, find the approximate probability that 15 or fewer contain missing pulses.

7.26 The capacitances of a certain type of capacitor are normally distributed with a mean of 53 μf and a standard deviation of 2 μf. If sixty-four such capacitors are to be used in an electronic system, approximate the probability that at least twelve of them will have capacitances below 50 μf.

7.27 The daily water demands for a city pumping station exceed 500,000 gallons with probability only .15. Over a 30-day period find the approximate probability that demand for over 500,000 gallons per day occurs no more than twice.

7.28 At a specific intersection vehicles entering from the east are equally likely to turn left, turn right, or proceed straight ahead. If 500 vehicles enter this intersection from the east tomorrow, what is the approximate probability that

(a) 150 or fewer turn right?

(b) at least 350 turn?

7.29 Waiting times at a service counter in a pharmacy are exponentially distributed with a mean of 10 minutes. If 100 customers come to the service counter in a day, approximate the probability that at least half of them must wait for more than 10 minutes.

7.30 A large construction firm has won 60 % of the jobs for which it has bid. Suppose this firm bids on 25 jobs next month.

(a) Approximate the probability that it will win at least 20 of these.

(b) Find the exact binomial probability that it will win at least 20 of these. Compare to your answer in part (a).

(c) What assumptions are necessary for your answers in parts (a) and (b) to be valid?

7.31 An auditor samples 100 of a firm's travel vouchers to check on how many of these vouchers are improperly documented. Find the approximate probability that more than 30 % of the sampled vouchers will show up as being improperly documented if, in fact, only 20 % of all the firm's vouchers are improperly documented.

7.5

COMBINATION OF CONVERGENCE IN PROBABILITY AND CONVERGENCE IN DISTRIBUTION

Many times we will be interested in the limiting behavior of the product or quotient of several functions of a set of random variables. The following theorem, which combines convergence in probability with convergence in distribution, applies to the quotient of two functions, X_n and Y_n.

Theorem 7.5 Suppose that X_n converges in distribution to a random variable X, and Y_n converges in probability to unity. Then X_n/Y_n converges in distribution to X.

Proof

The proof of the theorem is beyond the scope of this text, but we illustrate its usefulness with the following example. ■

Example 7.11 Suppose that $X_1, \ldots, X_n$ are independent and identically distributed random variables with $E(X_i) = \mu$ and $V(X_i) = \sigma^2$. Define S'^2 as

$$S'^2 = \frac{1}{n} \sum_{i=1}^{n} (X_i - \bar{X})^2.$$

Show that

$$\sqrt{n}\frac{\bar{X} - \mu}{S'}$$

converges in distribution to a standard normal random variable.

Solution

In Example 7.2, we showed that S'^2 converges in probability to σ^2. Hence it follows from Theorem 7.2, parts (c) and (d), that S'^2/σ^2 (and hence S'/σ) converges in probability to 1. We also know from Theorem 7.4 that

$$\sqrt{n}\frac{\bar{X} - \mu}{\sigma}$$

converges in distribution to a standard normal random variable. Therefore,

$$\sqrt{n}\frac{\bar{X} - \mu}{S'} = \sqrt{n}\frac{\bar{X} - \mu}{\sigma}\bigg/\frac{S'}{\sigma}$$

converges in distribution to a standard normal random variable by Theorem 7.5.

7.6

RENEWAL THEORY: RELIABILITY APPLICATIONS

Reliability theory deals with probabilistic models that govern the behavior of systems of components and allows one to formulate optimum maintenance policies, including replacement and repair of components, and to measure operating characteristics of the system, such as the proportion of downtime, the expected life, and the probability of survival beyond a specified age. The "system" could be a complex electronic system (such as a computer), an automobile, a factory, or even a human body. In the remainder of this section the system will be simplified to a single component that has a random life length but may be replaced before it fails, such as a light bulb in a socket.

For a Poisson process, as described in Section 3.11 (pages 110–113), the time between any two successive occurrences of the event being observed has an exponential distribution with mean $1/\lambda$, where λ is the expected number of occurrences per unit time (see Exercise 4.36). A more general stochastic process can be obtained by considering interarrival times to be nonnegative random variables, identically distributed, but not necessarily exponential.

Suppose that $X_1, X_2, \ldots, X_n$ is a sequence of independent, identically distributed random variables, each with distribution function $F(x)$ and probability density function $f(x)$. Also, assume that $E(X_i) = \mu$ and $V(X_i) = \sigma^2 < \infty$. The values of X_i could represent lifetimes of identically constructed electronic components, for example. If a system operates by inserting a new component, letting it burn continually until it fails, and then repeating the process with an identical new component assumed to operate independently, then the occurrences of interest are in-service failures of components, and X_i is the interarrival time between failure number $(i - 1)$ and failure number i. A random variable of interest in such problems is N_t, the number of in-service failures in the time interval $(0, t)$. Note that

$$N_t = \text{maximum integer } k, \text{ such that } \sum_{i=1}^{k} X_i \le t,$$

with $N_t = 0$ if $X_1 > t$.

Let us now consider the probability distribution of N_t. If X_i is an exponential random variable, then N_t has a Poisson distribution. However, if X_i is not exponential, the distribution of N_t may be difficult or impossible to obtain. Thus we shall investigate the asymptotic behavior of N_t as t tends to infinity.

Note that, with $S_r = \sum_{i=1}^{r} X_i$,

$$P(N_t \ge r) = P(S_r \le t).$$

The limiting probabilities will be unity unless $r \to \infty$, and some restrictions must be placed on the relationship between r and t. Thus, suppose that $r \to \infty$ and $t \to \infty$ in such a way that

$$\frac{t - r\mu}{\sqrt{r}\sigma} \to c,$$

for some constant c. This implies that $(\mu r/t) \to 1$ as $r \to \infty$, $t \to \infty$. Now the preceding probability equality can be written as

$$P\left[\frac{N_t - (t/\mu)}{t^{1/2}\mu^{-3/2}\sigma} \ge \frac{r - (t/\mu)}{t^{1/2}\mu^{-3/2}\sigma}\right] = P\left(\frac{S_r - r\mu}{r^{1/2}\sigma} \le \frac{t - r\mu}{r^{1/2}\sigma}\right).$$

We have

$$\frac{r - (t/\mu)}{t^{1/2}\mu^{-3/2}\sigma} = \frac{\mu r - t}{r^{1/2}\sigma}\left(\frac{\mu r}{t}\right)^{1/2},$$

and this quantity tends to $-c$ as $t \to \infty$ and $r \to \infty$. It follows that

$$P\left[\frac{N_t - (t/\mu)}{t^{1/2}\mu^{-3/2}\sigma} \ge -c\right] = P\left(\frac{S_r - r\mu}{r^{1/2}\sigma} \le c\right).$$

However,

$$\frac{S_r - r\mu}{r^{1/2}\sigma} = \sqrt{r}\,\frac{\bar{X} - \mu}{\sigma},$$

which has, approximately, a standard normal distribution for large r. Letting $\Phi(x)$ denote the standard normal distribution function, we can conclude that

$$P\left[\frac{N_t - (t/\mu)}{t^{1/2}\mu^{-3/2}\sigma} \geq -c\right] \to \Phi(c)$$

as $t \to \infty$ and $r \to \infty$, or

$$P\left[\frac{N_t - (t/\mu)}{t^{1/2}\mu^{-3/2}\sigma} \leq -c\right] \to [1 - \Phi(c)] = \Phi(-c).$$

Hence, N_t can be regarded as approximately normally distributed, for large t, with a mean of t/μ and a variance of $t\sigma^2/\mu^3$.

Example 7.12 A fuse in an electronic system is replaced with an identical fuse upon failure. It is known that these fuses have a mean life of 10 hours with a variance of 2.5. Beginning with a new fuse, the system is to operate for 400 hours.

(a) If the cost of a fuse is \$5, find the expected amount to be paid for fuses over the 400-hour period.

(b) If 42 replacement fuses are in stock, find the probability that they will all be used in the 400-hour period.

Solution

(a) No distribution of lifetimes for the fuses is given, so asymptotic results must be employed. (Note that $t = 400$ is large as compared to $\mu = 10$.) Letting N_t denote the number of replacements, we have

$$E(N_t) \doteq \frac{t}{\mu} = \frac{400}{10} = 40.$$

We would expect to use forty replacements, plus the one initially placed into the system. Thus, the expected cost of fuses is

$$(5)(41) = \$205.$$

(b) N_t is approximately normal with mean $t/\mu = 40$ and variance

$$\frac{t\sigma^2}{\mu^3} = \frac{400(2.5)}{1000} = 1.$$

Thus,

$$P(N_t \geq 42) = P\left[\frac{N_t - (t/\mu)}{t^{1/2}\mu^{-3/2}\sigma} \geq \frac{42 - 40}{1}\right]$$
$$= P(Z \geq 2) = .0228.$$

where Z denotes a standard normal random variable.

To minimize the number of in-service failures, components often are replaced at age T or at failure, whichever comes first. Here, age refers to length of time in service and T is a constant. Under this age replacement scheme, the interarrival times are given by

$$Y_i = \min(X_i, T).$$

Consider an age replacement policy in which c_1 represents the cost of replacing a component that has failed in service and c_2 the cost of replacing a component (which has not failed) at age T. (Usually, $c_1 > c_2$.) To find the total replacement costs up to time t, we must count the total number of replacements, N_t, and the number of in-service failures. Let

$$W_i = \begin{cases} 1 & \text{if } X_i < T \\ 0 & \text{if } X_i \geq T. \end{cases}$$

Then $\sum_{i=1}^{N_t} W_i$ denotes the total number of in-service failures. Thus, the total replacement cost up to time t is

$$C_t = c_1 \sum_{i=1}^{N_t} W_i + c_2\left(N_t - \sum_{i=1}^{N_t} W_i\right).$$

The expected replacement cost per unit time is frequently of interest, and this can be approximated on observing that, under the conditions stated in this section,

$$E\left(\sum_{i=1}^{N_t} W_i\right) = E(N_t)E(W_i).$$

Thus,

$$E(C_t) = c_1 E(N_t)E(W_i) + c_2 E(N_t)[1 - E(W_i)],$$

which, for large t, becomes

$$\frac{t}{E(Y_i)}[c_1 P(X_i < T) + c_2 P(X_i \geq T)].$$

It follows that

$$\frac{1}{t}E(C_t) = \frac{1}{E(Y_i)}\{c_1 F(T) + c_2[1 - F(T)]\},$$

where $F(x)$ is still the distribution function of X_i.

An optimum replacement policy can be found by choosing T to minimize the expected cost per unit time. This concept is continued in Exercise 7.44.

SUPPLEMENTARY EXERCISES

7.32 A large industry has an average hourly wage of \$4.00 per hour with a standard deviation of \$.50. A certain ethnic group consisting of sixty-four workers has an average wage of \$3.90 per hour. Is it reasonable to assume that the ethnic group is a random sample of workers from the industry? (Calculate the probability of obtaining a sample mean less than or equal to \$3.90 per hour.)

7.33 An anthropologist wishes to estimate the average height of men for a certain race of people. If the population standard deviation is assumed to be 2.5 inches and if she randomly samples 100 men, find the probability that the difference between the sample mean and the true population mean will not exceed .5 inch.

7.34 Suppose that the anthropologist of Exercise 7.9 wants the difference between the sample mean and the population mean to be less than .4 inch with probability .95. How many men should she sample to achieve this objective?

7.35 A machine is shut down for repairs if a random sample of 100 items selected from the daily output of the machine reveals at least 15% defectives. (Assume that the daily output is a large number of items.) If the machine, in fact, is producing only 10% defective items, find the probability that it will be shut down on a given day.

7.36 A pollster believes that 20% of the voters in a certain area favor a bond issue. If sixty-four voters are randomly sampled from the large number of voters in this area, approximate the probability that the sampled fraction of voters favoring the bond issue will not differ from the true fraction by more than .06.

7.37 Twenty-five heat lamps are connected in a greenhouse so that when one lamp fails, another takes over immediately. (Only one lamp is turned on at any time.) The lamps operate independently, and each has mean life of 50 hours and standard deviation of 4 hours. If the greenhouse is not checked for 1300 hours after the lamp system is turned on, what is the probability that a lamp will be burning at the end of the 1300-hour period?

7.38 Suppose that $X_1, \ldots, X_n$ are independent random variables, each with a mean of μ_1 and a variance of σ_1^2. Suppose, also, that $Y_1, \ldots, Y_n$ are independent random variables, each with a mean of μ_2 and a variance of σ_2^2. Show that the random variable

$$\frac{(\bar{X} - \bar{Y}) - (\mu_1 - \mu_2)}{\sqrt{(\sigma_1^2 + \sigma_2^2)/n}}$$

converges in distribution, as $n \to \infty$, to a standard normal random variable.

7.39 An experimenter is counting bacteria of two types, A and B, in a certain liquid. Let X denote the number of type A bacteria per cubic centimeter and let Y denote the number of type B bacteria per cubic centimeter. X and Y are assumed to have Poisson distributions with means λ_1 and λ_2, respectively. The experimenter plans to observe a number of independent values for X and Y. What function of the observed values of X

and Y would you suggest as an estimator of the ratio

$$\frac{\lambda_1}{\lambda_1 + \lambda_2}?$$

Why?

7.40 Let Y have a χ^2 distribution with n degrees of freedom. That is, Y has the density function

$$f(y) = \begin{cases} \dfrac{1}{2^{1/2}\Gamma(n/2)} y^{(n/2)-1} e^{-y/2} & y \geq 0 \\ 0 & \text{elsewhere.} \end{cases}$$

Show that the random variable

$$\frac{Y - n}{\sqrt{2n}}$$

is asymptotically standard normal in distribution, as $n \to \infty$.

7.41 A machine in a heavy-equipment factory produces steel rods of length Y, where Y is a normal random variable with a mean μ of 6 inches and a variance of .2. The cost, C, of repairing a rod that is not exacty 6 inches in length is proportional to the square of the error and is given in dollars by

$$C = 4(Y - \mu)^2.$$

If 50 rods with independent lengths are produced in a given day, approximate the probability that the total cost for repairs for that day exceeds \$48.

7.42 Let Y denote a random variable with distribution function $F(y)$, such that $F(0) = 0$. If $X = \min(Y, T)$ for a constant, T, show that

$$E(X) = \int_0^T [1 - F(x)]\,dx.$$

7.43 Probability distributions used to model length of life are sometimes characterized by their failure rate functions, $r(t)$, where

$$r(t) = \frac{f(t)}{1 - F(t)}.$$

(a) For $f(y) = (1/\theta)e^{-y/\theta}$, $y > 0$, show that $r(t) = 1/\theta$,

(b) For $f(y) = (1/\alpha)my^{m-1}e^{-y^m/\alpha}$, $y \geq 0$, show that $r(t) = (m/\alpha)y^{m-1}$. (Note that this is the Weibull density.)

7.44 Referring to Section 7.6, the optimum age replacement interval, T_0, is defined as the one that minimizes the expected cost per unit time, as given in Section 7.6.

(a) Find T_0 if the components have exponential life lengths with mean θ.

(b) Show that a finite T_0 always exists if $r(t)$ is strictly increasing to infinity. (Note that this will be the case for Weibull life lengths if $m > 1$.)

REFERENCES

Baily, N. T. J. 1964. *The Elements of Stochastic Processes with Applications to the Natural Sciences*. New York: John Wiley and Sons.

Barlow, R. E., and F. Proschan. 1965. *Mathematical Theory of Reliability*. New York: John Wiley and Sons.

Cohen, A. C., B. J. Whitten, and Y. Ding. 1984. Modified Moment Estimation for the Three-Parameter Weibull Distribution. *Journal of Quality Technology* 16, no. 3.

Feller, W. 1968. *An Introduction to Probability Theory and Its Applications,* vol. 1, 3d ed. New York: John Wiley and Sons.

Ferrani, P., et al. 1984. A Behavioral Approach to the Measurement of Motorway Circulation, Comfort and Safety. *Transportation Research,* 18A.

Goranson, U. G., and J. Hall. 1980. Airworthiness of Long-Life Jet Transport Structures. *Aeronautical Journal* (November).

Kamerad, D. B. 1983. The 55MPH Speed Limit: Costs, Benefits, and Implied Trade-Offs. *Transportation Research* 17A, no. 1: 51–64.

Karlin, S. 1968. *A First Course in Stochastic Processes*. New York: Academic Press.

Kemeny, J. G., A. Schleifer, J. L. Snell, and G. L. Thompson. 1962. *Finite Mathematics with Business Applications*. Englewood Cliffs, N.J.: Prentice-Hall.

Kennedy, Jr., W. J., and J. E. Gentle. 1980. *Statistical Computing*. New York: Marcel Dekker.

Larsen, R. J, and M. L. Marx. 1985. *An Introduction to Probability and Its Applications*. Englewood Cliffs, N.J.: Prentice-Hall.

Meyer, P. L. 1970. *Introductory Probability and Statistical Applications,* 2d ed. Reading, Mass.: Addison-Wesley Publishing Co.

Mosteller, F., R. E. K. Rourke, and G. B. Thomas. 1970. *Probability with Statistical Applications,* 2d ed. Reading, Mass.: Addison-Wesley Publishing Co.

Nelson, W. 1967. The Truncated Normal Distribution with Applications to Component Sorting. *Industrial Quality Control:* 261–68.

Parzen, E., 1964. *Modern Probability Theory and Its Applications*. New York: John Wiley and Sons.

Perruzzi, J. J., and E. J. Hilliard. 1984. Modeling Time-Delay Measurement Errors Using a Generalized Beta Density Function. *Journal of the Acoustical Society of America* 75, no. 1: 197–201.

Pyke, R. 1965. Spacings. *Journal of the Royal Statistical Society* B, no. 27: 395–436.

Ross, S. 1988. *A First Course in Probability,* 3d ed. New York: Macmillan Publishing Co.

Sullivan, B. E. 1984. Some Observations on the Present and Future Performance of Surface Public Transportation in the U.S. and Canada. *Transportation Research* 18A, no. 2.

Yang, M. C. K., and D. H. Robinson. 1986. *Understanding and Learning Statistics by Computer.* Philadelphia: World Scientific Publishing Co.

Zamurs, J. 1984. Assessing the Effect of Transportation Control Strategies on Carbon Monoxide Concentrations. *Air Pollution Control Association Journal* 34, no. 6.

Zimmels, Y. 1983. Theory of Hindered Sedimentation of Polydisperse Mixtures. *AIChE Journal* 29, no. 4: 669–76.

NOTES ON COMPUTER SIMULATIONS

The simulations presented in the computer activities sections were written in the program language, Turbo Pascal. Instead of presenting the actual Pascal code, brief descriptions will be given for how each simulation was obtained.

CHAPTER 3

Page 121

1. Use the algorithm for generating a binomial random variable to obtain the variable X, which is binomial with parameters n_1, p_1, and to obtain the variable Y, which is binomial with parameters n_2, p_2.
2. Let $W = X + Y$.
3. Repeat steps 1 and 2 for the desired number of simulated values for W.
4. Compute the sample mean and sample standard deviation for W.
5. Construct a histogram for all values of W simulated.

Pages 121–122

1. Declare the number of distinct coupons desired, denoted as N.
2. Generate a random number R, which is $U(0, 1)$.
3. Take the integer part of $N * R + 1$.
4. Repeat steps 2 and 3 until each distinct integer $1, 2, \ldots, N$ has appeared at least once.
5. Let X = number of integers generated to satisfy step 4.
6. Repeat steps 2–5 for the desired number of simulated values for X.
7. Compute the sample mean and sample standard deviation for X.
8. Construct a histogram for all values of X simulated.

CHAPTER 4

Pages 190–191

1. Use the algorithm for generating Poisson random variables to obtain X, which is a Poisson variable with parameter λ_1, and to obtain Y, which is a Poisson variable with a parameter of λ_2.

2. Let $W = X/Y$.
3. Repeat steps 1 and 2 for the desired number of simulated values for W.
4. Compute the sample mean and sample standard deviation for W.
5. Construct a histogram for the simulated values of W.

Page 191

1. Use the algorithm for generating exponential random variables to obtain X, which is an exponential variable with mean θ_x and to obtain Y, which is an exponential variable with a mean of θ_y.
2. Let $W = \max\{X, Y\}$.
3. Repeat steps 1 and 2 for the desired number of simulated values for W.
4. Compute the sample mean and sample standard deviation for W.
5. Construct a histogram for the simulated values of W.

Page 192

1. Use one of the algorithms for simulating normal random variables to obtain X, which is a normal variable with parameters μ_x and σ_x^2 and to obtain Y, which is a normal variable with parameters μ_y and σ_y^2.
2. Let $W = X/(X + Y)$.
3. Repeat steps 1 and 2 for the desired numbers of simulated values for W.
4. Compute the sample mean and sample standard deviation for W.
5. Construct a histogram for the simulated values of W.

APPENDIX TABLES

TABLE 1 Random numbers

Row	Column 1	2	3	4	5	6	7	8	9	10	11	12	13	14
1	10480	15011	01536	02011	81647	91646	69179	14194	62590	36207	20969	99570	91291	90700
2	22368	46573	25595	85393	30995	89198	27982	53402	93965	34095	52666	19174	39615	99505
3	24130	48360	22527	97265	76393	64809	15179	24830	49340	32081	30680	19655	63348	58629
4	42167	93093	06243	61680	07856	16376	39440	53537	71341	57004	00849	74917	97758	16379
5	37570	39975	81837	16656	06121	91782	60468	81305	49684	60672	14110	06927	01263	54613
6	77921	06907	11008	42751	27756	53498	18602	70659	90655	15053	21916	81825	44394	42880
7	99562	72905	56420	69994	98872	31016	71194	18738	44013	48840	63213	21069	10634	12952
8	96301	91977	05463	07972	18876	20922	94595	56869	69014	60045	18425	84903	42508	32307
9	89579	14342	63661	10281	17453	18103	57740	84378	25331	12566	58678	44947	05585	56941
10	85475	36857	53342	53988	53060	59533	38867	62300	08158	17983	16439	11458	18593	64952
11	28918	69578	88231	33276	70997	79936	56865	05859	90106	31595	01547	85590	91610	78188
12	63553	40961	48235	03427	49626	69445	18663	72695	52180	20847	12234	90511	33703	90322
13	09429	93969	52636	92737	88974	33488	36320	17617	30015	08272	84115	27156	30613	74952
14	10365	61129	87529	85689	48237	52267	67689	93394	01511	26358	85104	20285	29975	89868
15	07119	97336	71048	08178	77233	13916	47564	81056	97735	85977	29372	74461	28551	90707
16	51085	12765	51821	51259	77452	16308	60756	92144	49442	53900	70960	63990	75601	40719
17	02368	21382	52404	60268	89368	19885	55322	44819	01188	65255	64835	44919	05944	55157
18	01011	54092	33362	94904	31273	04146	18594	29852	71585	85030	51132	01915	92747	64951
19	52162	53916	46369	58586	23216	14513	83149	98736	23495	64350	94738	17752	35156	35749
20	07056	97628	33787	09998	42698	06691	76988	13602	51851	46104	88916	19509	25625	58104
21	48663	91245	85828	14346	09172	30168	90229	04734	59193	22178	30421	61666	99904	32812
22	54164	58492	22421	74103	47070	25306	76468	26384	58151	06646	21524	15227	96909	44592
23	32639	32363	05597	24200	13363	38005	94342	28728	35806	06912	17012	64161	18296	22851
24	29334	27001	87637	87308	58731	00256	45834	15398	46557	41135	10367	07684	36188	18510
25	02488	33062	28834	07351	19731	92420	60952	61280	50001	67658	32586	86679	50720	94953
26	81525	72295	04839	96423	24878	82651	66566	14778	76797	14780	13300	87074	79666	95725
27	29676	20591	68086	26432	46901	20849	89768	81536	86645	12659	92259	57102	80428	25280
28	00742	57392	39064	66432	84673	40027	32832	61362	98947	96067	64760	64584	96096	98253
29	05366	04213	25669	26422	44407	44048	37937	63904	45766	66134	75470	66520	34693	90449
30	91921	26418	64117	94305	26766	25940	39972	22209	71500	64568	91402	42416	07884	69618
31	00582	04711	87917	77341	42206	35126	74087	99547	81817	42607	43808	76655	62028	76630

***TABLE 1,** continued*

Row	**Column** 1	2	3	4	5	6	7	8	9	10	11	12	13	14
32	00725	69884	62797	56170	86324	88072	76222	36086	84637	93161	76038	65855	77919	88006
33	69011	65795	95876	55293	18988	27354	26575	08625	40801	59920	29841	80150	12777	48501
34	25976	57948	29888	88604	67917	48708	18912	82271	65424	69774	33611	54262	85963	03547
35	09763	83473	73577	12908	30883	18317	28290	35797	05998	41688	34952	37888	38917	88050
36	91576	42595	27958	30134	04024	86385	29880	99730	55536	84855	29080	09250	79656	73211
37	17955	56349	90999	49127	20044	59931	06115	20542	18059	02008	73708	83517	36103	42791
38	46503	18584	18845	49618	02304	51038	20655	58727	28168	15475	56942	53389	20562	87338
39	92157	89634	94824	78171	84610	82834	09922	25417	44137	48413	25555	21246	35509	20468
40	14577	62765	35605	81263	39667	47358	56873	56307	61607	49518	89656	20103	77490	18062
41	98427	07523	33362	64270	01638	92477	66969	98420	04880	45585	46565	04102	46880	45709
42	34914	63976	88720	82765	34476	17032	87589	40836	32427	70002	70663	88863	77775	69348
43	70060	28277	39475	46473	23219	53416	94970	25832	69975	94884	19661	72828	00102	66794
44	53976	54914	06990	67245	68350	82948	11398	42878	80287	88267	47363	46634	06541	97809
45	76072	29515	40980	07391	58745	25774	22987	80059	39911	96189	41151	14222	60697	59583
46	90725	52210	83974	29992	65831	38857	50490	83765	55657	14361	31720	57375	56228	41546
47	64364	67412	33339	31926	14883	24413	59744	92351	97473	89286	35931	04110	23726	51900
48	08962	00358	31662	25388	61642	34072	81249	35648	56891	69352	48373	45578	78547	81788
49	95012	68379	93526	70765	10592	04542	76463	54328	02349	17247	28865	14777	62730	92277
50	15664	10493	20492	38391	91132	21999	59516	81652	27195	48223	46751	22923	32261	85653
51	16408	81899	04153	53381	79401	21438	83035	92350	36693	31238	59649	91754	72772	02338
52	18629	81953	05520	91962	04739	13092	97662	24822	94730	06496	35090	04822	86774	98289
53	73115	35101	47498	87637	99016	71060	88824	71013	18735	20286	23153	72924	35165	43040
54	57491	16703	23167	49323	45021	33132	12544	41035	80780	45393	44812	12515	98931	91202
55	30405	83946	23792	14422	15059	45799	22716	19792	09983	74353	68668	30429	70735	25499
56	16631	35006	85900	98275	32388	52390	16815	69298	82732	38480	73817	32523	41961	44437
57	96773	20206	42559	78985	05300	22164	24369	54224	35083	19687	11052	91491	60383	19746
58	38935	64202	14349	82674	66523	44133	00697	35552	35970	19124	63318	29686	03387	59846
59	31624	76384	17403	53363	44167	64486	64758	75366	76554	31601	12614	33072	60332	92325
60	78919	19474	23632	27889	47914	02584	37680	20801	72152	39339	34806	08930	85001	87820
61	03931	33309	57047	74211	63445	17361	62825	39908	05607	91284	68833	25570	38818	46920
62	74426	33278	43972	10119	89917	15665	52872	73823	73144	88662	88970	74492	51805	99378
63	09066	00903	20795	95452	92648	45454	09552	88815	16553	51125	79375	97596	16296	66092
64	42238	12426	87025	14267	20979	04508	64535	31355	86064	29472	47689	05974	52468	16834
65	16153	08002	26504	41744	81959	65642	74240	56302	00033	67107	77510	70625	28725	34191
66	21457	40742	29820	96783	29400	21840	15035	34527	33310	06116	95240	15957	16572	06004
67	21581	57802	02050	89728	17937	37621	47075	42080	97403	48626	68995	43805	33386	21597
68	55612	78095	83197	33732	05810	24813	86902	60397	16489	03264	88525	42786	05269	92532
69	44657	66999	99324	51281	84463	60563	79312	93454	68876	25471	93911	25650	12682	73572
70	91340	84979	46949	81973	37949	61023	43997	15263	80644	43942	89203	71795	99533	50501
71	91227	21199	31935	27022	84067	05462	35216	14486	29891	68607	41867	14951	91696	85065
72	50001	38140	66321	19924	72163	09538	12151	06878	91903	18749	34405	56087	82790	70925
73	65390	05224	72958	28609	81406	39147	25549	48542	42627	45233	57202	94617	23772	07896
74	27504	96131	83944	41575	10573	08619	64482	73923	36152	05184	94142	25299	84387	34925
75	37169	94851	39117	89632	00959	16487	65536	49071	39782	17095	02330	74301	00275	48280
76	11508	70225	51111	38351	19444	66499	71945	05422	13442	78675	84081	66938	93654	59894
77	37449	30362	06694	54690	04052	53115	62757	95348	78662	11163	81651	50245	34971	52924
78	46515	70331	85922	38329	57015	15765	97161	17869	45349	61796	66345	81073	49106	79860

Source: Reprinted with permission from W. H. Beyer, ed., *CRC Standard Mathematical Tables,* 24th ed. (abridged). Copyright CRC Press, Inc., Boca Raton, FL, 1976.

Binomial probabilities

Tabulated values are $\sum_{x=0}^{k} p(x)$. (Computations are rounded at the third decimal place.)

Distribution function

(a) $n = 5$

k \ p	.01	.05	.10	.20	.30	.40	.50	.60	.70	.80	.90	.95	.99
0	.951	.774	.590	.328	.168	.078	.031	.010	.002	.000	.000	.000	.000
1	.999	.977	.919	.737	.528	.337	.188	.087	.031	.007	.000	.000	.000
2	1.000	.999	.991	.942	.837	.683	.500	.317	.163	.058	.009	.001	.000
3	1.000	1.000	1.000	.993	.969	.913	.812	.663	.472	.263	.081	.023	.001
4	1.000	1.000	1.000	1.000	.998	.990	.969	.922	.832	.672	.410	.226	.049

(b) $n = 10$

k \ p	.01	.05	.10	.20	.30	.40	.50	.60	.70	.80	.90	.95	.99
0	.904	.599	.349	.107	.028	.006	.001	.000	.000	.000	.000	.000	.000
1	.996	.914	.736	.376	.149	.046	.011	.002	.000	.000	.000	.000	.000
2	1.000	.988	.930	.678	.383	.167	.055	.012	.002	.000	.000	.000	.000
3	1.000	.999	.987	.879	.650	.382	.172	.055	.011	.001	.000	.000	.000
4	1.000	1.000	.998	.967	.850	.633	.377	.166	.047	.006	.000	.000	.000
5	1.000	1.000	1.000	.994	.953	.834	.623	.367	.150	.033	.002	.000	.000
6	1.000	1.000	1.000	.999	.989	.945	.828	.618	.350	.121	.013	.001	.000
7	1.000	1.000	1.000	1.000	.998	.988	.945	.833	.617	.322	.070	.012	.000
8	1.000	1.000	1.000	1.000	1.000	.998	.989	.954	.851	.624	.264	.086	.004
9	1.000	1.000	1.000	1.000	1.000	1.000	.999	.994	.972	.893	.651	.401	.096

(c) $n = 15$

k \ p	.01	.05	.10	.20	.30	.40	.50	.60	.70	.80	.90	.95	.99
0	.860	.463	.206	.035	.005	.000	.000	.000	.000	.000	.000	.000	.000
1	.990	.829	.549	.167	.035	.005	.000	.000	.000	.000	.000	.000	.000
2	1.000	.964	.816	.398	.127	.027	.004	.000	.000	.000	.000	.000	.000
3	1.000	.995	.944	.648	.297	.091	.018	.002	.000	.000	.000	.000	.000
4	1.000	.999	.987	.836	.515	.217	.059	.009	.001	.000	.000	.000	.000
5	1.000	1.000	.998	.939	.722	.403	.151	.034	.004	.000	.000	.000	.000
6	1.000	1.000	1.000	.982	.869	.610	.304	.095	.015	.001	.000	.000	.000
7	1.000	1.000	1.000	.996	.950	.787	.500	.213	.050	.004	.000	.000	.000
8	1.000	1.000	1.000	.999	.985	.905	.696	.390	.131	.018	.000	.000	.000
9	1.000	1.000	1.000	1.000	.996	.966	.849	.597	.278	.061	.002	.000	.000
10	1.000	1.000	1.000	1.000	.999	.991	.941	.783	.485	.164	.013	.001	.000
11	1.000	1.000	1.000	1.000	1.000	.998	.982	.909	.703	.352	.056	.005	.000
12	1.000	1.000	1.000	1.000	1.000	1.000	.996	.973	.873	.602	.184	.036	.000
13	1.000	1.000	1.000	1.000	1.000	1.000	1.000	.995	.965	.833	.451	.171	.010
14	1.000	1.000	1.000	1.000	1.000	1.000	1.000	1.000	.995	.965	.794	.537	.140

***TABLE 2**, continued*

(*d*) $n = 20$

k \ p	.01	.05	.10	.20	.30	.40	.50	.60	.70	.80	.90	.95	.99
0	.818	.358	.122	.002	.001	.000	.000	.000	.000	.000	.000	.000	.000
1	.983	.736	.392	.069	.008	.001	.000	.000	.000	.000	.000	.000	.000
2	.999	.925	.677	.206	.035	.004	.000	.000	.000	.000	.000	.000	.000
3	1.000	.984	.867	.411	.107	.016	.001	.000	.000	.000	.000	.000	.000
4	1.000	.997	.957	.630	.238	.051	.006	.000	.000	.000	.000	.000	.000
5	1.000	1.000	.989	.804	.416	.126	.021	.002	.000	.000	.000	.000	.000
6	1.000	1.000	.998	.913	.608	.250	.058	.006	.000	.000	.000	.000	.000
7	1.000	1.000	1.000	.968	.772	.416	.132	.021	.001	.000	.000	.000	.000
8	1.000	1.000	1.000	.990	.887	.596	.252	.057	.005	.000	.000	.000	.000
9	1.000	1.000	1.000	.997	.952	.755	.412	.128	.017	.001	.000	.000	.000
10	1.000	1.000	1.000	.999	.983	.872	.588	.245	.048	.003	.000	.000	.000
11	1.000	1.000	1.000	1.000	.995	.943	.748	.404	.113	.010	.000	.000	.000
12	1.000	1.000	1.000	1.000	.999	.979	.868	.584	.228	.032	.000	.000	.000
13	1.000	1.000	1.000	1.000	1.000	.994	.942	.750	.392	.087	.002	.000	.000
14	1.000	1.000	1.000	1.000	1.000	.998	.979	.874	.584	.196	.011	.000	.000
15	1.000	1.000	1.000	1.000	1.000	1.000	.994	.949	.762	.370	.043	.003	.000
16	1.000	1.000	1.000	1.000	1.000	1.000	.999	.984	.893	.589	.133	.016	.000
17	1.000	1.000	1.000	1.000	1.000	1.000	1.000	.996	.965	.794	.323	.075	.001
18	1.000	1.000	1.000	1.000	1.000	1.000	1.000	.999	.992	.931	.608	.264	.017
19	1.000	1.000	1.000	1.000	1.000	1.000	1.000	1.000	.999	.988	.878	.642	.182

(*e*) $n = 25$

k \ p	.01	.05	.10	.20	.30	.40	.50	.60	.70	.80	.90	.95	.99
0	.778	.277	.072	.004	.000	.000	.000	.000	.000	.000	.000	.000	.000
1	.974	.642	.271	.027	.002	.000	.000	.000	.000	.000	.000	.000	.000
2	.998	.873	.537	.098	.009	.000	.000	.000	.000	.000	.000	.000	.000
3	1.000	.966	.764	.234	.033	.002	.000	.000	.000	.000	.000	.000	.000
4	1.000	.993	.902	.421	.090	.009	.000	.000	.000	.000	.000	.000	.000
5	1.000	.999	.967	.617	.193	.029	.002	.000	.000	.000	.000	.000	.000
6	1.000	1.000	.991	.780	.341	.074	.007	.000	.000	.000	.000	.000	.000
7	1.000	1.000	.998	.891	.512	.154	.022	.001	.000	.000	.000	.000	.000
8	1.000	1.000	1.000	.953	.677	.274	.054	.004	.000	.000	.000	.000	.000
9	1.000	1.000	1.000	.983	.811	.425	.115	.013	.000	.000	.000	.000	.000
10	1.000	1.000	1.000	.994	.902	.586	.212	.034	.002	.000	.000	.000	.000
11	1.000	1.000	1.000	.998	.956	.732	.345	.078	.006	.000	.000	.000	.000
12	1.000	1.000	1.000	1.000	.983	.846	.500	.154	.017	.000	.000	.000	.000
13	1.000	1.000	1.000	1.000	.994	.922	.655	.268	.044	.002	.000	.000	.000
14	1.000	1.000	1.000	1.000	.998	.966	.788	.414	.098	.006	.000	.000	.000
15	1.000	1.000	1.000	1.000	1.000	.987	.885	.575	.189	.017	.000	.000	.000
16	1.000	1.000	1.000	1.000	1.000	.996	.946	.726	.323	.047	.000	.000	.000
17	1.000	1.000	1.000	1.000	1.000	.999	.978	.846	.488	.109	.002	.000	.000
18	1.000	1.000	1.000	1.000	1.000	1.000	.993	.926	.659	.220	.009	.000	.000
19	1.000	1.000	1.000	1.000	1.000	1.000	.998	.971	.807	.383	.033	.001	.000
20	1.000	1.000	1.000	1.000	1.000	1.000	1.000	.991	.910	.579	.098	.007	.000
21	1.000	1.000	1.000	1.000	1.000	1.000	1.000	.998	.967	.766	.236	.034	.000
22	1.000	1.000	1.000	1.000	1.000	1.000	1.000	1.000	.991	.902	.463	.127	.002
23	1.000	1.000	1.000	1.000	1.000	1.000	1.000	1.000	.998	.973	.729	.358	.026
24	1.000	1.000	1.000	1.000	1.000	1.000	1.000	1.000	1.000	.996	.928	.723	.222

TABLE 3 **Poisson distribution function**

$$F(x, \lambda) = \sum_{k=0}^{x} e^{-\lambda} \frac{\lambda^k}{k!}$$

λ \ x	0	1	2	3	4	5	6	7	8	9
.02	.980	1.000								
.04	.961	.999	1.000							
.06	.942	.998	1.000							
.08	.923	.997	1.000							
.10	.905	.995	1.000							
.15	.861	.990	.999	1.000						
.20	.819	.982	.999	1.000						
.25	.779	.974	.998	1.000						
.30	.741	.963	.996	1.000						
.35	.705	.951	.994	1.000						
.40	.670	.938	.992	.999	1.000					
.45	.638	.925	.989	.999	1.000					
.50	.607	.910	.986	.998	1.000					
.55	.577	.894	.982	.998	1.000					
.60	.549	.878	.977	.997	1.000					
.65	.522	.861	.972	.996	.999	1.000				
.70	.497	.844	.966	.994	.999	1.000				
.75	.472	.827	.959	.993	.999	1.000				
.80	.449	.809	.953	.991	.999	1.000				
.85	.427	.791	.945	.989	.998	1.000				
.90	.407	.772	.937	.987	.998	1.000				
.95	.387	.754	.929	.981	.997	1.000				
1.00	.368	.736	.920	.981	.996	.999	1.000			
1.1	.333	.699	.900	.974	.995	.999	1.000			
1.2	.301	.663	.879	.966	.992	.998	1.000			
1.3	.273	.627	.857	.957	.989	.998	1.000			
1.4	.247	.592	.833	.946	.986	.997	.999	1.000		
1.5	.223	.558	.809	.934	.981	.996	.999	1.000		
1.6	.202	.525	.783	.921	.976	.994	.999	1.000		
1.7	.183	.493	.757	.907	.970	.992	.998	1.000		
1.8	.165	.463	.731	.891	.964	.990	.997	.999	1.000	
1.9	.150	.434	.704	.875	.956	.987	.997	.999	1.000	
2.0	.135	.406	.677	.857	.947	.983	.995	.999	1.000	
2.2	.111	.355	.623	.819	.928	.975	.993	.998	1.000	
2.4	.091	.308	.570	.779	.904	.964	.988	.997	.999	1.000
2.6	.074	.267	.518	.736	.877	.951	.983	.995	.999	1.000
2.8	.061	.231	.469	.692	.848	.935	.976	.992	.998	.999
3.0	.050	.199	.423	.647	.815	.916	.966	.998	.996	.999
3.2	.041	.171	.380	.603	.781	.895	.955	.983	.994	.998
3.4	.033	.147	.340	.558	.744	.871	.942	.977	.992	.997
3.6	.027	.126	.303	.515	.706	.844	.927	.969	.988	.996
3.8	.022	.107	.269	.473	.668	.816	.909	.960	.984	.994
4.0	.018	.092	.238	.433	.629	.785	.889	.949	.979	.992
4.2	.015	.078	.210	.395	.590	.753	.867	.936	.972	.989
4.4	.012	.066	.185	.359	.551	.720	.844	.921	.964	.985
4.6	.010	.056	.163	.326	.513	.686	.818	.905	.955	.980

***TABLE 3**, continued*

λ \ x	0	1	2	3	4	5	6	7	8	9
4.8	.008	.048	.143	.294	.476	.651	.791	.887	.944	.975
5.0	.007	.040	.125	.265	.440	.616	.762	.867	.932	.968
5.2	.006	.034	.109	.238	.406	.581	.732	.845	.918	.960
5.4	.005	.029	.095	.213	.373	.546	.702	.822	.903	.951
5.6	.004	.024	.082	.191	.342	.512	.670	.797	.886	.941
5.8	.003	.021	.072	.170	.313	.478	.638	.771	.867	.929
6.0	.002	.017	.062	.151	.285	.446	.606	.744	.847	.916

	10	11	12	13	14	15	16
2.8	1.000						
3.0	1.000						
3.2	1.000						
3.4	.999	1.000					
3.6	.999	1.000					
3.8	.998	.999	1.000				
4.0	.997	.999	1.000				
4.2	.996	.999	1.000				
4.4	.994	.998	.999	1.000			
4.6	.992	.997	.999	1.000			
4.8	.990	.996	.999	1.000			
5.0	.986	.995	.998	.999	1.000		
5.2	.982	.993	.997	.999	1.000		
5.4	.977	.990	.996	.999	1.000		
5.6	.972	.988	.995	.998	.999	1.000	
5.8	.965	.984	.993	.997	.999	1.000	
6.0	.957	.980	.991	.996	.999	.999	1.000

	0	1	2	3	4	5	6	7	8	9
6.2	.002	.015	.054	.134	.259	.414	.574	.716	.826	.902
6.4	.002	.012	.046	.119	.235	.384	.542	.687	.803	.886
6.6	.001	.010	.040	.105	.213	.355	.511	.658	.780	.869
6.8	.001	.009	.034	.093	.192	.327	.480	.628	.755	.850
7.0	.001	.007	.030	.082	.173	.301	.450	.599	.729	.830
7.2	.001	.006	.025	.072	.156	.276	.420	.569	.703	.810
7.4	.001	.005	.022	.063	.140	.253	.392	.539	.676	.788
7.6	.001	.004	.019	.055	.125	.231	.365	.510	.648	.765
7.8	.000	.004	.016	.048	.112	.210	.338	.481	.620	.741
8.0	.000	.003	.014	.042	.100	.191	.313	.453	.593	.717
8.5	.000	.002	.009	.030	.074	.150	.256	.386	.523	.653
9.0	.000	.001	.006	.021	.055	.116	.207	.324	.456	.587
9.5	.000	.001	.004	.015	.040	.089	.165	.269	.392	.522
10.0	.000	.000	.003	.010	.029	.067	.130	.220	.333	.458

TABLE 3, continued

λ \ x	10	11	12	13	14	15	16	17	18	19
6.2	.949	.975	.989	.995	.998	.999	1.000			
6.4	.939	.969	.986	.994	.997	.999	1.000			
6.6	.927	.963	.982	.992	.997	.999	.999	1.000		
6.8	.915	.955	.978	.990	.996	.998	.999	1.000		
7.0	.901	.947	.973	.987	.994	.998	.999	1.000		
7.2	.887	.937	.967	.984	.993	.997	.999	.999	1.000	
7.4	.871	.926	.961	.980	.991	.996	.998	.999	1.000	
7.6	.854	.915	.954	.976	.989	.995	.998	.999	1.000	
7.8	.835	.902	.945	.971	.986	.993	.997	.999	1.000	
8.0	.816	.888	.936	.966	.983	.992	.996	.998	.999	1.000
8.5	.763	.849	.909	.949	.973	.986	.993	.997	.999	.999
9.0	.706	.803	.876	.926	.959	.978	.989	.995	.998	.999
9.5	.645	.752	.836	.898	.940	.967	.982	.991	.996	.998
10.0	.583	.697	.792	.864	.917	.951	.973	.986	.993	.997

	20	21	22
8.5	1.000		
9.0	1.000		
9.5	.999	1.000	
10.0	.998	.999	1.000

	0	1	2	3	4	5	6	7	8	9
10.5	.000	.000	.002	.007	.021	.050	.102	.179	.279	.397
11.0	.000	.000	.001	.005	.015	.038	.079	.143	.232	.341
11.5	.000	.000	.001	.003	.011	.028	.060	.114	.191	.289
12.0	.000	.000	.001	.002	.008	.020	.046	.090	.155	.242
12.5	.000	.000	.000	.002	.005	.015	.035	.070	.125	.201
13.0	.000	.000	.000	.001	.004	.011	.026	.054	.100	.166
13.5	.000	.000	.000	.001	.003	.008	.019	.041	.079	.135
14.0	.000	.000	.000	.000	.002	.006	.014	.032	.062	.109
14.5	.000	.000	.000	.000	.001	.004	.010	.024	.048	.088
15.0	.000	.000	.000	.000	.001	.003	.008	.018	.037	.070

	10	11	12	13	14	15	16	17	18	19
10.5	.521	.639	.742	.825	.888	.932	.960	.978	.988	.994
11.0	.460	.579	.689	.781	.854	.907	.944	.968	.982	.991
11.5	.402	.520	.633	.733	.815	.878	.924	.954	.974	.986
12.0	.347	.462	.576	.682	.772	.844	.899	.937	.963	.979
12.5	.297	.406	.519	.628	.725	.806	.869	.916	.948	.969
13.0	.252	.353	.463	.573	.675	.764	.835	.890	.930	.957
13.5	.211	.304	.409	.518	.623	.718	.798	.861	.908	.942
14.0	.176	.260	.358	.464	.570	.669	.756	.827	.883	.923
14.5	.145	.220	.311	.413	.518	.619	.711	.790	.853	.901
15.0	.118	.185	.268	.363	.466	.568	.664	.749	.819	.875

TABLE 3, *continued*

λ \ x	20	21	22	23	24	25	26	27	28	29
10.5	.997	.999	.999	1.000						
11.0	.995	.998	.999	1.000						
11.5	.992	.996	.998	.999	1.000					
12.0	.988	.994	.997	.999	.999	1.000				
12.5	.983	.991	.995	.998	.999	.999	1.000			
13.0	.975	.986	.992	.996	.998	.999	1.000			
13.5	.965	.980	.989	.994	.997	.998	.999	1.000		
14.0	.952	.971	.983	.991	.995	.997	.999	.999	1.000	
14.5	.936	.960	.976	.986	.992	.996	.998	.999	.999	1.000
15.0	.917	.947	.967	.981	.989	.994	.997	.998	.999	1.000

	4	5	6	7	8	9	10	11	12	13
16	.000	.001	.004	.010	.022	.043	.077	.127	.193	.275
17	.000	.001	.002	.005	.013	.026	.049	.085	.135	.201
18	.000	.000	.001	.003	.007	.015	.030	.055	.092	.143
19	.000	.000	.001	.002	.004	.009	.018	.035	.061	.098
20	.000	.000	.000	.001	.002	.005	.011	.021	.039	.066
21	.000	.000	.000	.000	.001	.003	.006	.013	.025	.043
22	.000	.000	.000	.000	.001	.002	.004	.008	.015	.028
23	.000	.000	.000	.000	.000	.001	.002	.004	.009	.017
24	.000	.000	.000	.000	.000	.000	.001	.003	.005	.011
25	.000	.000	.000	.000	.000	.000	.001	.001	.003	.006

	14	15	16	17	18	19	20	21	22	23
16	.368	.467	.566	.659	.742	.812	.868	.911	.942	.963
17	.281	.371	.468	.564	.655	.736	805	.861	.905	.937
18	.208	.287	.375	.469	.562	.651	.731	.799	.855	.899
19	.150	.215	.292	.378	.469	.561	.647	.725	.793	.849
20	.105	.157	.221	.297	.381	.470	.559	.644	.721	.787
21	.072	.111	.163	.227	.302	.384	.471	.558	.640	.716
22	.048	.077	.117	.169	.232	.306	.387	.472	.556	.637
23	.031	.052	.082	.123	.175	.238	.310	.389	.472	.555
24	.020	.034	.056	.087	.128	.180	.243	.314	.392	.473
25	.012	.022	.038	.060	.092	.134	.185	.247	.318	.394

	24	25	26	27	28	29	30	31	32	33
16	.978	.987	.993	.996	.998	.999	.999	1.000		
17	.959	.975	.985	.991	.995	.997	.999	.999	1.000	
18	.932	.955	.972	.983	.990	.994	.997	.998	.999	1.000
19	.893	.927	.951	.969	.980	.988	.993	.996	.998	.999
20	.843	.888	.922	.948	.966	.978	.987	.992	.995	.997
21	.782	.838	.883	.917	.944	.963	.976	.985	.991	.994
22	.712	.777	.832	.877	.913	.940	.959	.973	.983	.989
23	.635	.708	.772	.827	.873	.908	.936	.956	.971	.981
24	.554	.632	.704	.768	.823	.868	.904	.932	.953	.969
25	.473	.553	.629	.700	.763	.818	.863	.900	.929	.950

***TABLE 3**, continued*

λ \ x	34	35	36	37	38	39	40	41	42	43
19	.999	1.000								
20	.999	.999	1.000							
21	.997	.998	.999	.999	1.000					
22	.994	.996	.998	.999	.999	1.000				
23	.988	.993	.996	.997	.999	.999	1.000			
24	.979	.987	.992	.995	.997	.998	.999	.999	1.000	
25	.966	.978	.985	.991	.995	.997	.998	.999	.999	1.000

Source: Reprinted by permission from E. C. Molina, *Poisson's Exponential Binomial Limit* (Princeton, N.J.: D. Van Nostrand Company, 1947).

TABLE 4 **Normal curve areas**

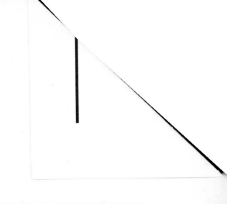

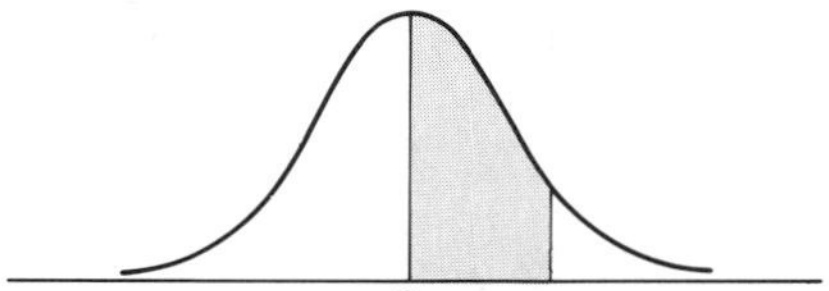

z	.00	.01	.02	.03	.04	.05	.06	.07	.08	.09
.0	.0000	.0040	.0080	.0120	.0160	.0199	.0239	.0279	.0319	.0359
.1	.0398	.0438	.0478	.0517	.0557	.0596	.0636	.0675	.0174	.0753
.2	.0793	.0832	.0871	.0910	.0948	.0987	.1026	.1064	.1103	.1141
.3	.1179	.1217	.1255	.1293	.1331	.1368	.1406	.1443	.1480	.1517
.4	.1554	.1591	.1628	.1664	.1700	.1736	.1772	.1808	.1844	.1879
.5	.1915	.1950	.1985	.2019	.2054	.2088	.2123	.2157	.2190	.2224
.6	.2257	.2291	.2324	.2357	.2389	.2422	.2454	.2486	.2517	.2549
.7	.2580	.2611	.2642	.2673	.2704	.2734	.2764	.2794	.2823	.2852
.8	.2881	.2910	.2939	.2967	.2995	.3023	.3051	.3078	.3106	.3133
.9	.3159	.3186	.3212	.3238	.3264	.3289	.3315	.3340	.3365	.3389
1.0	.3413	.3438	.3461	.3485	.3508	.3531	.3554	.3577	.3599	.3621
1.1	.3643	.3665	.3686	.3708	.3729	.3749	.3770	.3790	.3810	.3830
1.2	.3849	.3869	.3888	.3907	.3925	.3944	.3962	.3980	.3997	.4015
1.3	.4032	.4049	.4066	.4082	.4099	.4115	.4131	.4147	.4162	.4177
1.4	.4192	.4207	.4222	.4236	.4251	.4265	.4279	.4292	.4306	.4319
1.5	.4332	.4345	.4357	.4370	.4382	.4394	.4406	.4418	.4429	.4441
1.6	.4452	.4463	.4474	.4484	.4495	.4505	.4515	.4525	.4535	.4545
1.7	.4554	.4564	.4573	.4582	.4591	.4599	.4608	.4616	.4625	.4633
1.8	.4641	.4649	.4656	.4664	.4671	.4678	.4686	.4693	.4699	.4706
1.9	.4713	.4719	.4726	.4732	.4738	.4744	.4750	.4756	.4761	.4767
2.0	.4772	.4778	.4783	.4788	.4793	.4798	.4803	.4808	.4812	.4817
2.1	.4821	.4826	.4830	.4834	.4838	.4842	.4846	.4850	.4854	.4857
2.2	.4861	.4864	.4868	.4871	.4875	.4878	.4881	.4884	.4887	.4890
2.3	.4893	.4896	.4898	.4901	.4904	.4906	.4909	.4911	.4913	.4916
2.4	.4918	.4920	.4922	.4925	.4927	.4929	.4931	.4932	.4934	.4936
2.5	.4938	.4940	.4941	.4943	.4945	.4946	.4948	.4949	.4951	.4952
2.6	.4953	.4955	.4956	.4957	.4959	.4960	.4961	.4962	.4963	.4964
2.7	.4965	.4966	.4967	.4968	.4969	.4970	.4971	.4972	.4973	.4974
2.8	.4974	.4975	.4976	.4977	.4977	.4978	.4979	.4979	.4980	.4981
2.9	.4981	.4982	.4982	.4983	.4984	.4984	.4985	.4985	.4986	.4986
3.0	.4987	.4987	.4987	.4988	.4988	.4989	.4989	.4989	.4990	.4990

Source: Abridged from Table I of A. Hald. *Statistical Tables and Formulas* (New York: John Wiley & Sons, 1952). Reproduced by permission of A. Hald and the publisher.

ANSWERS TO SELECTED EXERCISES

CHAPTER 2

2.1 **a.** 3

b. 13

c. 7

d. 9

2.2 **a.** $JD, JS, DM, DN, MN, JM, JN, DS, MS, SN$

b. $A = \{JD, JM, JS, JN, DM, DS, DN\}$

c. $B = \{JM, JS, JN, DM, DS, DN\}$

d. $\bar{A} = \{MS, MN, SN\}$

e. $\bar{A} = \{MS, MN, SN\}$

$AB = B$

$A \cup B = A$

$\overline{AB} = \{JD, MS, MN, SN\}$

2.5 **a.** L = left turn, R = right turn, S = straight

b. $P(L) = P(R) = P(S) = 1/3$

c. 2/3

2.6 **a.** .4

b. .9

c. .4

d. .1

e. .5

f. .1

g. .1

2.7 **a.** 1/3

b. 6/15

c. 19/48

d. 2/3

2.8 **a.** .43

b. .05

c. .59

d. .74

e. .91

2.9 **a.** .08

b. .16

c. .14

d. .84

2.10 **a.** *SS, SR, SL, RS, RR, RL, LS, LR, LL*

b. 5/9

c. 5/9

2.11 **a.** (I, I), (I, II), (II, I), (II, II)

b. $P[(\text{I}, \text{I})] = 1/4$

$P[(\text{I}, \text{I})] + P[(\text{II}, \text{II})] = 1/2$

2.12 **a.** (I, I), (I, II), (I, III), (II, I), (II, II), (II, III), (III, I), (III, II), (III, III)

b. 1/3

c. 5/9

2.13 3/10

2.14 **a.** 42

b. 21

2.15 1/5

2.16 5040

2.17 **a.** 168

b. 1/8

2.19 **a.** 3/5

b. 2/5

c. 3/10

2.20 ***a.*** .09375
b. .140625
c. .578125
d. .5625

2.21 ***a.*** .48
b. .04
c. .48

2.22 ***a.*** 24
b. 1/2

2.23 ***a.*** 1680
b. 1/12

2.24 5/8

2.25 ***a.*** 2/3
b. 1/9
c. 1/3

2.26 ***a.*** .4979
b. .5144
c. .7443
d. .4399
e. .6894

2.27 ***a.*** .16
b. 1 − .16 = .84
c. .96
d. No

2.28 ***a.*** .10
b. .03
c. .06
d. .018

2.29 ***a.*** .6316
b. .9684
c. .6522

2.30 ***a.*** .8053
b. .9947
c. .8096

2.31 **a.** 1/10
b. 7/10
c. 1/7

2.32 **a.** .7225
b. .19

2.33 **a.** .999
b. .9009

2.34 Series: .81
Parallel: .99

2.35 .40

2.36 .8235

2.37 .0833

2.38 **a.** 3/5
b. 2/5
c. Yes
d. Yes

2.41 .5073

2.42 **a.** $HHHH$ $THHH$
$HHHT$ $THHT$
$HHTH$ $THTH$
$HHTT$ $THTT$
$HTHH$ $TTHH$
$HTHT$ $TTHT$
$HTTH$ $TTTH$
$HTTT$ $TTTT$
b. $A = \{HHHT, HHTH, HTHH, THHH\}$
c. 1/4

2.43 **a.** N_1N_2 N_2N_3 N_3N_4 N_4D_1 D_1D_2
N_1N_3 N_2N_4 N_3D_1 N_4D_2 D_1D_3
N_1N_4 N_2D_1 N_3D_2 N_4D_3 D_2D_3
N_1D_1 N_2D_2 N_3D_3 N_1D_2 N_2D_3
N_1D_3
b. $A = \{N_1N_2, N_1N_3, N_1N_4, N_2N_3, N_2N_4, N_3N_4\}$
c. 2/7

2.44 **a.**

Year	Desired Probability
1973	.07
1974	.32
1975	.29

b.

Year	Desired Probability
1973	.23
1974	.43
1975	.39

c.

Year	Desired Probability
1973	.59
1974	.75
1975	.65

2.45 **a.** 36

b. 1/6

2.47 **a.** .57

b. .18

c. .9

d. .3158

2.48 120

2.49 9,000,000

2.50 720

2.51 18

2.52 40,320

2.53 **a.** .0362

b. .000495, .001981

2.54 **a.** .216

b. .936

c. .648

2.55 **a.** 1/8

b. 1/64

c. No

2.56 1/6

2.57 .5952

2.58 **a.** 5/16

b. $27(1/2)^{10}$

2.59 **a.** $(1/4)^7$

b. $9(1/4)^3$

2.60 No

2.61 .8704

2.62 .0625

2.65 No

2.66 1/2

2.67 1/7

2.69 A

2.70 **a.** .00892

b. .9890

CHAPTER 3

3.1

X	$p(X)$
0	1/30
1	3/10
2	1/2
3	1/6

3.2 **a.**

Y	$p(Y)$
0	1/16
1	1/4
2	3/8
3	1/4
4	1/16

b. No

3.3

X	$p(X)$
0	.2585
1	.4419
2	.2518
3	.0478

Assumes independence

3.4

X	$p(X)$
0	1/8
1	3/8
2	3/8
3	1/8

3.5 **a.**

X	$p(X)$
0	.0053
1	.0575
2	.2331
3	.4201
4	.2840

b. .9947

3.6

X	$p(X)$
0	.1296
1	.3456
2	.3456
3	.1536
4	.0256

3.7

x	$p(x)$
0	8/27
1	12/27
2	6/27
3	1/27

x	$p(y)$
0	2744/3375
1	588/3375
2	42/3375
3	1/3375

$x + y$	$p(x + y)$
0	.24090
1	.41295
2	.26179
3	.07445
4	.00935
5	.00053
6	.00003

3.8 **a.**

x	$p(x)$
0	.49
1	.28
2	.18
3	.04
4	.01

b. .51

3.9 **a.**

x	$p(x)$
0	1/6
1	2/3
2	1/6

b.

x	$p(x)$
0	1/2
1	1/2

c. $p(0) = 1$

3.10

y	$p(y)$
0	.019
1	.252
2	.729

3.11 **a.** 0, 2/3

b. 0, 12/5

3.12 **a.** $\mu = 313.22$

$\sigma = 27.62$

b. Yes

3.13 $\mu = .4$

$\sigma = .6633$

3.14 $\mu = .2$

$\sigma^2 = .18$

3.15 \$60,000 for firm I

\$120,000 for both firms

3.16 $\mu = 1$

$\sigma^2 = 1/2$

3.17 **a.** (1.4702, 6.5298)

b. Yes

3.18 41

3.19 **a.** (84.1886, 115.8114)

b. No

3.20 **a.** $\mu = \$200$

$\sigma = \$80$

b. \$360

3.21 **a.** .1536
b. .1808
c. .9728
d. .8
e. .64

3.22 **a.** .250
b. .057
c. .180

3.23 **a.** .537
b. .098

3.24 **a.** .109
b. .999
c. .589

3.25 **a.** 16
b. 3.2

3.26 **a.** .672
b. .672

3.27 8

3.28 **a.** 1/16
b. 1/4

3.29 .5931

3.30 **a.** Independence
b. .7379

3.31 **a.** .99
b. .9999

3.32 2

3.33 **a.** .1536
b. .9728

3.34 **a.** $400,000
b. $474,342

3.35 3.96

3.36 **a.** 22
b. 45

3.37 840

3.38 **a.** .9

b. $P(Y > 4 \mid Y > 2) = P(Y > 2) = (1 - p)^2$

3.39 **a.** .648

b. 1

3.40 .09

3.41 **a.** .04374

b. .99144

3.42 .1

3.43 **a.** $\mu = 10/9$

$\sigma^2 = .1235$

b. $\mu = 10/3$

$\sigma^2 = .3704$

3.44 .0645

3.45 $\mu = 150$

$\sigma^2 = 4500$

No

3.46 **a.** .09877

b. .14815

3.47 **a.** .128

b. .03072

3.48 $\mu = 15$

$\sigma^2 = 60$

3.49 **a.** .06561

b. $\mu = 4.4444$

$\sigma^2 = .4938$

3.50 **a.**

Try	Probability
1	.4
2	.24
3	.144

b. .1728

3.51 **a.** 3/16

b. 3/16

c. 1/8

d. 1/2

3.52 **a.** .0902

b. .143

c. .857

d. .2407

3.53 **a.** .0183

b. .908

c. .997

3.54 **a.** .905

b. .005

c. .819

3.55 **a.** .467

b. .188

3.56 **a.** .022

b. .866

3.57 **a.** .8187

b. .5488

3.58 .353

3.59 **a.** .140

b. .042

c. .997

3.60 $\mu = 80$

$\sigma^2 = 800$

No

3.61 **a.** 1.44×10^{-5}

b. 1.44×10^{-5}

3.62 **a.** .9817

b. $1 - e^{-12}$

3.63 $\mu = 320$

$\sigma = 56.5685$

3.64 **a.** .9997

b. 2

3.65 $E[Y(Y - 1)] = \lambda^2$

3.66 47.5 hours

3.67 **a.** .001
b. .000017

3.68 **a.** 4/7
b. 6/7
c. 1/3
d. 4/7

3.69 1/42

3.70 $\mu = 100$
$\sigma^2 = 5000/3$
$P(\text{cost} < \$222.48) = 8/9$

3.71 **a.** 4/5
b. 1/5

3.72 1/30, probably not random

3.73 **a.**

y	$p(y)$
0	7/15
1	7/15
2	1/15

b.

y	$p(y)$
0	1/6
1	1/2
2	3/10
3	1/30

3.74 **a.** 1/14
b. 13/14
c. 1/14

3.75 3/14

3.76 **a.** 41/42
b. 11/42
c. 1/42
d. ϕ

3.77 **a.** 9/14
b. 13/14
c. 1

3.78

y	$p(y)$
0	1/15
1	8/15
2	6/15

3.79 **a.** 1
b. 1
c. 18/19
d. 49/57
e. 728/969

3.80 **a.** 1
b. 1
c. 1
d. 113/114
e. 938/969

3.81 10/21

3.82 $P(t) = [pt + 1 - p]$

3.87 $P(t) = [pt + (1 - p)]^n$

3.88 **a.** .5037
b. .3679

3.90 $E(\text{duration}) = 3$
Probability A wins = 1/4

3.93 $P(Y = 4) = .32805$
$P(Y \geq 1) = .99999$

3.94 **a.** .9606
b. .9994

3.95 **a.** 1
b. .5905
c. .1681
d. .03125
e. 0

3.96 (a, b, c)

p	$p(Y \leq 0)$	$p(Y \leq 1)$	$p(Y \leq 2)$
0	1.0000	1.0000	1.0000
.05	.5987	.9139	.9885
.10	.3487	.7361	.9298
.30	.0283	.1493	.3828
.50	.0010	.0107	.0597
1.00	.0000	.0000	.0000

3.97 **a.** $n = 25, a = 5$

b. $n = 25, a = 5$

3.98 $E(C) = \$80$

$\sqrt{V(C)} = \$9.49$

(\$61.02, \$98.98)

3.99 **a.** .758

b. $\mu = 12, \sigma = \sqrt{12}$

c. (5, 19)

3.100 **a.** .083

b. .895

3.101 **a.** 5

b. .007

c. .384

3.102 **a.** $ke^{\lambda t}$

b. 3.2974

3.103 .993

3.104 .04096

3.106 .837

3.107 .18522

3.108 **a.** .081

b. .81

3.109 $P(Y = 5) = .01536$

$P(Y \geq 5) = .0256$

3.110 **a.** .019

b. .1745

3.111 $E(Y) = 900$

$V(Y) = 90$

$P(81.026 < Y < 918.974) \geq .75$

3.112

y	$p(y)$ (a) Binomial	$p(y)$ (b) Poisson
0	.358	.368
1	.378	.368
2	.189	.184
3	.059	.061
4	.013	.015

3.113 **a.**

y	$p(y)$
0	$(2/3)^4$
1	$2(2/3)^4$
2	$(2/3)^3$
3	$1/3(2/3)^3$
4	$(1/3)^4$

b. 1/9

c. 4/3

d. 8/9

3.115 **a.** .1192

b. .117

Yes

3.116 \$149.09

3.117 3

3.118 **a.** $\frac{N}{K}(1 + K[1 - (.95)^K])$

b. 5

c. $(.5738)N$

3.119 Number of combinations = 6,760,000

$E(\text{winnings}) = \$.0311$

No

CHAPTER 4

4.2 **a.** 6

b. .648

c. .3929

d. $F(b) = \begin{cases} 0 & b < 0 \\ 3b^2 - 2b^3 & 0 \le b \le 1 \\ 1 & b > 1 \end{cases}$

4.3 **a.** .84375

b. 4

4.4 **a.** $F(x) = \begin{cases} 0 & x < 0 \\ \dfrac{x^3}{256}(16 - 3x) & 0 \le x \le 4 \\ 1 & x > 4 \end{cases}$

b. 11/16

c. 67/256

d. 3.4298

4.5 **b.** $F(x) = \begin{cases} 0 & x < 5 \\ \dfrac{(x - 7)^3}{8} + 1 & 5 \le x \le 7 \\ 1 & x > 7 \end{cases}$

c. 7/8

d. 37/56

4.6 **a.** 1/2

b. 1/4

4.7 **a.** 3/4

b. 4/5

c. 1

d. $F(x) = \begin{cases} 0 & x < 0 \\ x^2 & 0 \le x \le 1 \\ 1 & x > 1 \end{cases}$

4.8 **a.** .08392

b. .99046

4.9 $E(X) = 60$

$V(X) = 1/3$

4.10 **a.** $E(X) = 2/3$

$V(X) = 1/18$

b. $E(Y) = 220/3$

$V(Y) = 20{,}000/9$

c. $(-20.9476, 167.6142)$

4.11 4

4.12 **a.** $E(X) = 2.4$

$V(X) = .64$

b. $E(\text{weekly costs}) = 480$

$V(\text{weekly costs}) = 25{,}600$

c. Yes; $P(Y > 600) = .2617$

4.13 **a.** $E(X) = 5.5$

$V(X) = .15$

b. (4.7254, 6.2746)

c. Yes; $P(X < 5.5) = .5781$

4.14 .7368

4.15 **a.** $F(x) = \begin{cases} 0 & x < a \\ \dfrac{x - a}{b - a} & a \le x \le b \\ 1 & x > b \end{cases}$

b. $\dfrac{b - c}{b - a}$

c. $\dfrac{b - d}{b - c}$

4.16 **a.** 2/5

b. 1/5

c. \$22,500

4.17 **a.** 1/20

b. 1/20

c. 1/2

4.18 .2

4.19 3/4

4.20 2/5

4.21 **a.** 1/8

b. 1/8

c. 1/4

4.22 **a.** 1/5

b. $E(X) = 0$

$V(X) = 1/1200$

4.23 **a.** 2/7

b. $E(X) = .015$

$V(X) = 49/120{,}000$

4.24 $E(\text{volume}) = 2.0 \times 10^{-5}$

$V(\text{volume}) = 3.4795 \times 10^{-10}$

4.25 1/6

4.26 1/4

4.27 **a.** 1/2

b. 1/4

4.28 3/8

4.29 **a.** $E(X) = 60$

$V(X) = 100/3$

b. 4

4.31 **a.** .2865

b. .1481

4.32 .7355

4.33 **a.** .1353

b. 460.52 cfs

4.34 **a.** $\mu = \dfrac{1}{2}$, $\sigma^2 = 1/4$

b. .9975

4.35 **a.** $E(C) = 1100$

$V(C) = 2{,}920{,}000$

b. No; $P(C > 2000) = .14$

4.36 **a.** .5057

b. 1936

4.37 **a.** .3679

b. .6065

4.38 .2636

4.39 **a.** .6065

b. .7788

c. No

4.40 **a.** .08208

b. .02732

4.41 **a.** .2865

b. .5091

4.42 **a.** .5353

b. .5353

4.43 **a.** .3653

b. .1903

c. ~~50.66 minutes~~ 101 min

4.44 $E(\text{area}) = 200\pi$

$V(\text{area}) = 200{,}000\pi^2$

4.45 **a.** $E(X) = 3.2$

$V(X) = 6.4$

b. (0, 8.26)

4.46 **a.** $E(X) = 30{,}000$

$V(X) = 1{,}500{,}000$

b. No; $P(X > 35{,}000) < .06$

4.47 **a.** $E(L) = 276$

$V(L) = 47{,}664$

b. (0, 930.963)

4.48 **a.** $E(Y) = 1$

$V(Y) = 1/2$

$$f(y) = \begin{cases} 4ye^{-2y} & y > 0 \\ 0 & y \le 0 \end{cases}$$

b. $E(Y) = 3/2$

$V(Y) = 3/4$

$$f(y) = \begin{cases} 4y^2e^{-2y} & y > 0 \\ 0 & y \le 0 \end{cases}$$

4.49 **a.** $E(Y) = 20$

$V(Y) = 200$

b. $E(A) = 10$

$V(A) = 50$

4.50 **a.** $E(Y) = 140$

$V(Y) = 280$

$$f(y) = \begin{cases} \frac{1}{2^{70}(69!)} y^{69} e^{-y/2} & y > 0 \\ 0 & y \leq 0 \end{cases}$$

b. 206.93

4.51 **a.** $E(Y) = 240$

$\sqrt{V(Y)} = 189.74$

b. (0, 809.21)

4.52 $E(Y) = 60$

$V(Y) = 900$

$$f(y) = \begin{cases} \frac{1}{303{,}750} y^{3} e^{-y/15} & y > 0 \\ 0 & y \leq 0 \end{cases}$$

4.53 $E(Y) = 9.6$

$V(Y) = 30.72$

$$f(x) = \begin{cases} \frac{1}{65.536} y^{2} e^{-y/3.2} & y > 0 \\ 0 & y \leq 0 \end{cases}$$

4.54 $$f(x) = \begin{cases} \frac{1}{4} x e^{-x/2} & y > 0 \\ 0 & x \leq 0 \end{cases}$$

4.55 **a.** .3849

b. .3159

c. .3227

d. .1586

e. .0366

4.56 **a.** 0

b. 1.15

c. 1.19

d. −.44

e. 1.645

f. 1.96

4.57 .0062

4.58 $425.60

4.59 .0730

4.60 $\mu = 1$

4.61 **a.** .9544

b. .8297

4.62 .5859

4.63 $P(|X| > 5) = .6170$

$P(|X| > 10) = .3174$

4.64 **a.** .1498

b. .0224

4.65 **a.** .0062

b. 225.6 hours

4.66 **a.** .0062

b. 1171 hours

4.67 13.67 ounces

4.68 .5102

4.69 **a.** 60

b. $E(X) = 4/7$

$V(X) = 3/98$

4.71 **a.** $E(C) = 17.33$

$V(C) = 29.96$

b. (6.387, 28.280)

4.72 **a.** .8208

b. $E(V) = 4.7$

$V(V) = .01$

4.73 $E(X) = 2/3$ corresponds to angle of 240°

4.74 **a.** .75

b. $E(X) = 1/3$

$\sqrt{V(X)} = \sqrt{1/18}$

4.75 **a.** $E(X) = 1/2$

$V(X) = 1/28$

b. $E(X) = 1/2$

$V(X) = 1/20$

c. $E(X) = 1/2$
$V(X) = 1/12$

d. a

4.76 **a.** 7/8
b. .002128

4.77 **a.** .6321
b. $\sqrt{\pi}$

4.78 **a.** .5547
b. .6966

4.79 .6576

4.80 .03091

4.81 .06573

4.82 .09813

4.83 **a.** .08209
b. .01855

4.84 $E(X) = 2.8025$
$V(X) = 2.1460$

4.85 42.9193

4.86 **a.** $2\sqrt{\frac{2KT}{m\pi}}$
b. $\frac{3}{2}KT$

4.88 $E(X^2) = 2\theta^2$

4.89 $M_2(t) = e^{t^2/2}$

4.90 $M_{z^2}(t) = (1 - 2t)^{-1/2}$,
gamma (1/2, 2)

4.91 **a.** 1/2
b. $F(y) = \begin{cases} 0 & y < 0 \\ y^2/4 & 0 \le y \le 2 \\ 1 & y > 2 \end{cases}$

d. 3/4

e. 3/4

4.92 **a.** $-3/8$

b. $F(y) = \begin{cases} 0 & y < 0 \\ \dfrac{y^2}{2} - \dfrac{y^3}{8} & 0 \leq y \leq 2 \\ 1 & y > 2 \end{cases}$

d. $F(-1) = 0$

$F(0) = 0$

$F(1) = 3/8$

e. $7/64$

f. $E(Y) = 7/6$

$V(Y) = 43/180$

4.93 **a.** $6/5$

b. $F(y) = \begin{cases} 0 & y \leq -1 \\ \dfrac{1}{5}(y + 1) & -1 < y \leq 0 \\ \dfrac{1}{5}(1 + y + 3y^2) & 0 < y \leq 1 \\ 1 & y > 1 \end{cases}$

d. $F(-1) = 0$

$F(0) = 1/5$

$F(1) = 1$

e. $1/4$

f. $E(Y) = 2/5$

$V(Y) = 41/150$

4.94 .1151

4.95 15.87 %

4.96 .0015

4.97 .073

4.98 .3155

4.99 **a.** $E(Y) = c/4$

$V(Y) = c(6 - c)/16$

b. $c/(2 - t)^2$

c. $c = 4$

4.100 $E(X^k) = \dfrac{\Gamma(\alpha + \beta)\Gamma(\alpha + k)}{\Gamma(\alpha + \beta + k)\Gamma(\alpha)}$

4.101 2.3833×10^{-7}

4.102 **a.** .9975

b. $f(y) = \begin{cases} \dfrac{1}{60{,}000} y^3 e^{-y/10} & y > 0 \\ 0 & y \leq 0 \end{cases}$

4.103 **a.** 105

b. $E(X) = .375$

$V(X) = .02604$

4.104 $1 - e^{-4}$

4.105 $\dfrac{29}{8} e^{-3/2}$

4.106 \$53.58

4.107 \$113.33

4.108 .736

4.109 .875

4.110 **a.** R has a Weibull distribution, $\gamma = 2$, $\theta = \dfrac{1}{\lambda\pi}$

b. $1/(2\sqrt{\lambda})$

4.111 **a.** .0045

b. .9726

4.112 **a.** $E(X) = e^{11} \times 10^{-2} g$

$V(X) = (e^{38} - e^{22}) \times 10^{-4} g^2$

b. (0, 3,570,245)

c. .3156

4.113 $M(t) = 1/(1 - t^2)$

$E(Y) = 0$

4.114 **a.** $F_X(x) = \begin{cases} 0 & x \leq 0 \\ 1 - e^{-x/100} & 0 < x < 200 \\ 1 & x \geq 200 \end{cases}$

b. 86.4665

4.119 .04999

CHAPTER 5

5.1 **a, b.**

		x_1		
		0	1	2
x_2	0	1/9	2/9	1/9
	1	2/9	2/9	0
	2	1/9	0	0
	$p(x_1)$	4/9	4/9	1/9

c. 1/2

5.2 **b.** .08

c. .12

5.3

		x_1			
		0	1	2	3
x_2	0	0	4/84	12/84	4/84
	1	3/84	24/84	18/84	0
	2	6/84	12/84	0	0
	3	1/84	0	0	0

5.4 **a.** $k = 1$

b. 2/3

5.5 **a.** $f_1(x_1) = 1 \quad 0 \le x_1 \le 1$

b. .5

c. Yes

5.6 **a.** $f(x_2) = 2(1 - x_2) \quad 0 \le x_2 \le 1$

b. .64

c. No

d. 1/2

5.7 **a.** 7/8

b. 1/2

c. 2/3

5.8 **a.** $f_1(x_1) = 2(1 - x_1) \quad 0 \le x_1 \le 1$

$f_2(x_2) = 2(1 - x_2) \quad 0 \le x_2 \le 1$

b. No

c. ~~2/3~~ 1/3

5.9 **a.** 21/64

b. 1/3

c. No

5.10 11/32

5.11 **a.** Yes

b. $\frac{3}{2e}$

5.12 7/32

5.13 11/36

5.14 1/4

5.15 **a.** 1/4

b. 23/144

5.16 1/4

5.17 **a.** $E(X_i) = .2$

$V(X_i) = .16$

$i = 1, 2$

b. $-.04$

c. $E(Y) = .4$

$V(Y) = .24$

5.18 **a.** $E(Y) = 1$

$V(Y) = 1/6$

b. (.1835, 1.8165)

5.19 **a.** $E(Y) = 2/3$

$V(Y) = 1/8$

b. (1/3, 1)

5.20 **a.** $E(Y) = 61$

$V(Y) = 20$

b. No, $P(Y > 75) \leq .1653$

5.21 **a.** $\frac{1}{e} - \frac{2}{e^2}$

b. 1/2

c. $1/e$

d. $f_1(y_1) = y_1 e^{-y_1} \qquad y_1 > 0$

$f_2(y_2) = e^{-y/2} \qquad y_2 > 0$

5.22 1/3

5.23 **a.** 1
b. 1
c. No; $P(Y_1 - Y_2 > 2) = e^{-2}$

5.24 1/2

5.26 .08953

5.27 66,960

5.28 **a.** .04594
b. .2262

5.29 .07776

5.30 .09352

5.31 **a.** 4/27
b. 1/27
c. 4/9

5.32 .40951

5.33 $\mu = 2.5,\ \sigma^2 = 4.875$

5.34 **a.** .2759
b. .80313

5.35 **a.** .08575
b. .7627
c. $E(Y_1) = 40$
$V(Y_1) = 24$

5.36 $M_Y(t) = \left(\dfrac{pe^t}{1 - (1 - p)e^t}\right)^r$

5.37 .05213

5.39 **a.** .1587
b. .3085

5.40 .0228

5.41 **a.** 4
b. $f_i(x_i) = 2x_i \qquad 0 \le x_i \le 1$
$i = 1, 2$

c. $F(x_1, x_2) = \begin{cases} 0 & x_1 < 0 \text{ or } x_2 < 0 \\ x_1^2 x_2^2 & 0 \le x_i \le 1, i = 1, 2 \\ x_1^2 & 0 \le x_1 \le 1, x_2 > 1 \\ x_2^2 & 0 \le x_2 \le 1, x_1 > 1 \\ 1 & x_1 > 1, x_2 > 1 \end{cases}$

d. 9/64

e. 1/4

5.42 **a.** $f_1(x_1) = 3x_1^2 \quad 0 \le x_1 \le 1$

$f_2(x_2) = \frac{3}{2}(1 - x_2^2) \quad 0 \le x_2 \le 1$

b. 23/64

c. 0

5.43 **a, b.**

		x_2: 0	1	2	3	$p(x_1)$
x_1	0	0	1/28	1/14	1/84	10/84
	1	1/21	2/7	1/7	0	10/21
	2	1/7	3/14	0	0	5/14
	3	1/21	0	0	0	1/21
	$p(x_2)$	5/21	15/28	3/14	1/84	

c. 9/16

5.44 $f(x_1 \mid X_2 = x_2) = 2x_1 \quad 0 \le x_1 \le 1$

Yes

5.45 **a.** $f(x_1 \mid x_2) = \dfrac{2x_1}{1 - x_2^2} \quad 0 \le x_2 \le x_1 \le 1$

b. $f(x_2 \mid x_1) = 1/x_1 \quad 0 \le x_2 \le x_1 \le 1$

c. $f(x_1 \mid x_2) \ne f_1(x_1)$

d. 5/12

5.46 **a.** $f(x_1, x_2) = 1/x_1 \quad 0 \le x_2 \le x_1 \le 1$

b. 1/2

c. ln 2/ln 4

5.47 **a.** $f(x_1) = \dfrac{2}{\pi}\sqrt{1 - x^2} \quad |x_1| \le 1$

b. 1/2

5.48 **a.** $f(x_i) = x_i + \dfrac{1}{2} \qquad 0 \le x_i \le 1 \qquad i = 1, 2$

b. No

c. $f(x_1 \mid x_2) = \dfrac{x_1 + x_2}{x_2 + \dfrac{1}{2}} \qquad 0 \le x_i \le 1 \qquad i = 1, 2$

5.49 **a.** 1

b. $f(x_1) = x/2 \qquad 0 \le x_1 \le 2$

$f(x_2) = 2(1 - x_2) \qquad 0 \le x_2 \le 1$

c. $f(x_1 \mid x_2) = \dfrac{1}{2(1 - x_2)} \qquad 0 \le 2x_2 \le x_1 \le 1$

d. $f(x_2 \mid x_1) = \dfrac{2}{x_1} \qquad 0 \le 2x_2 \le x_1 \le 1$

e. 1/2

f. 8/9

5.50 **a.** $f(x_2) = 2(1 - x_2) \qquad 0 \le x_2 \le 1$

b. $f(x_1) = 1 - |x_1| \qquad |x_1| \le 1$

c. 1/4

5.51 **a.** 2/3

b. 1/18

c. 0

5.52 3/160

5.53 **a.** $-.3333$

b, c. $E(X_1 + X_2) = 7/3$

$V(X_1 + X_2) = 7/18$

5.54 **a.** $-1/144$

b. 7/12

c. 1.0764

5.55 **a.** 2

b. 2/3

5.56 1/4

5.57 $f(x) = \left(\dfrac{1}{2}\right)^{(x+1)} \qquad x = 0, 1, 2, \ldots$

5.58 $f(x_1, x_2, x_3) = \left(\frac{1}{\theta}\right)^3 e^{(-1/\theta)(x_1+x_2+x_3)}$

$\theta, x_i > 0, i = 1, 2, 3$

5.59 $E(G) = 42$

$V(G) = 26$

No

5.60 $E(Y) = np$

$V(Y) = np(1 - p)$

5.61 **a.** $x_1/2$

b, c. 3/8

5.62 **a, b.** 1

5.63 3/8

5.64 300

5.66 **a.** $(p_1e^{t_1} + p_2e^{t_2} + p_3e^{t_3})^n$

b. $-np_1p_2$

5.70 $-\sqrt{\dfrac{p_1p_2}{(1 - p_1)(1 - p_2)}}$

5.75 **a.** $f(x_1, x_2) = \frac{1}{9}e^{-(x_1+x_2)/3} \qquad x_1 > 0 \qquad x_2 > 0$

b. .0446

5.76 **a.** .0868

b. $1 - e^{-5}$

CHAPTER 6

6.1 **a.** $f_{u_1}(u) = \frac{1}{2}(1 - u) \qquad |u| \leq 1$

b. $f_{u_2}(u) = \frac{1}{2}(1 + u) \qquad |u| \leq 1$

c. $f_{u_3}(u) = \dfrac{1 - \sqrt{u}}{\sqrt{u}} \qquad 0 < u \leq 1$

6.2 **a.** $f_{u_1}(u) = \dfrac{u^2}{18} \qquad |u| \leq 3$

b. $f_{u_2}(u) = \frac{3}{2}(3 - u)^2 \qquad 2 \leq u \leq 4$

$$f_{u_3}(u) = \frac{3\sqrt{u}}{2} \qquad 0 \le u \le 1$$

6.3 **a.** $f_u(u) = \begin{cases} \dfrac{u+4}{100} & -4 \le u \le 6 \\ \dfrac{1}{10} & 6 < u \le 11 \\ 0 & \text{otherwise} \end{cases}$

b, c. 67/12

6.4 $f_u(u) = \dfrac{1}{8\sqrt{2(u-3)}} \qquad 5 \le u \le 53$

6.5 **a.** $f_u(u) = \begin{cases} u & 0 \le u \le 1 \\ 2 - u & 1 < u \le 2 \\ 0 & \text{otherwise} \end{cases}$

b. 1

6.6 **a.** $f_u(u) = e^{-u} \qquad u > 0$

b. $E(U) = 1$

$V(U) = 1$

6.7 **a.** $f_u(u) = 2u \qquad 0 \le u \le 1$

b, c. 2/3

6.8 $f_u(u) = 1 \qquad 0 \le u \le 1$

6.9 **a.** $f_{u_1}(u) = \dfrac{1-u}{2} \qquad |u| \le 1$

b. $f_{u_2}(u) = \dfrac{1+u}{2} \qquad |u| \le 1$

c. $f_{u_3}(u) = \dfrac{1-\sqrt{u}}{\sqrt{u}} \qquad 0 < u \le 1$

6.10 **a.** $f_{u_1}(x) = \dfrac{u^2}{18} \qquad |x| \le 3$

b. $f_{u_2}(u) = \dfrac{3}{2}(3-u)^2 \qquad 2 \le u \le 4$

c. $f_{u_3}(u) = \dfrac{3\sqrt{u}}{2} \qquad 0 \le u \le 1$

6.11 $f_u(u) = 18u(1 - u \pm u \ln u)$

6.12 **a.** $f_u(u) = \dfrac{1}{\theta} e^{-u/\theta} \qquad u > 0$

b. $E(u) = \theta$
$V(u) = \theta^2$

6.13 **a.** $f_u(u) = \dfrac{1}{\alpha} e^{-u/\alpha} \qquad u > 0$

b. $\Gamma\left(\dfrac{k}{m} + 1\right)\alpha^{k/m}$

6.14 **a.** $f_u(u) = 2(1 - u) \qquad 0 \le u \le 1$

b. $f_u(u) = 2u \qquad 0 \le u \le 1$

6.15 $g_1(x) = e^{-(x-4)} \qquad x \ge 4$

6.16 **a.** $g_1(x) = ne^{-n(x-\theta)} \qquad x > \theta$

b. $E(X_{(1)}) = \dfrac{1}{n} + \theta$

6.21 $1 - [1 + 200(1 - e^{-.5}) \exp[-200(1 - e^{-.5})]$

6.22 $\left(1 - \dfrac{d}{t}\right)^2$

6.23 **a.** $f_{u_1}(u) = \dfrac{1}{2\sqrt{u}} \qquad 0 < u \le 1$

b. $f_{u_2}(u) = \begin{cases} 1/2 & 0 \le u \le 1 \\ \dfrac{1}{2u^2} & u > 1 \\ 0 & \text{otherwise} \end{cases}$

c. $f_{u_3}(u) = ue^{-u} \qquad u > 0$

d. $f_{u_4}(u) = -\ln u \qquad 0 < u \le 1$

6.24 **a.** $p_u(u) = \dfrac{e^{-(\lambda_1+\lambda_2)}(\lambda_1 + \lambda_2)^u}{u!} \qquad u = 0, 1, 2, \ldots$

b. $P(X_1 = u \mid X_2 + X_1 = m) = \dbinom{m}{u}\left(\dfrac{\lambda_1}{\lambda_1 + \lambda_2}\right)^u\left(\dfrac{\lambda_2}{\lambda_1 + \lambda_2}\right)^{m-u} \qquad u = 0, 1, 2, \ldots$

6.25 **a, b.** $E(U_1) = -1/3$
$E(U_2) = 1/3$
$E(U_3) = 1/6$

6.26 $f_u(u) = 1 \qquad 0 \le u \le 1$

6.27 **a.** $f_u(u) = 2(1 - u) \qquad 0 \le u \le 1$

b. 1/3

c. 1/18

6.28 $f_u(u) = \dfrac{1}{\Gamma(4)} u^3 e^{-u} \qquad u > 0$

6.29 $f_u(u) = 1 \qquad 0 \le u \le 1$

6.30 3/4

6.31 $f_u(u) = \dfrac{1}{\pi(1 + u^2)} \qquad -\infty < u < \infty$

6.32 $f_u(u) = \begin{cases} \dfrac{1}{4\sqrt{u}} & |u| \le 1 \\ \dfrac{1}{8\sqrt{u}} & 1 < u \le 9 \end{cases}$

6.33 $[1 - F(x)]^3[1 + F(x)]$

6.34 $f_R(u) = n(n - 1)(u^{n-2} - u^{n-1})$

6.37 $f_u(u) = ue^{-u^2/2} \qquad u > 0$

6.39 $f_u(u) = \dfrac{1}{2\pi}\left(\dfrac{3u}{4\pi}\right)^{-1/3} \qquad 0 \le u \le \dfrac{4}{3}\pi$

6.40 $f_E(u) = 4\sqrt{\dfrac{2}{\pi}}\left(\dfrac{b}{m}\right)^{3/2} \sqrt{u}\, e^{-2bu/m} \qquad u > 0$

6.41 Poisson with mean $\theta\lambda$

CHAPTER 7

7.1 $\mu = 3/4$

7.2 $\mu = ab$

7.3 **a.** $\mu = \theta/2$

7.4 No

7.5 **a.** $1 - 2e^{-1}$

7.6 .0668

7.7 .6826

7.8 385

7.9 .9090

7.10 153

7.11 **a.** .5328

b. .9772

c. Independence and random sampling

7.12 .0132

7.13 0

7.14 ***a.*** 1
b. .1230
c. Independence

7.15 ***a.*** .9938
b. 110

7.16 4.4653

7.17 .0013

7.18 88

7.19 .9876

7.20 51

7.21 .0062

7.22 .0287

7.23 ***a.*** .5948
b. .0559
c. .0017

7.24 .3936

7.25 .1292

7.26 .0

7.27 .1539

7.28 ***a.*** .0630
b. .0630

7.29 .0041

7.30 ***a.*** .0329
b. .029
c. Independence

7.31 .0043

7.32 .0548

7.33 .9544

7.34 151

7.35 .0668

7.36 .7698

7.37 .0071

7.39 $\bar{X}/(\bar{X} + \bar{Y})$

7.41 .1587

7.44 **a.** $T_o \to \infty$

INDEX